CMOS Cookbook

Second Edition

Don Lancaster

Revised by
Howard M. Berlin

BPB PUBLICATIONS

B-14, CONNAUGHT PLACE, NEW DELHI-110001

FIRST EDITION 1997, REPRINTED 2013

Distributors:

MICRO BOOK CENTRE
2, City Centre, CG Road,
Near Swastic Char Rasta,
AHMEDABAD-380009 Phone: 26421611

COMPUTER BOOK CENTRE
12, Shrungar Shopping Centre, M.G. Road,
BANGALORE-560001 Phone: 25587923, 25584641

MICRO BOOKS
Shanti Niketan Building, 8, Camac Street,
KOLKATTA-700017 Phone: 2826518, 2826519

BUSINESS PROMOTION BUREAU
8/1, Ritchie Street, Mount Road,
CHENNAI-600002 Phone: 28534796, 28550491

DECCAN AGENCIES
4-3-329, Bank Street,
HYDERABAD-500195 Phone: 24756400, 24756967

MICRO MEDIA
Shop No. 5, Mahendra Chambers, 150 D.N. Roa
Next to Capital Cinema V.T. (C.S.T.) Station,
MUMBAI-400001 Ph.: 22078296, 22078297, 2200

BPB PUBLICATIONS
B-14, Connaught Place, **NEW DELHI-11000**
Phone: 23325760, 23723393, 23737742

INFOTECH
G-2, Sidhartha Building, 96 Nehru Place,
NEW DELHI-110019
Phone: 26438245, 26415092, 26234208

INFOTECH
Shop No. 2, F-38, South Extension Part-1
NEW DELHI-110049
Phone: 24691288, 24641941

BPB BOOK CENTRE
376, Old Lajpat Rai Market,
DELHI-110006 PHONE: 23861747

Printed in India by arrangement with
Howard W. Sams & Co. Inc., USA.

ISBN 81-7029-795-8

Published by Manish Jain for BPB Publications, B-14, Connaught Place
New Delhi-110 001 and Printed by him at Pressworks, Delhi.

Contents

Preface

CMOS has been called the first "hassle-free" digital-logic family. It is ultralow in cost and is available in hundreds of devices from a dozen major manufacturers. It works over a very wide, noncritical, power-supply range, and it uses *zero* power when the inputs aren't changing and very little power when they are. Its inputs are essentially open circuits, and its outputs swing the whole range between supply limits. As an added advantage, output drive to other CMOS packages is virtually unlimited.

CMOS logic is very forgiving of system noise and doesn't generate much noise of its own. It is easily converted to linear operation and offers dozens of options towards high-performance, low-parts-count timers, oscillators, and pulse sources.

But most important, CMOS is the first digital-logic family that is genuinely fun to work with. It is extremely tolerant of the usual rat's-nest breadboards and poor power supplies that are typical of experimenter, student, and industrial lash-ups. Very often, CMOS turns out to be the top choice for digital-logic design, particularly in portable, low-cost, low-frequency applications. These applications include digital instruments, voltmeters, frequency counters, displays, video games, tv typewriters, microprocessors and their peripherals, electronic music, alarms, remote controls, and much, much more. And CMOS is almost certainly the best choice for teaching and learning digital logic, since it lets you concentrate on what the logic is supposed to be doing.

The *CMOS Cookbook* will tell you all you need to know to understand and profitably use CMOS. It will show you all the basics of working with digital logic and many of its end applications along the way. Regardless of whether you are a newcomer to logic and electronics or a senior design engineer, you will find valuable help and inside information here. You can use this book as a self-learning guide, as a reference handbook, as a project idea book, or as a text for teaching others digital logic on the high-school through university levels.

While similar in organization to our older *TTL Cookbook* (Catalog No. 21035), very little material is repeated. We have kept the math-free, informal coverage, the attention to detail, the user-orientation, and the extensive application examples of the earlier texts. This new edition also covers the newer 74C, 74HC, and 74HCT CMOS subfamilies, which permit easier interfacing between TTL and CMOS, as well as making the transition from working with TTL devices to working with CMOS devices virtually painless.

We begin in Chapter 1 with some basics—what CMOS is, who makes it, and how the basic transistors, inverters, logic gates, and transmission gates work. We follow this information with CMOS usage rules, power-supply-design examples, information on breadboards, state testing, tools, and interface. Chapter 2 is a minicatalog of approximately 160 CMOS devices. It shows their pinouts and gives detailed descriptions of how they work and how to use them. Unlike industry catalogs, this chapter contains only what you need to know, is an industry-wide reference, and, most important, puts the hangups, tricks, and use restrictions out front where they belong.

The next chapter is on logic, starting with the basic gate fundamentals, then covering TRI-STATE® logic, and ending with the new *redundant* design methods, using data selectors, ROMs, PLAs, and microprocessors. These new techniques offer single-package solutions to virtually any logic design and have the advantages of easy changes, minimum cost, and virtually instant design. They totally obsolete the so-called logical-minimum design tehcniques of the 1950s and 1960s. We end up with some little-known guidelines and philosophy of logic design, along with some ASCII-coded computer peripheral examples.

Multivibrators are covered in Chapter 4, with most of the astable, monostable, bistable, and linear techniques explored in depth. Clocked-logic designs and the extensive applications of JK and D flip-flops follow in the next chapter. Chapter 6 takes a detailed look at counter and register techniques. Digital sine-wave generators are also discussed in this chapter.

Things that you can do *only* with CMOS are the topic of Chapter 7—CMOS linear operational amplifiers; wide-range, micropower phase-locked loops; bidirectional analog and digital switches; and some other neat tricks that simply aren't possible with any other digital-logic family. Chapter 8 covers how CMOS devices are connected to single LEDs, 7-segment LED displays, and liquid-crystal displays.

The final chapter, as usual, is on applications, where we will be taking a close look at CMOS digital instruments, voltmeters, counters, video games, wristwatches, tv typewriters, cassette systems, polytonic electronic-music synthesizers, and a number of other real-world, exciting, and useful devices.

DON LANCASTER

x

Some Basics

A *digital logic family* is a group of compatible building blocks that have inputs and outputs. These blocks make simple *yes-no* decisions. They output yes-no decisions based on the presence of yeses and nos on all their inputs. Often, they will also use internal *memory* or storage to take into account the history of yeses and nos that have been input previously.

Sometimes, we need only a single yes-no output. An intrusion alarm, an auto "lights-on" warning, or an industrial "the tank is empty" signal are typical examples. But, more generally, we want to combine the outputs of many different logical building blocks into related groups of yes-no decisions. We can also call these decisions ON-OFF decisions, or "one" and "zero" states.

Four yes-no decisions together can be a decimal digit. Six or seven equal an alphanumeric character. Ten or twelve can be converted into a continuous, or *analog*, output such as a musical note or a process control signal.

Four hundred decisions can build you a better-grade hand calculator. Four thousand groups, or *words*, of eight decisions, or *bits*, per word will build you the memory for a microprocessor, mini-computer, or hobbyist computer. Four million words of sixteen bits per word will get you up into the large-computer main-frame class, while around four billion bits can approximate the human brain.

We can make our "1"-"0" decisions into just about anything—a musical note, a test waveform, a measured and displayed value, a video presentation, a calculation, a clock, a game, an industrial control, a toy, a microcomputer, an art form, a community information access service, or just about anything else you can dream up. All it takes to do the job is the right number of properly connected logic blocks.

There are lots of digital logic families available. TTL or T²L (Transistor-Transistor Logic), RTL (Resistor-Transistor Logic), and ECL (Emitter-Coupled Logic) are examples of older general-purpose logic families.

A newer logic family is called CMOS, short for Complementary Metal-Oxide Semiconductor. CMOS has some very important advantages over earlier logic families. As we'll see in detail later on, these benefits include very low cost (from 3 cents per gate and up), ultralow and noncritical power needs, wide logic swings, "down-the-middle" transfer characteristics, open-circuit inputs, lots of "fan-out" drive, good noise performance, and lots of different devices available from many highly competitive sources.

We also gain some system-level benefits with CMOS. These include the ability to swallow, rather than perpetuate, system noise, the generation of little power-line noise during output changes, and the use of essentially zero supply power when the logic blocks aren't changing. What is even nicer about CMOS is that we can now do new things with a general-purpose logic family that simply was not possible before with the older families. These new features include analog switching, simple and effective oscillators and pulse shapers, wide-range phase-locked loops, excellent linear techniques, *bilateral* digital logic that has interchangeable inputs and outputs, and effective digital sine-wave generators.

But, what's best of all about CMOS is that it is fun to work with. It usually tries to help you rather than fight you. It is far more forgiving and far less critical to get on line than *any* earlier logic family. It is also far more tolerant of rat's-nest breadboarding, poor power supplies, and the usual things that go with school-lab or kitchen-table experimental lash-ups.

Of course, CMOS isn't perfect. Most CMOS is limited to five million or less input changes per second, which is equivalent to a maximum *clock*, or *data rate*, of five megabits per second. Newer and premium versions of CMOS with ten times this speed limit are showing up. CMOS outputs are also sensitive to outside-world loading, particularly capacitance. Thus, you have to be somewhat more careful with a CMOS *interface* (connections), to other logic or the rest of the world. As we'll see, there are a few usage rules that must be strictly obeyed. But even these rules are simple and easy to live with. Today, CMOS often ends up as the top choice for many electronic systems, particularly if low cost, portability, low supply power, easy design, simplicity, and good noise performance are important.

2

THE CMOS PROCESS

Let's see what is involved in building two basic types of MOS, or *Metal-Oxide Semiconductor*, transistors. We can then connect these basic transistor types together to form the simplest possible CMOS building block—the *inverter*. From there, we'll look at other simple CMOS building blocks and then see how they are combined into more useful circuits. These more elaborate circuits include the SSI (Small-Scale Integration) gates, flip-flops, and monostables; MSI (Medium-Scale Integration) counters, analog switches, registers, phase-locked loops, and arithmetic units; and LSI (Large-Scale Integration) circuits that include thousand-bit memories, microprocessors, multiple decade counters, and complete clock and stopwatch circuits.

Fig. 1-1 shows how we might build a transistor called an *n-channel enhancement-mode MOS device*. We start with a bar of p-type silicon. P-type silicon is ultrapure, single-crystal silicon (derived from ordinary beach sand) with just enough of an impurity introduced that there are too few electrons to go around. The absence of an electron, where an electron is expected to be, is called a *hole*. We say that p-type silicon has an excess of holes. A hole has an equivalent positive charge that equals and offsets the negative charge of an electron.

Now, we *diffuse* two *junctions* into our silicon block, or *substrate*. This builds two n-type silicon regions. The n-type silicon regions

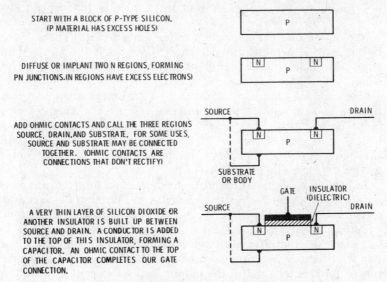

START WITH A BLOCK OF P-TYPE SILICON. (P MATERIAL HAS EXCESS HOLES)

DIFFUSE OR IMPLANT TWO N REGIONS, FORMING PN JUNCTIONS.(N REGIONS HAVE EXCESS ELECTRONS)

ADD OHMIC CONTACTS AND CALL THE THREE REGIONS SOURCE, DRAIN, AND SUBSTRATE. FOR SOME USES, SOURCE AND SUBSTRATE MAY BE CONNECTED TOGETHER. (OHMIC CONTACTS ARE CONNECTIONS THAT DON'T RECTIFY)

A VERY THIN LAYER OF SILICON DIOXIDE OR ANOTHER INSULATOR IS BUILT UP BETWEEN SOURCE AND DRAIN. A CONDUCTOR IS ADDED TO THE TOP OF THIS INSULATOR, FORMING A CAPACITOR. AN OHMIC CONTACT TO THE TOP OF THE CAPACITOR COMPLETES OUR GATE CONNECTION.

Fig. 1-1. Building an n-channel MOS transistor.

have an excess of electrons in them. These two regions are introduced by *diffusing,* or *ion-implanting,* additional impurities that are carefully selected to tip the balance towards an excess of electrons in these new regions.

The contact area between the n and p regions is called a *pn junction.* It will conduct current when p is positive with respect to n. It blocks current when n is positive with respect to p. Simple pn junctions are also used in small-signal and power silicon diodes and rectifiers.

Next, we make some physical or *ohmic* contacts to the three regions in our silicon bar. By an ohmic contact, we mean some type of direct connection (like a solder blob) that does not rectify. We call the lead that connects to our original silicon bar the *substrate,* or body. We call one of the n regions a *source* and one of the n regions a *drain.* Sometimes we may externally connect the substrate to the source terminal.

So far, all we have done is build two back-to-back silicon diodes and then optionally shorted one of them out. To get this device to "transist," we have to be able somehow to *control* the current flow between our source/substrate terminal and our drain terminal. To do this, we build a very special capacitor between the source and drain on the surface of the transistor.

We begin building our capacitor with a *dielectric,* or insulating, layer of silicon dioxide, glass, or some other insulator. This dielectric layer is extremely thin. As in any good capacitor, the dielectric does not pass a dc current. It is an insulator. On top of the insulator, we place a new conductor which we call a *gate.* An ohmic contact brings out an external connection called the *gate lead.* The gate can be made of either metal or silicon, so long as it conducts. Metal gates are older and simpler, but silicon gates are more sensitive, faster, and smaller.

This completes our basic transistor. To get it to do something useful, we have to apply external voltages or currents. This is known as properly *biasing* the device. Fig. 1-2 shows how we bias our transistor. We usually use the gate as an input. Most often, the drain is used as an output, and the substrate and source are connected to ground or some other voltage.

In Fig. 1-2A, we have grounded our source and substrate and have connected the drain to a positive voltage, +V, through a *load resistor* or some other current source. We have also grounded our gate input.

Since our input is grounded, we have zero voltage difference on our gate capacitor. The capacitor won't charge and so behaves as if it wasn't there. The outside world sees the transistor as a reverse-biased pn junction, and so the output lead goes positive, pulled up to +V by the load resistor. A grounded input thus gives us a posi-

4

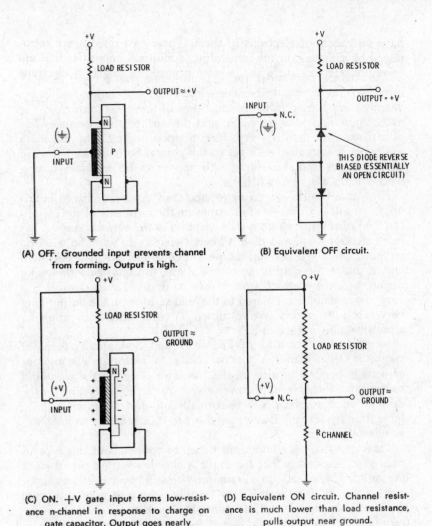

(A) OFF. Grounded input prevents channel from forming. Output is high.

(B) Equivalent OFF circuit.

(C) ON. +V gate input forms low-resistance n-channel in response to charge on gate capacitor. Output goes nearly to ground.

(D) Equivalent ON circuit. Channel resistance is much lower than load resistance, pulls output near ground.

Fig. 1-2. Biasing an n-channel MOS transistor.

tive output. The equivalent circuit of our input is an open circuit to dc and a very small capacitance to ac signals. The equivalent circuit, looking in from the output to ground is the very high impedance of a reverse-biased diode, completely swamped by a load resistor that pulls the output to +V. And, since we've done absolutely nothing, our transistor still isn't "transisting."

But, suppose we connect the gate input to the +V positive source, as shown in Fig. 1-2C. The left side of the gate capacitor now has positive charges, or holes, placed on it. When it charges, it must end up with extra electrons on the right side. It can get these electrons from our substrate-to-ground connection.

The surface just under the gate insulator was p-type material, meaning that it was deficient in electrons. As a charge builds up, electrons are gained from ground. Some of these electrons cancel or even *recombine* with the holes, and the surface state becomes *less* of a p material than it was before. Suppose we pick up still more electrons. Eventually, we'll get to the point where all the holes are cancelled out and our silicon will appear to be *intrinsic*, having neither extra electrons nor holes.

How about adding even more electrons? As more charge builds up, the result is an excess of electrons on the right side of the capacitor. *The silicon turns into n-type material at the surface,* since it has more electrons on hand than it knows what to do with. So what do we have now? We have an n-type source connected by a narrow n region under the gate to an n-type drain, forming a *continuous* n region, or an *n channel,* from source to drain. The n channel is a very low resistance compared to the load resistor, so the output goes very nearly to ground. We've turned on our transistor by applying a positive gate voltage to it.

So, to leave our transistor OFF, we ground the input. To turn our transistor ON, we apply a positive voltage to the input. The amount of voltage needed to turn on the transistor is called the *threshold voltage* and varies from just over a volt upwards, depending on the device. Since our transistor is normally off, and since we have to force it on by actively doing something to it, it is called an *enhancement-mode* device.

There are two very important things to notice about this type of transistor. The first is that *the input is always an open circuit since the gate lead goes only to a capacitor.* We will never need any input current, except for the brief instant when we charge or discharge the small gate capacitor. The second thing is that *when the transistor turns on, it is simply a solid bar of material,* or *a plain old resistor.* There are no saturation voltages, saturated junctions, stored charges, or other voltage offsets. The only voltage drop you will get is a plain old Ohm's-law current drop across the channel resistance. Very handy, if we watch our biasing; *this resistor conducts equally well in either direction, or is a bidirectional resistor.* We will see later that this bidirectional resistor lets us freely interchange inputs and outputs on *some* digital logic blocks, as well as switch analog signals simply and effectively.

Fig. 1-3 shows the symbol and connections for an n-channel MOS transistor *inverter.* If we ground the input, we get a positive output. If we make the input positive, we get a grounded output. If we call a positive voltage a "1" and a grounded voltage a "0," a 1 in produces a 0 out and vice versa. Since the inputs of other inverters will be open circuits, we can freely connect this inverter's output to lots of other inverter inputs without any loading problems.

This looks like a great device; in fact, it is so good that a whole integrated-circuit technology has built up around it. These devices are called *n-channel MOS integrated circuits*. Typical available examples include random-access memories, keyboard encoders, microprocessors, character generators, shift registers, code converters, and many other circuits. These devices are widely available and low in cost.

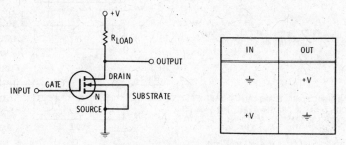

Fig. 1-3. Inverter made from n-channel MOS transistor.

But, they are not CMOS, nor are they available as a general-purpose logic family. One major limitation of n-channel devices is that power-supply current is needed continuously in the ON state. So, even if our logic is sitting still, or *quiescent,* we're going to need supply current, and probably a bunch of it in a complex circuit. A second limitation of n-channel technology is that things seem to be unbalanced. Our transistor has to pull down harder than our load resistor can pull up, and this can lead to problems involving loading and speed of response.

What would really be nice is to find a "mirror" technology so that our equivalent load resistor would be there only in the output-high state and would be disconnected, "off," or otherwise invisible in the low state. This mirror technology is called *p-channel MOS transistor* technology. We'll see shortly how pairs of n and p *complementary* transistors form the key to our CMOS logic blocks.

Fig. 1-4 shows how we build a p-channel MOS transistor, again an enhancement-mode device. All we do is switch the n and p regions around. We start with a block of n-type silicon and diffuse two p-type regions into it. We make the same substrate, source, and drain ohmic connections. Then, we build a similar gate dielectric and gate lead. The biasing is shown in Fig. 1-5.

Everything is similar only upside down. This time, we connect our substrate and source to +V and our load resistor goes to ground. A

+V input now puts zero charge on our gate capacitor, and it stays off, giving us a grounded output. A ground input now puts electrons on the left side of the gate capacitor, so a positive charge, obtained from the +V supply, has to build up on the right-hand side of our gate capacitor. These positive charges, or holes, first make the material less of an n-type, then make it intrinsic, and finally convert it into a temporary p-type material. This forms a continuous p channel from input to output.

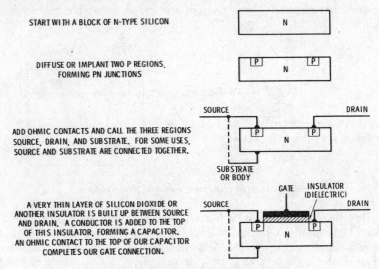

START WITH A BLOCK OF N-TYPE SILICON

DIFFUSE OR IMPLANT TWO P REGIONS, FORMING PN JUNCTIONS

ADD OHMIC CONTACTS AND CALL THE THREE REGIONS SOURCE, DRAIN, AND SUBSTRATE. FOR SOME USES, SOURCE AND SUBSTRATE ARE CONNECTED TOGETHER.

A VERY THIN LAYER OF SILICON DIOXIDE OR ANOTHER INSULATOR IS BUILT UP BETWEEN SOURCE AND DRAIN. A CONDUCTOR IS ADDED TO THE TOP OF THIS INSULATOR, FORMING A CAPACITOR. AN OHMIC CONTACT TO THE TOP OF OUR CAPACITOR COMPLETES OUR GATE CONNECTION.

Fig. 1-4. Building a p-channel MOS transistor.

A p-channel MOS inverter appears in Fig. 1-6. As with the n-channel inverter, a "0" at the input gets you a "1" at the output and vice versa. P-channel technology by itself has just about as many problems as n-channel does. While p-channel ICs are available, many of them are dated and being replaced by equivalent n-channel devices that have speed, cost, and power-supply advantages. So, when used by themselves, both p- and n-channel systems have problems, but amazing things happen when the two technologies are used together.

Fig. 1-7 shows a CMOS, or *complementary* MOS, inverter. Here, an n-channel transistor and a p-channel transistor form each other's load resistors. When the input is low, the p-channel device is ON and the n-channel device is OFF. The output sees a low resistance to +V. If we don't load the output, there is no supply current needed. When the input is high, the p-channel device is OFF and the n-channel one is ON. The output sees a low resistance to ground. Again, no supply current is needed. The only time you need supply current is when you change the input state because it will take some energy to charge the gate capacitor. In addition, both transistors will be partially on during the transition between states.

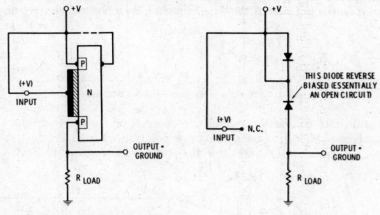

(A) OFF. Input at +V prevents channel from forming. Output is grounded.

(B) Equivalent OFF circuit.

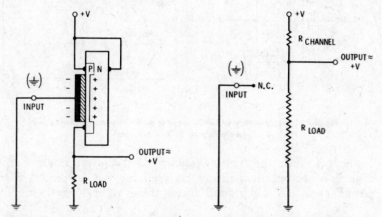

(C) ON. Grounded gate input forms low-resistance p-channel in response to charge on gate. Output goes nearly to +V.

(D) Equivalent ON circuit. Channel resistance is much lower than load resistance, pulls output near +V.

Fig. 1-5. Biasing a p-channel MOS transistor.

The problem of needing different types of substrates for n and p devices is handled by the IC people in several ways. Two popular techniques are to build a *p tub* or use thin layers of *dielectric isolation* to separate the two types of devices. Note that the arrows on our p- and n-channel transistors are "backwards" compared with the bipolar npn and pnp transistor arrows, since the arrows represent

9

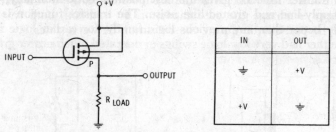

IN	OUT
⏚	+V
+V	⏚

Fig. 1-6. Inverter made from p-channel MOS transistor.

substrate connections which are reverse- or back-biased. On a p-channel MOS transistor, the arrow points *away* from the device. On an n-channel MOS transistor, the arrow points *toward* the device.

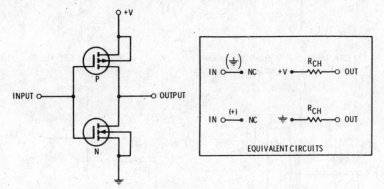

Fig. 1-7. Complementary or CMOS inverter made from one n-channel and one p-channel MOS transistor.

Fig. 1-8 shows the *transfer function* of the inverter. The output changes state exactly halfway between the two extremes of the power supply +V and ground, slicing things right down the middle.

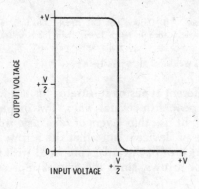

Fig. 1-8. Response or transfer function of a CMOS inverter.

This transfer function gives the best possible noise immunity, both to supply-line and ground-line noise. The transfer function is also vastly better than any previous logic family, for earlier logic families either had very low logic swings or operated much nearer ground or in a segmented manner.

CMOS FEATURES

There are several unique things about our CMOS logic family:

- The inputs to all devices are open circuits and thus very easy to drive.

- No power-supply current is needed except during input logic changes. Operating current is extremely low, particularly at low frequencies.

- The logic changes from high to low exactly halfway up to the supply voltage, giving good noise immunity.

- The circuits will work over a very wide power-supply voltage range, typically +3 to +15 volts.

- The unloaded output logic swing goes from ground to the positive supply, the full range of the available supply voltage.

- CMOS output stages do not generate large current spikes on the power-supply lines, thus they create very little noise of their own.

- CMOS logic is usually designed so that the transition (output change) times are longer than the propagation (input to output delay) times. This makes them tend to swallow glitches and system noise, rather than legitimizing noise into unwanted, but apparently valid, output signals.

LOGIC AND TRANSMISSION GATES

We'll see later on how inverters are used to change the definition of a 1 and a 0; to introduce small amounts of delay; to build things like contact conditioners, pulsers, and oscillators; to interface with other logic; and to increase the available drive of a point in a digital circuit. But to do logically more useful things, we have to connect groups of inverters and inverter-like combinations of p- and n-channel transistors into basic building blocks. Three of the most important elemental connections are the two-input NOR gate, the two-input NAND gate, and the transmission gate, as shown in Figs. 1-9, 1-10, and 1-11.

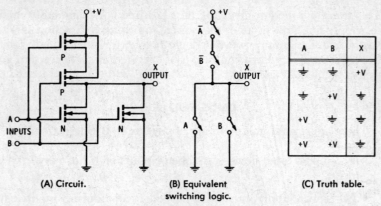

A	B	X
⏚	⏚	+V
⏚	+V	⏚
+V	⏚	⏚
+V	+V	⏚

(A) Circuit.

(B) Equivalent switching logic.

(C) Truth table.

Fig. 1-9. CMOS two-input NOR gate.

In Fig. 1-9A, we have arranged two n-channel MOS transistors in parallel so that either can pull the output to ground for a positive input. We have also arranged two p-channel transistors in series so that both must work together to pull the output to +V for grounded inputs. If either A or B is positive, the path to +V will be disconnected and the path to ground will be completed. The output will go to +V only if both A and B are grounded.

The equivalent switching is shown in Fig. 1-9B. Switch A closes when A input is positive. The *complement* of A, or $\overline{A}$, closes when the A input is grounded. Similarly, B closes for a positive B input and $\overline{B}$ closes for a grounded B input.

This is called a positive logic NOR circuit and follows the *truth table* of Fig. 1-9C. The circuit is used as a basic building block whenever you want either input positive to do something. Two of these blocks may be cross-coupled to produce a simple memory or *set-reset flip-flop*.

In Fig. 1-10A, we've put the n-channel transistors in series to ground and the p-channel transistors in parallel to +V, ending up with the switching shown in Fig. 1-10B. This time, either input grounded pulls the output high, but only both inputs simultaneously positive allows the output to go low, following the truth table of Fig. 1-10C. This is a positive-logic NAND circuit and is used whenever you want the coincidence of two inputs to do something useful.

The inverter, NOR, and NAND logic building blocks are the cornerstones of most digital logic families. From these basic building blocks, we can build just about any traditional logic block or system of any complexity we want. All we have to do is suitably connect enough of the basic blocks together.

Fig. 1-11 shows a new way to connect our MOS transistors. It builds a brand-new and unique-to-CMOS logic block called a *trans-*

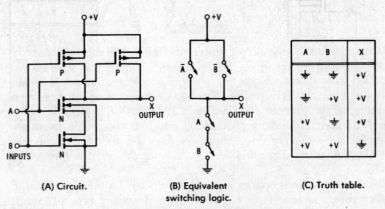

| (A) Circuit. | (B) Equivalent switching logic. | (C) Truth table. |

Fig. 1-10. CMOS two-input NAND gate.

mission gate. The transmission gate connects terminal X to terminal Y when A is positive. It disconnects terminal X from terminal Y when A is grounded. *X and Y can be either inputs or outputs. They can control digital or analog signals in either direction.* The only restriction is that the voltage present on X or Y must never be more than +V or less than ground.

We use transmission gates for digital or analog switching, or to design digital logic blocks that have interchangeable inputs and outputs. For instance, the same transmission gate can be used as a *data selector*, a *data distributor*, a *multiplexer*, or a *demultiplexer*. When used inside a package, transmission gates greatly simplify design of clocked logic, flip-flops, register, and memory stages. We will be looking at the transmission gate in more detail in Chapter 7.

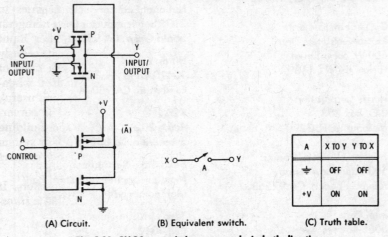

| (A) Circuit. | (B) Equivalent switch. | (C) Truth table. |

Fig. 1-11. CMOS transmission gate works in both directions.

SOURCES

Today, many hundreds of different CMOS circuits are readily available, and more are being introduced. Chapter 2 will give a detailed look at some of the more popular devices. The mainstream of available CMOS is called the "4000 Series" and represents a merging of the pioneer 4000 Series of RCA with the 4500 Series of *Motorola* and a few of the MC14400 numbers thrown in by other manufacturers. Today, most suppliers offer the majority of the circuits in the series, regardless of who first introduced the particular device.

The series is designed as a *general-purpose* digital logic family. This means you can use it by itself to build up any type of circuit or system you like, or else you can use it as the "glue" to interface and hold together a microprocessor or other complex system using large-scale integration.

Chart 1-1. Some CMOS Sources

American Microsystems
3800 Homestead Road
Santa Clara, CA 95051

Analog Devices
Box280
Norwood, MA 02062

Fairchild Semiconductor
464 Ellis Street
Mountain View, CA 94042

General Instruments
Mocroelectronics Div.
600 W. John Street
Hicksville, NY 11802

Harris Corporation
P.O. Box 883
Melbourne, FL 32901

Hughes Microelectronics
500 Superior Avenue
Newport Beach, CA 92663

Intel Corp.
3065 Bowers Avenue
Santa Clara, CA 95051

Intersil Inc.
10710 N. Tantau Avenue
Cupertino, CA 95014

Mostek
1215 W. Crosby Road
Carrollton, TX 75006

Motorola Semiconductor
Box 20912
Phoenix, AZ 85036

National Semiconductor Corp.
2900 Semiconductor Drive
Santa Clara, CA 95052

Nitron Corp.
10420 Bubb Road
Cupertino, CA 95014

RCA
Route 202
Somerville, NJ 08876

Solid State Scientific
Montgomeryville Industrial Park
Montgomeryville, PA 18936

Texas Instruments
Box 5012
Dallas, TX 75222

Chart 1-1 lists the major CMOS sources. You can buy CMOS from just about any electronics distributor. Extensive mail order ads appear regularly in *Popular Electronics, Radio Electronics,* and the various ham and computer hobbyist magazines.

Many of the manufacturers also have good data books that give very detailed information on their particular CMOS devices. These books range in price from free to several dollars. It is a good idea to get as many of them as you can. Several of the better or more available ones are shown in Chart 1-2. You can usually get specific data sheets from most of the suppliers simply by asking for them. You'll get the fastest and best results by asking for only one or two specific sheets at a time and by typing your request in the most professional form you can, preferably on business letterhead paper.

Chart 1-2. CMOS Data Books

Isoplanar CMOS Data Book	Fairchild Semiconductor 464 Ellis Street Mountain View, CA 94042
MOS Data Books	Motorola Semiconductor Products Inc. Box 20912 Phoenix, AZ 85036-0924
High-Speed Logic and Data Book Series	National Semiconductor Corp. 2900 Semiconductor Drive Santa Clara, CA 95052
RCA Cosmos Databook	RCA Solid State Box 3200 Somerville, NJ 08876

PINS AND PACKAGES

Most CMOS devices are available in the standard 14-, 16-, and 24-pin, dual in-line, plastic packages numbered as shown in Fig. 1-12. A code notch and dot usually identify pin one, to the left and toward you, when *viewed from the top.* The numbering is then counterclockwise. Pins are normally on 0.1-inch centers. The 14- and 16-pin packages are usually 0.3 inch wide, while the 24-pin version is usually 0.6 inch wide. *Note that the pin numbering reverses for a bottom-side printed-circuit (pc) layout or breadboard wiring.*

Supply power is almost always applied to diagonally opposite pins, with pin 7 being the most negative (often ground) and pin 14 being the most positive voltage (+V) on a 14-pin package. Pins 8 and 16 are ground and +V, respectively, on a 16-pin package. The

24-pin packages may have pin 12 as ground and pin 24 as +V, or they may not, depending on who made the IC and how it fits into a system; so be sure to check. Supply-pin connections of memory ICs may also be different.

You'll also find a few special packages like 28- and 40-pin packages for CMOS microprocessors and other special LSI devices, and

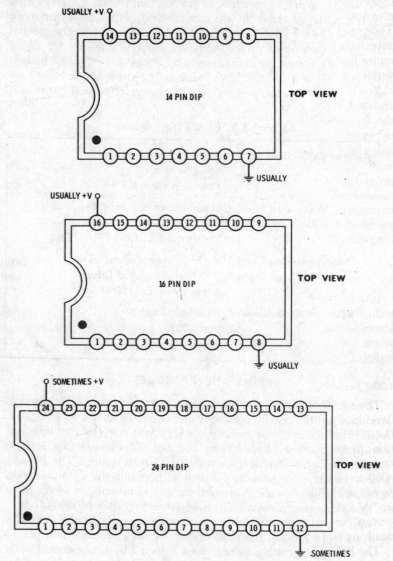

Fig. 1-12. CMOS pinouts and packages.

several subminiature and "bare-chip" packages for digital watches and other timing circuits.

Quite often, you'll get several different functions in one package. For instance, a quad two-input NOR gate like the 4001 or the 74C02 gives you four independent NOR gates in one package. Each gate has two inputs. The dual four-input NOR gates like the 4002 or the 74C25 give you two separate gates in the package, each with four inputs. These may be used together or in totally separate circuits. All that the gates have in common are the supply-voltage and ground pins. Other multiple packages include dual flip-flops, triple gates, and hex inverters, with the latter having six inverters per package.

You'll also find CMOS available in several different package materials and operating-temperature ranges. For practically all uses, the limited-temperature plastic package is cheapest and best. Typical examples include the RCA -AE package, the Motorola -P package, and the Fairchild -C package. All of these operate over at least a −40° to +85° C temperature range. There are also available some premium ceramic packages with wider temperature ranges at much higher costs. When you buy from a major distributor, it is extremely important to spell out that you want the plastic package, for these premium ceramic devices can be ridiculously expensive and can have long delivery times as well.

CMOS FAMILIES

Like the evolution of the many TTL subfamilies or series, CMOS technology also has evolved to the point where there now exists several series, each with improved characteristics over the previous series. Currently there are the 4000, 74C00, 74HC00, and 74HCT00 series of CMOS devices in common use.

4000 Series

The 4000 series of CMOS devices is the original full-line family as developed by RCA. It is much slower than the standard TTL family, but it offered a significant advantage in reducing power consumption. Several years after their introduction, an improved, higher-voltage, 4000 series was developed. To differentiate this improved 4000-series family of devices from the earlier version, a "B" (for *buffered*) suffix was added. The older-type devices were marked using an "A" suffix. Internal buffers, which usually consist of a pair of inverters, were added to all outputs as shown in Fig. 1-13, with the resulting response improvements shown in Fig. 1-14.

The buffers do several things. They make the output drive in both directions a uniform and standard value. This makes the rise and fall times more nearly identical. The additional gain makes up for

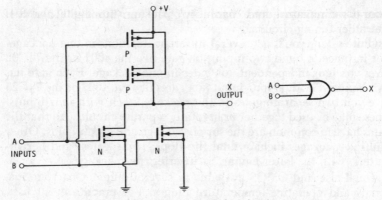

(A) Unbuffered, conventional, or A-series.

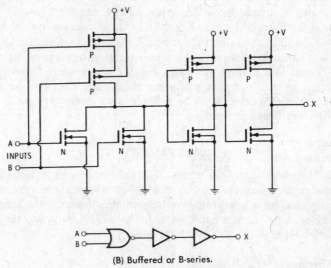

(B) Buffered or B-series.

Fig. 1-13. Two types of CMOS.

slowly changing input signals and gives sharp rise and fall times, particularly with resistor-capacitor timing and pulsing circuits. While the buffers add some delay to the response, the circuits can now be built with smaller internal transistors. Therefore, buffered devices are often slightly *faster* than their A-series equivalents.

The B-series also lets manufacturers quietly fix some of the really bad surprises that were built into the original A-series. These included (1) the 4009 and 4010 inverters (replaced by 4049 and 4050 types, respectively) that would self-destruct if the supplies were turned on in the wrong sequence, (2) the 4030 (replaced by the 4070) that had a *low* input impedance and wouldn't work in pulse circuits, (3) the

18

poor one-directional drive capability of the four-input 4002 and 4012 gates, (5) the very limited drive on some outputs of the 4018 counter, and so on.

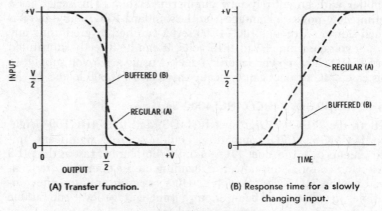

(A) Transfer function.

(B) Response time for a slowly changing input.

Fig. 1-14. A-series vs. B-series.

Generally, the B-series is the better choice by far and should be used, particularly if you are using resistor-capacitor pulsing networks or are working at lower voltages or higher speeds. One exception where you might still use the A-series is in interfaces. We'll see later how you can double and quadruple the available current-sinking output of a 4001 or a 4002 simply by paralleling inputs on an A-series device. Since B-series devices have constant source and sink currents, this dodge won't work.

Although the 4000 series can be powered using a negative supply voltage ($-V_{SS}$) and a positive supply voltage ($+V_{DD}$), the popular convention of positive logic is to assign logic 0 to ground ($V_{SS} = 0$). The recommended positive supply voltage can range anywhere from +3 to +18 volts. The minimum voltage for a logic 1 at the input to a 4000-series device is 70% V_{DD}, while the maximum logic 0 level is 30% V_{DD}. In general, the 4000 series are not pin-for-pin compatible with the equivalent TTL functions. As discussed later in this chapter, any 4000-series device has to be properly interfaced with a 7400-series TTL device.

The 4000 series of CMOS devices are listed using the numbering system introduced by RCA. Other manufacturers generally follow the same numbering system, but may use different prefixes. For example, Motorola's CMOS devices belong to the *MC14000* series, while those manufactured by Fairchild Semiconductor belong to the 34000 series.

74C00 Series

The 74C00 series was developed to be pin-for-pin compatible with the 7400 and 74LS00 series. This allowed users of TTL devices to make the transition from TTL to CMOS without having to become familiar with an entirely new numbering system. This series has a definite low-power advantage over the standard TTL family, but it is much slower. Although the 74C00 series can be powered using any supply voltage from +3 to +15 volts, it can be directly substituted for 74LS00 (Low-power Schottky) devices using a +5-volt supply. In this case, 74C devices can typically drive two 74LS00 loads.

74HC/HCT00 and 74HC/74HCT4000 Series

Both the 74HC00 (High-speed CMOS) and the 74HCT00 (High-speed CMOS, TTL compatible) families offer significant improvements over the older 74C00 family. Both are as fast as the 74LS series and consume less power, depending on the operating frequency. Like the 74C series, the 74HC/HCT series are pin-for-pin compatible, in addition to being input/output voltage-level compatible with the TTL family when operated from a +5-volt supply. The 74HC/74HCT4000 series is identical with the original 4000-series devices, but are TTL voltage-level compatible when using a +5-volt supply.

INPUT PROTECTION

The inputs to MOS transistors are essentially open circuits. But, at the same time, the dielectric insulator on the gate capacitor is extremely thin. If static electricity is allowed to reach the gate dielectric, extreme field strengths can result, and the gate can be permanently and instantly destroyed. Without some kind of protection network, the static generated by scuffing across a carpet or by pushing a device into a styrofoam carrier generates more than enough voltage to destroy the circuit.

To get around these problems, all CMOS integrated circuits have input-gate protection built in, along with other forms of static protection. The particular protection arrangement varies with the device and manufacturer. Four typical methods are shown in Fig. 1-15.

In Fig. 1-15A, diodes are placed from input to +V and from input to ground. One or the other of the diodes conducts if the input goes above +V or below ground. In Fig. 1-15B, a single zener diode to ground is used. This diode conducts directly for below-ground voltages and conducts as a zener diode for positive voltages above 30 volts or so. This allows us to apply gate voltages more positive than the supply voltage, and is handy for translating from a 10- or 15-volt

CMOS system to a 5-volt CMOS and TTL system. The 4049 and 4050 use this type of input network.

In Fig. 1-15C, we use a zener connected to +V. This diode conducts in the forward direction for voltages above +V and conducts as a zener diode for voltages 50 volts or so negative with respect to +V. Thus, input gate voltages below ground are allowed with this combination. Many Motorola and a few National Semiconductor circuits use this method.

In Fig. 1-15D, a 200-ohm resistor is in series with the input, followed by protection diodes to +V and ground. Input voltages can range from very slightly above +V to very slightly below ground. Many Fairchild Semiconductor devices use this combination.

Any and all of these methods give good *in-circuit* protection against static problems. They also will give you good out-of-circuit protection, provided you are reasonably careful with your handling of the devices. Most lists of rules for handling MOS devices are overdone and introduce more problems than they solve. To take care of MOS circuits, keep them in *conductive* foam or metal carriers before use, and solder them in place with an ordinary small soldering iron. Make sure your circuits have some sort of input resistor for MOS leads going off the pc board. And that's about all the precautions you'll really need.

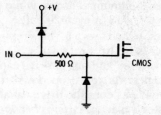

(A) Double diode. Input must stay between +V and ground.

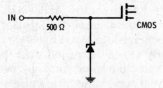

(B) Zener to ground. Input must stay between +30 volts and ground.

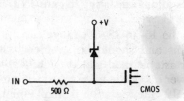

(C) Zener to +V. Input must stay between +V and −20 volts.

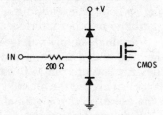

(D) Improved double diode. Input must stay between +V and ground.

Fig. 1-15. Internal CMOS static protection circuits vary with device and manufacturer. Current-limiting to 10 mA or less must be provided if input voltages exceed ranges shown.

The protection networks do a fine job against static, but they introduce a new problem—you now have to protect the protection against *current* overloads.

It is extremely important to always limit the input current through any protection diode to 10 milliamperes or less. Ideally, you should keep your input swings between ground and +V so that these diodes never conduct. Some interface circuits, and some combinations of powered inputs and unpowered CMOS, can forward-bias these diodes; therefore, the current must be limited.

These protection diodes can also do strange things to the astable, monostable, and pulse circuits described in Chapter 4. The zener circuits (Figs. 1-15B and C) in particular can give you a one-way "dc restorer" that will wildly change your RC time constants and that can introduce supply and temperature variations, along with recovery problems. Often a series input resistor or an extra diode will minimize this problem.

POWER SUPPLIES

CMOS is inherently a low-power logic family, and its power needs are far simpler and easier to live with than most other logic families. Most CMOS devices work over a 3- to 15-volt range and are reasonably forgiving of voltage ripple and poor regulation. Some watch circuits operate on even lower minimum voltages, while some B-series devices will work properly up to 13 or even 20 volts.

Except for the 74HC/74HCT series, which uses a 5-volt supply, the *optimum* supply voltage for CMOS is 9, 10, or 12 volts. At these voltage levels, you'll get very fast operation and lots of drive capability along with good noise performance. As you drop down to +5 volts, the speed drops by over one half, as does the drive capability and noise immunity. Pulse and resistor-capacitor circuits become less well defined and more voltage and temperature dependent. So *while CMOS does work on a 5-volt supply, it is far better to operate at a higher voltage*, even if this means some translating to gain TTL compatibility or other interface.

Very high supply voltages in the 15- to 18-volt range can lead to heating and dissipation problems and should be avoided. CMOS devices that are run in linear or astable modes at these higher voltages can exceed the allowed power dissipation if you are not careful. The extra speed you gain above 12-volt operation is small compared with the major improvement in performance you get when you go from 5 to 12 volts.

How much current do we need? As a general rule, practically all your supply current will go to external things like LED displays, interfaces to TTL, and other outside-world connections. And practically all the remaining current will be used by one or two stages that are running faster than the rest of the system. So, the first question we have to ask is how much current does everything in the circuit *but* the CMOS need? Then we add the CMOS current to it.

Fig. 1-16 shows how a CMOS *gate* varies its supply current with frequency. Except for micropower systems (powered by hearing-aid batteries, etc.), the current is utterly negligible at audio and sub-audio frequencies. It rises in proportion to frequency. When a CMOS device gets up in the 2- to 5-megahertz region, it draws about as much current as low-power Schottky TTL, and a single gate by itself offers no power savings.

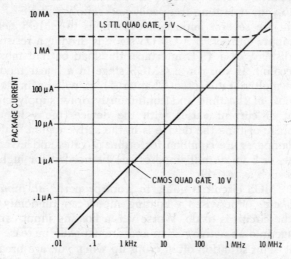

Fig. 1-16. CMOS operating current versus frequency.

Note that this frequency "break-even" point between CMOS and LS TTL applies only to single gates driving identical off-chip loads. In the real world, when we start using MSI and LSI devices, CMOS will offer a 10:1 to 100:1 power advantage over LS TTL for operating frequencies in the 2- to 5-megahertz region. There are two reasons for this. The first is that TTL gates draw power whether or not they change, while CMOS gates draw power only while changing. In a MSI package, only a few of the gates normally change at the input-frequency rate. This is particularly true of counters and arithmetic units. Secondly, at these frequencies, the CMOS current is proportional to the load capacitance, while the LS TTL is not. Since most of the CMOS gates drive very low internal capacitances, they do not need nearly the current that the CMOS output devices do.

23

CMOS supply current doubles with doubling supply voltage, so you will need twice the current and *four* times the power at 10 volts that you need at 5.

Chapter 2 gives the supply currents needed at 1 megahertz for many CMOS devices. The currents are specified at both 5 and 10 volts. For instance, a 4001 at 1 megahertz with a 10-volt supply needs 0.8 milliampere, or 800 microamperes, of supply current. The same package needs 80 microamperes at 100 kHz, 8 microamperes at 10 kHz, and only 0.8 microampere at 1 kHz. So, for most slow CMOS systems, practically all of the supply current goes to the one or two circuits that are continuously running fast. You can calculate your system current needs by using the currents in Chapter 2 and scaling them to your operating frequency and supply voltage.

There are several important exceptions to CMOS needing very little supply current. If a CMOS stage is driving a resistor load, an LED display, or a TTL interface, this load current must be taken into account. If you run a CMOS stage in a linear mode or as an astable multivibrator (see Chapter 4), it will be in its active region most of the time and continuously draw supply current. The amount of current varies with the device, the voltage, and the percentage of time the device is in the active region. Values of 1 to 3 milliamperes are common for ordinary gates and inverters, while currents of 5 to 20 milliamperes can be needed for high-power inverters such as a 4049.

If a CMOS circuit is going to work properly, *all inputs must go somewhere*. Otherwise, a floating input can randomly determine what the circuit is to do. Worse yet, a floating input can often put the stage into an active region and dramatically increase the supply current. This situation often crops up when you are breadboarding circuits. You need an inverter, so you put a 4049 or 74C04 hex inverter into your circuit and carefully use one stage. Then, for some reason, your supply current skyrockets beyond a hundred milliamperes. Why? Because all five unused and unconnected inverters bias themselves into linear operation. The cure is simple. Be sure to disable *all* unused CMOS stages when you *first* put them into your circuit.

If your CMOS inputs only go near ground and near +V instead of directly to ground and to +V, your circuits will probably still work. There will be a slight loss of noise immunity. In addition, your supposedly "off" transistors inside the CMOS package will be slightly conductive, and this will raise your supply-current needs somewhat. This situation often crops up in interface circuits, where a CMOS stage drives an outside load or a current-sourcing TTL input. Because of the output current, the output voltage swing can't get clear to ground or clear to +V. Any other CMOS stage inputs connected to this point will be biased slightly active (depending on the internal logic) and will tend to raise your supply current.

So, if micropower operation is absolutely essential, do not use CMOS in a linear mode. Don't use resistive output loads. Keep all unused gates out of their linear region. Minimize capacitive loading of all outputs. And make sure all your input-logic swings go completely from ground to +V.

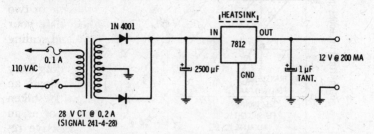

Fig. 1-17. Line-operated CMOS power supply. Unregulated supplies may also be used.

Fig. 1-17 shows a general-purpose regulated 12-volt power supply that gives up to 200 milliamperes of current and is more than adequate for most larger CMOS systems. For the 74HC/74HCT series of devices, a 7805 5-volt regulator is used instead, along with a 12-V CT transformer. Fig. 1-18 shows a typical battery supply. You can estimate battery life by using the graph of Fig. 1-19 for various sizes and styles of batteries.

For CMOS-only breadboards, ordinary 9-volt transistor radio batteries are ideal. If you're into displays or other higher-current loads, use four or six series-connected C- or D-size flashlight cells.

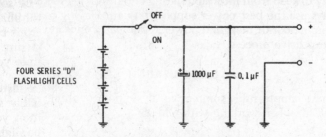

Fig. 1-18. CMOS battery power supply.

Some power-supply bypassing and decoupling are needed, particularly with battery supplies. A large electrolytic capacitor or a good regulator should serve the entire circuit, while a 0.1-μF bypass capacitor should be used for every six packages or so. Even though CMOS is very forgiving about sloppy power-supply design, it pays

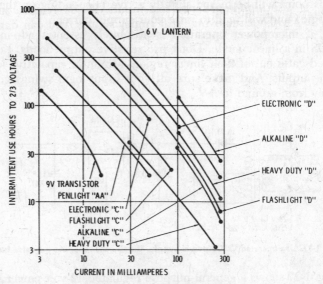

Fig. 1-19. Estimating battery life.

to use the best power supply you can. Sometimes the non-CMOS components in your circuit will demand much more in the way of a good power supply than your CMOS will. For instance, if you are interfacing TTL, you will need a tightly regulated +5-volt supply, low-impedance (wide foil) supply runs, and lots of despiking capacitors. If your CMOS is generating musical notes (see Chapter 10), chances are you will need a hum- and noise-free regulated power supply to keep from intermodulating and distorting the output notes. Always use the best power supply you can. Having to go back and beef up a poor or marginal power supply later on is always a painful and expensive process.

CMOS USAGE RULES

These simple rules summarize the important things you have to watch for when working with CMOS:

- **All inputs must go somewhere.** A solid connection to an input signal, to +V, or to ground *must* be provided for *all* inputs, particularly when breadboarding.

- **Protect the protection.** Avoid ever having the input-protection diodes conduct. If you must use diode current, limit the current to 10 milliamperes or less. Watch for the effects of diodes on time constants and other shaping circuits.

- **Use high-impedance test inputs.** If you remove supply power without removing "stiff" input signals, you can damage the input protection or latch up the package.

- **Avoid static during handling.** Store the ICs in conductive foam or metal carriers. Leave them on foil or on a cookie sheet during benchwork. Don't use a soldering gun. Make sure any inputs going off board have a load resistor (1 megohm) across them.

- **Condition all mechanical inputs going to clocked logic.** Push buttons, switches, and keyboard contacts must be debounced with contact conditioning (Chapter 4) to make them noise- and bounce-free.

- **Use fast rise and fall clocks.** The rise and fall times on the clock input of clocked logic blocks must be faster than 5 microseconds. Otherwise, erratic operation caused by *clock skew* can result.

Failing to follow one of these rules can cause subtle errors which can really get to you. For instance, an unconnected and unused input lead can have a time constant of half an hour or more. The first time you try it, things work perfectly. Later on, response is erratic or just plain wrong. Unconnected pins can also "track" neighboring pins most of the time, but not all of the time. A CMOS package with its supply-power pin disconnected (this can happen if the pin bends and misses the socket or connector) may seem to work perfectly, only to turn out slow and limited in drive power. Why? Because the chip is getting its supply power from the *inputs* through the protection diodes. As long as at least one input on the package is high, the chip gets some supply power. Watch these usage details very carefully, particularly when you are breadboarding.

MOUNTING AND BREADBOARDING

CMOS is far and away the most forgiving logic family for sloppy breadboards, improper power supplies, and rat's-nest wiring. Nevertheless, it pays to keep your breadboard and development circuits as orderly and as neat as possible, if for no other reason than that it makes them easier to trace, troubleshoot, and verify. A neatly organized circuit is also easier to expand and presents far fewer problems in going to a final design.

There are several important things to watch for in any breadboarding or testing technique. The most important of these is to *continuously* keep a record of what you have. You can do this with schematics, log books, wiring lists, or whatever works best for you. Be sure to save records of older "nonworking" circuits, for more often than not, you'll probably want to refer back to them. And don't *ever* take apart an "I-won't-need-this-anymore" circuit without first carefully making some sort of record of it.

Another important thing to do is to pick a breadboard or testing method that is much larger than you think you're going to need. Most circuits have a way of using up 120% of the available space and 200% of the available power-supply current. Provisions for resistors, capacitors, and adjustable controls in particular are almost always far underdone, as are mountings for switches and connectors.

Solderless Blocks

Plastic-block solderless breadboarding systems (Fig. 1-20) are ideal for preliminary testing of a CMOS circuit, particularly smaller circuits using ten ICs or less. The integrated circuits snap in place without any solder, and other components and wire leads are easily added. A medium-sized bare board costs under $20 and is usually well worth the price. Everything from small blocks up to complete and elaborate test stations that include power supplies and test generators is available. Chart 1-3 shows some of the more popular sources.

There are one or two tricks that make these breadboards much easier to work with. The first of these is to have precut and color-

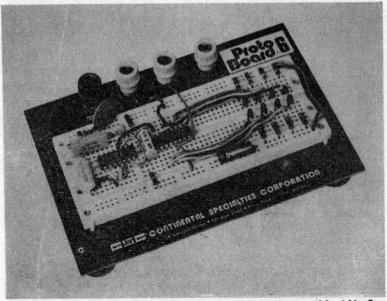

Courtesy Continental Specialties Corp.

Fig. 1-20. Solderless breadboarding system.

coded snap-in wires to go with the system. You can make these yourself, or you can use the wire-kit sources also shown in Chart 1-3. A second trick is to get some sort of pins-and-header combination that lets you get input and output wires and, in particular, trimmers onto and off of the board. All you really need is something with a lot of pins on 0.1-inch centers that is solidly held together by some sort of spacer. Electronics surplus stores often have something you can

Chart 1-3. Sources of Breadboarding Systems

AP Products/3M 1359 West Jackson St. Painesville, OH 44077	Heath Dept. 180 Benton Harbor, MI 49022
Continental Specialties 44 Kendall Street New Haven, CT 06509	Hewlett Packard 1501 Page Mill Road Palo Alto, CA 94304
E & L Instruments 61 First Street Derby, CT 06418	Solid State Systems Box 617 Columbia, MD 65201

Sources of Jumper Wire Kits

Fancort Industries 2436 McDonald Avenue Brooklyn, NY 11223	Squires Electronics 8900 S.W. Burnham Road Tigard, OR 97223

carve up for about a nickel each. You then cut these headers to the lengths you need to mount pots, supply leads, input cables, and so on. If you add a Velcro strip (that fuzzy zipper stuff) to the back of these blocks, several projects or several students can share a common power-supply base or test module.

Wire Wrapping

Wire wrapping is a premium interconnect method that works well whenever you need only a few copies of a complex circuit, or whenever what you are building is going to need lots of field changes or will be customized for different users. Wire wrapping is best used in digital systems that have a minimum number of controls, discrete components, and related analog circuitry.

In a typical wire-wrapping system, long pins extend back from wire-wrapping sockets or panels, and stripped wires are twisted eight times or so around these pins. The pins are usually .025 inch square and long enough for at least three connections. Special tools are used for wrapping, and the wrapping sequence usually follows a wiring list or table rather than a schematic.

Most wire wrapping is expensive. A five-bay Augat panel will run $300, and $100 is normal for a battery-powered Gardner-Denver wrapping gun. Nevertheless, when only a few systems of premium quality have to be delivered, particularly in a short time, wire wrapping is often the cheapest possible route.

Some reasonably priced wire-wrapping components and tools are available. Cambion and Vector are two sources, and bargain pre-wrapped panels are also available as surplus items for as little as a dollar each. It's a simple matter to remove all the old wires and reuse the panels.

Sockets

There are all kinds of sockets available for mounting ICs to printed-circuit boards, for hand-wiring breadboards, and for wire-wrapping systems, and the prices of many sockets are at realistic (1¢ per contact) levels. Today, sockets are strongly recommended, particularly for experimental setups from which you will want to recover the ICs later on, and for hobby kits that may be constructed by persons with limited ability. A disadvantage of the cheaper, "low-profile" sockets is that it may be hard to use a glomper clip, scope probe, or other test lead with them.

The Molex Soldercon (Fig. 1-21) is an interesting socket substitute. Soldercons consist of a strip of socket pins on a carrier. They are soldered in place on a pc board, and then the carrier is broken

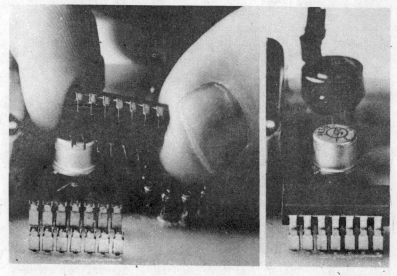

Courtesy Molex Inc.

Fig. 1-21. The Soldercon system is an economical socket alternative.

off. This leaves you with a series of isolated contacts that hold the IC. Their cost is less than a penny per contact. The strips are available from many different mail-order sources. Optional nylon "nests" are also available to protect the pins and make them look like real sockets.

Soldercons are ideal for many pc-based test and breadboard systems, but there are a few places where they should not be used. These include any system in which the ICs are to be changed very often, and any double-sided pc system that does not use plated-through holes, particularly if it has a through-the-pins lead routing. Be very careful when inserting and removing ICs from Soldercons, for it is very easy to bend a pin over or have the pin make intermittent contact on the outside or inside of the Soldercon. Also, make sure that an IC is in place when you do any soldering because the pins by themselves can rotate or fall loose when new connections are being made.

Wiring Pencils

Wiring pencils (Fig. 1-22), such as the Vector P173, offer a compact, fast, and supercheap way to breadboard circuits. The concept of the system is simple. Fine-gauge wire with heat-meltable insulation is used to make point-to-point connections. You don't have to prestrip the wire since ordinary soldering temperatures are high

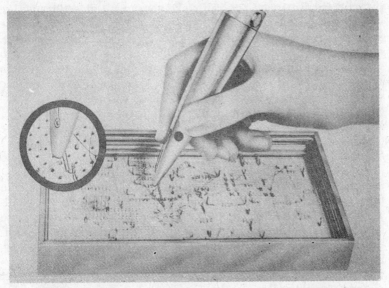

Courtesy Vector Electronics Co.

Fig. 1-22. Vector P-173 wiring pencil.

31

enough to melt through the insulation. Better yet, you can use a single wire for multiple or common connections without having to cut and strip each individual connection. Wiring pencils are especially attractive for modification and correction of pc boards. Several colors and several gauges of wire are available, as are support systems that include sockets, guides, and terminals.

The wiring pencils cost from $5 to $10, or you can make your own version for even less. They will work with just about any breadboarding system that can accept soldered point-to-point wiring. Some of the workable systems include the Vector perforated pc panels and terminals, Circuit-Stik copper IC patterns, self-stick modules such as the Christansen or Calectro Mini-Mounts, rubber-stamped pc resist patterns like the Stamp-It-Etch-It system, and special pc pad-isolating mini-drills such as those by AFS Precision Tool Systems.

PC Boards

Printed circuits are usually the simplest and most economical way to mount and use ICs. There is quite a bit of initial overhead and time involved in artwork generation and so on, so pc boards are best used for high-volume applications.

One type of printed circuit is the "universal" layout. Here, a dozen or more IC patterns already have their supply, ground, and bypassing runs provided for you. Premium units often use double-sided, plated-through boards. You can make your own connections with a wiring pencil or other point-to-point method. These are quite usable as breadboards and may be reused, but they do tend to be expensive.

Many of the electronics hobby magazines offer pc boards to go with their construction projects. These are generally a real bargain since all the layout and testing costs have been amortized and since your board is usually ready to go as soon as you get it.

You can build your own pc boards by using any of a number of kits and techniques. Three popular printed-circuit techniques are the *direct* method, the *photographic* method, and the *silk-screen* method.

With the direct method, you put resist inks and tape-and-dot combinations directly on the pc board. You then etch the board in ferric chloride, which eats away all the unprotected copper. The direct method takes a long time and is limited in detail. It is prone to undercutting, particularly at the pads, and to thumbprints. In addition, an *individual* layout is needed for each and every board copy. Besides the usual tape-and-dot systems (Bishop, Brady, Datak, Circuit Aides, etc.), there are several products that make the direct method somewhat less of a hassle. These include rubber stamps in the shape of integrated-circuit "footprints" and other patterns, and transfer emulsion systems like the Calectro Lift-It that will transfer a *printed* 1:1 layout pattern onto your board. The ordinary Sharpie "marks on anything" pens at office supply stores are an economical resist source.

The photographic method gives you far more control and is generally much better. You start by generating a 1:1 artwork negative. This is usually done by placing double-sized (2:1) tape, dots, callouts, and IC patterns on a gridded Mylar® sheet. This master artwork is then taken to a photolithographer and a reduced (1:1) negative is generated. Cost of the negative is a few dollars. Almost any printer has or knows of a litho camera if you can't find a local litho service.

After you have the 1:1 negative, you can make as many printed-circuit boards as you want. You take a piece of pc material and chemically clean it by scouring it with a bleach-type cleanser. Then you spray on a thin layer of photoresist, such as Dynachem DCR-3140A. Next, the board is *gently* heated to 120 °F for 15 minutes to drive off any solvent vapors and is allowed to cool in a dark place. You then put the negative and the board together, emulsion to emulsion, and insert them in a contact printer or place them under weighted glass. The board is now exposed to ultraviolet light, either from the direct sun (5 minutes) or a quartz photoflood lamp (3 minutes).

Next, the board is put in a developer solution where it is gently sloshed around for a minute. Developers with a built-in dye are best since you can see the results, but they are extremely messy to work with. Incidentally, both the resist and the dye are soluble in lacquer thinner, and a mixture of a lacquer thinner and sand is handy for cleanup, particularly if you're late in getting to this detail.

The best etchant to use is ammonium persulfate, heated to 110 °F. You support the board upside down in the solution and gently shake or rock the Pyrex dish or whatever you are holding the etchant in. You can speed things up by adding bubbles to the solution from an aquarium pump. After the etching is completed, scour the board with cleanser or steel wool to take off the remaining resist.

With the silk-screen method, you convert the 1:1 negative into a positive by using litho contact-printing "PMT" materials or Scotch Color-Key.

A silk screen is cut using a photo process somewhat like the photographic method. Silk-screening a pc board is much cheaper and much faster than the photographic method, but you will have a somewhat less resolution and extra front-end expenses.

The screen is used to directly print resist ink onto the pc material, which is ready to etch as soon as the resist dries. Screens also let you put trademarks, component locations, and connector pinouts on the topside of your board. The break-even point between screens and the photographic method is somewhere around ten boards; for more boards than this, or where callouts are extra important, the silk-screen method is far quicker and cheaper. The screen costs $15 to $30.

If you do your own layouts and pc work, there are some important things that can save you time and effort. Cleanliness and attention to detail are essential in any photo or etch process.

It is extremely important that you have a complete schematic and parts list before you start any layout. Be sure there is enough room around your ICs to use a glomper clip or other test connection. Your ICs should be arranged orderly, and should all point in the same direction. Components should be logically grouped where possible. A minimum line width of 25 mils, with between-the-lines spacing of 25 mils (1:1), is about as close as you want to get.

If at all possible, stick with a single-sided board layout. Besides being far faster and cheaper, single-sided boards look better where component-overlay topside lettering is used and the production yields are much higher, particularly with home or school-lab etching setups. Jumpers are one way to get around the interconnect problem. A neater trick is to use Eldre or Rodgers Buss-Strip or your own double-sided pc bus strips to tie down common supply and other runs and get them off the board foil (Fig. 1-23).

Double-sided pc boards without plated-through holes result in alignment problems and an extremely difficult assembly. The ICs have to be soldered on both the top and bottom sides when double-sided boards are used, making missed connections extra easy. Most Soldercons and sockets can't be used with double-sided boards without plate-through holes, although you can sometimes use wire-wrapping sockets placed slightly off the board. "Through-the-pins" lead routing is an unmitigated disaster if hand soldering has to be done on both sides of the board. Premium industrial pc processes with plated-through holes eliminate many of these problems but require quite a bit in the way of special equipment and add much to the board cost.

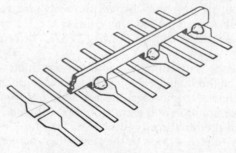

Fig. 1-23. Bus strips of double-sided pc material are a good way to multiply the connections available on a single-sided pc layout.

TESTING AND MONITORING STATES

The process of troubleshooting digital circuits is usually called *debugging* or *state checking*. When state checking, you monitor the output states of each logic block. This can be done either *statically* on a one-state-at-a-time basis, or *dynamically* in which the system runs at normal speeds.

One nearly essential debugging tool is a *glomper clip* (Fig. 1-24). This device is a snap-on or lock-on clip that brings all 14 or all 16 pins out for convenient testing and attaching of leads and scope probes. Glomper clips are available from API, Continental Specialties, Guest International, Micro Electronic Systems, Pomona, and several others. Cost is typically $5.

Two *state monitors* are shown in Figs. 1-25 and 1-26. The *logic pen* in Fig. 1-25 lights its light-emitting diode (LED) for a one and leaves it off for a zero. Internal monostable circuits can also be added to stretch out brief pulses long enough to be seen. A similar circuit repeated 16 times and built onto a glomper clip forms the state monitor in Fig. 1-26. These monitors can include output test pins, masks for individual circuits, automatic supply and ground location, and various other features. They are available from a number of sources with prices ranging from $20 to $120.

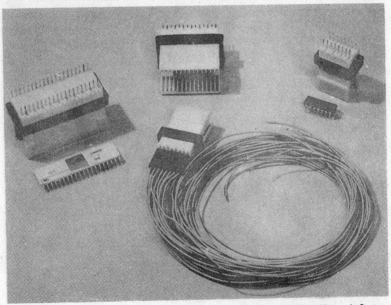

Courtesy Micro Electronic Systems

Fig. 1-24. Glomper clips ease testing and debugging.

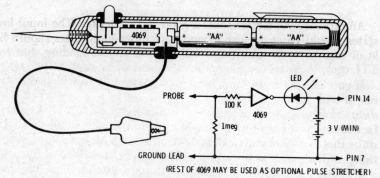

Fig. 1-25. CMOS logic pen.

A *logic injector* is the opposite of a state monitor and compares roughly to the signal injector used for radio troubleshooting. Logic injectors provide a way of force-feeding a one or a zero pulse onto a logic input. The pulse has a very low impedance and completely swamps whatever is normally driving the point under test. The pulse width is kept short enough (usually a few microseconds) that no damage is done by the force-feeding.

Logic injectors can be built with TTL-style buffers or analog line drivers, and are usually made to automatically pulse a normally high line in the low direction and vice versa. Hewlett Packard is one commercial source for logic injectors.

A *bounceless push button* (see Chapter 4) is a switch or push button that has *contact conditioning* added to it so that its ultimate output is noise free and has one, and only one, pulse per switch action. These devices are essential in testing any clocked CMOS logic where contact noise and bounce can be read as apparently valid input signals.

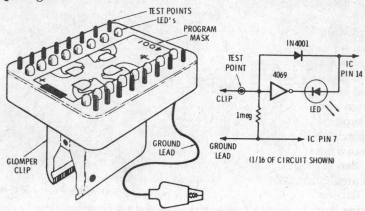

Fig. 1-26. CMOS state monitor.

An ordinary voltmeter makes a good static checker. The input-low states should be at or near ground, and input-high states should be at or near +V. Output states will depend on the loading, but for TTL compatibility, the output-low state must be less than half a volt.

If you use a voltmeter on a dynamic or rapidly changing logic signal, it will respond to the average value of the signal and reflect its *duty cycle*. This is the principle behind the operation of analog tachometers and frequency meters. For instance, a binary-divider stage that is running fast will be up half the time and down half the time, and the voltmeter will read one half the supply voltage, or +V/2. Many logic pens and state monitors will also light to half brightness on a signal with a 50-50 duty cycle.

A good oscilloscope is almost essential for full dynamic logic testing. The scope MUST have triggered sweep and at least a 5-mega·hertz bandwidth. Ideally, it should also have such features as two channels, a vertical delay, a sweep delay, and a 50-megahertz bandwidth. Low-cost units are available from Heath, Hewlett Packard, Tektronix, Telequipment, and Phillips.

INTERFACE

Interface is the art of interconnecting CMOS logic to other logic systems and to the outside world. The key to any interface scheme is to make the ones and zeros look like they "belong" to the systems being interfaced.

At the input to CMOS, this means we want a voltage very near ground for a zero and near +V for a one. Recommended limits are below 1 volt and above 4 volts for a 5-volt system, and below 2 volts and above 8 volts for a 10-volt system. Since CMOS inputs are essentially an open circuit, we don't need much in the way of drive current. But, we do have to make the source impedance low enough that we can reliably charge and discharge the stray and input capacitances at the highest speed of interest. Somewhere along the way, we will probably have to sharpen the rise time and provide debouncing and noise immunity to slowly changing input signals, but this is most often done by the first CMOS stage, following the techniques in Chapter 4. If we are using clocked CMOS, its logic inputs must be bounceless and noise free, and the rise and fall times must be better than 5 microseconds. A final restriction on our CMOS input interface is that it must protect the protection by providing current limiting at any input voltages above +V or below ground.

On the output side, CMOS can actually provide quite a bit of drive current. Most CMOS stages are short-circuit–protected at the output, except for large inverters and high supply voltages. Once

37

again, "force-fed" output voltages below ground and above +V must be current-limited.

There are two things a CMOS output can't do: it can't provide a lot of current and still be within the half a volt of ground that TTL needs for a logic zero, and it can't drive much capacitance without slowing down a great deal. So if we are going to interface TTL, we have to be careful just how we do it. And, if we are going to run at any speed at all, we have to somehow isolate our CMOS outputs from heavy capacitive loading.

Fig. 1-27 shows the typical output current capability of a B-series CMOS gate. With the B-series, the source and sink currents are just

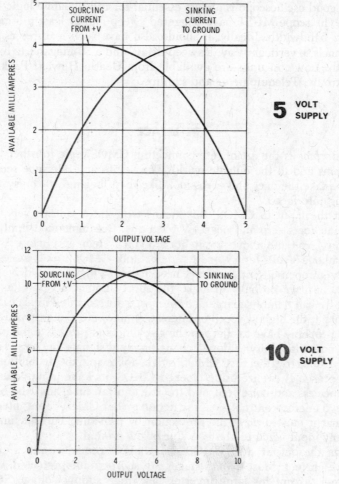

Fig. 1-27. Typical B-series output characteristics.

Chart 1-4. Typical TTL and CMOS Voltage and Current Levels for a 5-Volt Supply

Family	Input Levels		Output Levels		Input Levels		Output Levels	
	V_{IL} max	V_{IH} min	V_{OL} max	V_{OH} min	I_{IL} max	I_{IH} min	I_{OL} max	I_{OH} min
7400	0.8V	2.0V	0.4 V	2.4 V	1.6mA	40.0µA	16.0mA	0.4mA
74LS00	0.8V	2.0V	0.4 V	2.7 V	0.4mA	20.0µA	8.0mA	0.4mA
4000	1.5V	3.5V	0.05V	4.95V	1.0µA	1.0µA	0.5mA	0.5mA
74C00	1.5V	3.5V	0.5 V	4.5 V	1.0µA	1.0µA	0.4mA	0.4mA
74HC00	1.0V	3.5V	0.1 V	4.9 V	1.0µA	1.0µA	4.0mA	4.0mA
74HCT00	0.8V	2.0V	0.1 V	4.9 V	1.0µA	1.0µA	4.0mA	4.0mA

Note:
V_{IL} max — maximum input voltage necessary to produce a logic 0 (low).
V_{IH} min — minimum input voltage necessary to produce a logic 1 (high).
V_{OL} max — maximum output voltage necessary to produce a logic 0 (low).
V_{OH} max — maximum output voltage necessary to produce a logic 1 (high).
I_{IL} max — maximum input current drain at a logic 0 (low).
I_{IH} min — minimum input current drain at a logic 1 (high).
I_{OL} max — maximum output current available at a logic 0 (low).
I_{OH} max — maximum output current available at a logic 1 (high).

about the same. With a 5-volt supply, we can get up to 4 milliamperes of output current, sourced from +5 volts or sunk to ground. At 10 volts of supply, we can get over 11 milliamperes in either direction. Higher supply voltages give us still more current, but we have to watch for dissipation limits. High output currents can also shorten IC life due to the limited size of some internal metallization paths.

Most problems arise when trying to interface either CMOS with TTL or one CMOS family with another. With the many CMOS and TTL families currently available, careful attention must be paid to the specific voltage levels and current requirements that correspond to logic 0 and 1 levels. Chart 1-4 summarizes the typical voltage and current levels for the 4000, 74C00, 74HC00, and 74HCT00 CMOS families, as well as the 7400 and 74LS00 TTL families operating from a 5-volt supply.

CMOS and TTL

Fig. 1-28 shows how we gain CMOS and TTL compatibility. When any CMOS family runs on the same +5-volt and ground supply as TTL (Fig. 1-28A), any 7400 or 74LS00 gate with a 2.2K pull-up resistor will drive any number of CMOS loads. The pull-up resistor is needed because the minimum output voltage from either a 7400 or 74LS00 gate that corresponds to a logic 1 (V_{OH} min) is less than the minimum input voltage for a logic 1 (V_{IH} min) for any CMOS load. The exception is that no pull-up resistor is needed when TTL is driving a 74HCT00 load. When driving TTL loads from CMOS, any 4000B or 74C00 CMOS gate will drive any *one* 74LS00 gate (because of input and output current requirements), but any

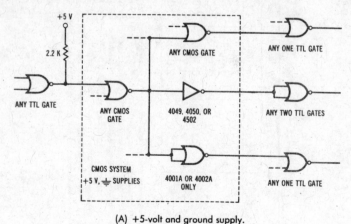

(A) +5-volt and ground supply.

Fig. 1-28. CMOS

40

74HC00 or 74HCT00 CMOS gate will drive either *two* 7400 or *ten* 74LS00 loads. The "artillery" CMOS buffers, the 4049, 4050, and 4502, can be used to drive two 7400 loads or six 74LS00 loads. We can also use 4000A-series CMOS gates and parallel transistors, doubling or quadrupling the current. Two 4001A (2-input NOR gate) in-

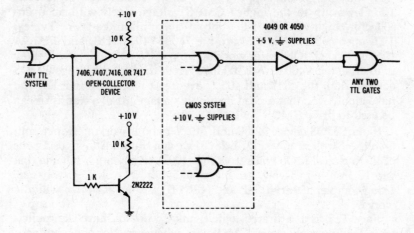

(B) +10-volt and ground supply.

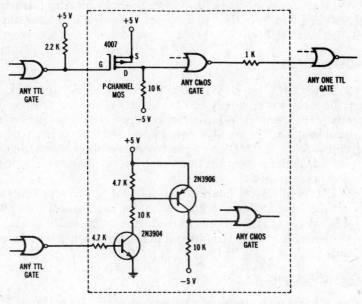

(C) +5-volt and −5-volt supply.

and TTL interface.

puts connected together will typically drive *one* 7400 TTL load. All four 4002A (4-input NOR gate) inputs tied together will drive one worst-case TTL load over a wide temperature range.

Our techniques change slightly if the CMOS device is working from a higher supply voltage. For example, if we are using a +10-volt and ground supply (Fig. 1-28B), we have to translate the TTL levels up to the higher CMOS voltage levels with an open-collector, high-voltage, TTL output and pull-up resistor. Typical TTL open-collector devices are the 7406, 7407, 7416, and 7417. A 4049 or 4050 CMOS buffer working from a *+5-volt* supply will allow us to drive *two* 7400 or 74LS00 loads from the higher-voltage CMOS devices. Both the 4049 and 4050 have input protection that lets you have input levels that exceed its supply voltage. Also, a transistor can be used to drive CMOS devices.

If our CMOS device (4000 and 74C00 only) is working from a split 5-volt, −5-volt supply (Fig. 1-28C), we can translate down using one of the p-chanel MOS transistors in a 4007, or any other scheme that gets us from a +5-volt, ground signal to a +5-volt, −5-volt one. Using complementary (PNP and NPN) transistors as shown will also work.

Since TTL is a saturated-logic family, it doesn't have the capacitance loading problems of CMOS. So, whenever a lot of capacitance (lots of inputs, long lines, off-board connections, etc.) must be driven, it often pays to use a low-power Schottky inverter (such as a 74LS04) as a CMOS buffer. Extra supply power will be needed for this high drive ability.

CMOS to CMOS

CMOS to CMOS interface is usually trivial. CMOS gates working from the same power supply can be directly connected together in essentially any quantity. A fan-out of less than 50 is recommended as the problem here is usually capacitance and not current drive capability. So, with CMOS, you can ignore completely any fan-in or fan-out calculations; they just don't exist for anything reasonable in the way of system size.

For CMOS to CMOS interface on different power supplies, you can use the same methods of Fig. 1-28, using a p-channel transistor to get below the negative supply of the higher-voltage gate and a 1K output resistor to protect the internal input diodes when translating upward.

CMOS to Other Logic

CMOS to other logic is usually easier than the TTL interfaces. The older RTL and DTL families are pretty much directly compatible, although you may not be able to simultaneously drive an RTL gate and a CMOS gate from the same CMOS output. PMOS and NMOS devices are usually easy to work with, particularly if they are already TTL compatible. Sometimes a pull-up or pull-down resistor is needed. ECL (emitter-coupled logic) to CMOS and back again is done with commercial translator integrated circuits.

NMOS devices, such as RAMs, can be operated at the same positive voltage levels as CMOS, and the NMOS inputs are CMOS compatible. Fig. 1-29A shows how CMOS interfaces with a 2102 NMOS 1K static RAM. The Data Output (DO) requires only a single 2.2K pull-up resistor. All address lines $(A_0 - A_7)$, and the Data Input (DI), Read/Write (R/W), and $\overline{\text{Chip Enable}}$ $(\overline{\text{CE}})$ inputs require no inter-

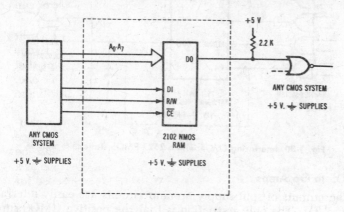

(A) Using a 2102 RAM.

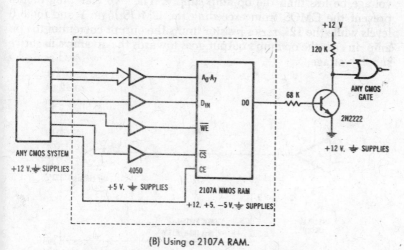

(B) Using a 2107A RAM.

Fig. 1-29. Interfacing CMOS with an NMOS RAM.

43

facing. NMOS devices, such as the 2107A 4K dynamic RAM, use positive and negative supplies, and are interfaced with CMOS using both 4050 and transistor buffers (Fig. 1-29B).

PMOS devices generally operate with a single positive ($+V_{DD}$) and negative ($-V_{GG}$) supply voltage. As with NMOS, PMOS devices are directly compatible with CMOS. Fig. 1-30 shows a 2521 PMOS 128-bit static shift register that uses only clamp diodes at the outputs to ground if CMOS is operated from a positive voltage and a ground supply.

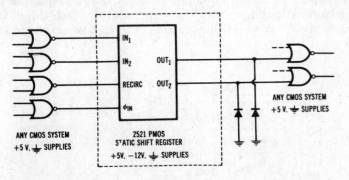

Fig. 1-30. Interfacing CMOS with a 2521 PMOS static shift register.

CMOS to Op Amps

The outputs of split-supply op-amp circuits are easily interfaced (Fig. 1-31). The only restriction is that the positive CMOS supply voltage be less than the op-amp supply. The two clamping diodes prevent the CMOS from exceeding the CMOS logic 1 and logic 0 levels while the 12K series resistor limits the output current of the op amp in case the op amp's output goes towards the op amp's negative supply voltage.

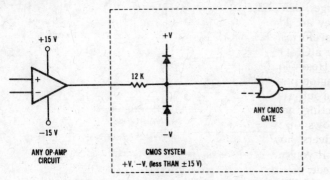

Fig. 1-31. CMOS and op amp interfacing.

CMOS to Lamps

CMOS to lamps or digital displays are fully covered in Chapter 8. Other applications, such as driving LEDs directly from either a 4511 decoder or from 4026/4033 decade counters, are shown in Chapter 6.

High-current loads, such as incandescent lamps, relays, SCRs, and similar devices, are best driven through interface transistors, either alone or in Darlington pairs, as shown in Fig. 1-32.

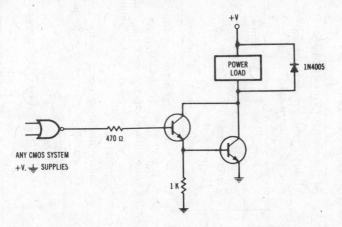

Fig. 1-32. Darlington high-power driver (High = on). Diode is needed only on inductive loads.

Other Inputs and Outputs

Chart 1-5 gives a set of interface guidelines that can tie CMOS into just about any electronic system. We can convert from CMOS to analog signals by placing weighted resistors on the CMOS output stages. Since the CMOS outputs swing the full supply-voltage range, they are ideal for this use. Sine-wave generators, electronic music waveforms, and digital-to-analog converters are typical applications for a CMOS interface.

Heavy loads are best driven using a high-gain transistor or a Darlington pair of transistors. Here the problem is usually to "amplify" the output current as needed to drive the load, at the same time protecting the CMOS from output transients. The same type of problem shows up when providing a high-voltage interface, and special drivers should be used in these applications.

If you have to interface over long distances or in a noisy environment, IC linear line drivers and receivers can be used. For total isolation, LED-phototransistor optical couplers can be used. These are low in cost and are particularly handy both for getting CMOS signals

Chart 1-5. CMOS Interface Guidelines

At the Input to CMOS:

* Provide these levels to the CMOS input stage:

Logic	0, 5-V Supply	0, 10-V Supply
Zero	0 to +1 Volt	0 to +2 Volts
One	+4 to +5 Volts	+8 to +10 Volts

* Protect the protection by limiting out-of-range input currents to 10 milliamperes.

* Keep input impedances low enough that they can rapidly charge stray and input capacitances at highest frequency of interest.

* If needed, use first CMOS stage to condition and debounce input signals. Clocked CMOS logic needs one noise-free pulse per desired event and rise and fall times better than 5 microseconds.

At the Output from CMOS:

* Don't try to get more current than this out of a B-series output stage:

Current	0, 5-V Supply	0, 10-V Supply
Within Half a Volt of Either Supply Rail	0.8 mA	1.6 mA
Short Circuit	4 mA	11 mA

(More current is available from 4049, 4050, and 4502 devices.)

* Protect the protection by never "force-feeding" more than 10 milliamperes into the output for out-of-range supplies.

* Satisfy the logic needs of the circuit being driven. Voltage translation and voltage or current gain might be needed.

* Avoid capacitive loading, particularly at high speeds. Use an LS TTL driver for this type of load.

onto and off of a hot-chassis system without having any electrical shock problems and for isolating grounds between systems.

TOOLS

All of the usual electronic hand tools are good for CMOS work. "Jeweler's-size" diagonal cutters and needle-nose pliers are essential, and a pair of traverse cutters is handy. So are an X-Acto knife, a wire stripper, and several sizes of screwdrivers. Optional, but handy, tools include an Adel nibbling tool; pin vises with 1/16-inch, No. 60, and No. 67 drills; some files; a high-intensity lamp; a magnifier; and a carbide-tipped scriber. The latter is particularly useful for opening pc runs in a way that can be resoldered.

A notebook or some other record system that works for you is absolutely essential.

A small soldering iron such as an Ungar handle with a 1235 element and a PL338 tip is a good choice for general IC work, as is a wiring pencil for breadboarding.

Your desoldering equipment should include a syringe-type solder sucker and a clean-out wire, along with several sizes of capillary solder braid (Fig. 1-33). A can of Dust-Off or other photographer's "air-in-a-can" is useful to clear plated-through holes. Fancier desoldering tools like the Ungar 7800 are also useful, and even a propane torch with a fan tip can be carefully used to salvage ICs from a pc board that is to be scrapped.

Courtesy Solder Removal Co.

Fig. 1-33. Solder braid is one good way to desolder integrated circuits.

47

There are a few premium tools you'll find essential if you get into advanced CMOS work. One of these is a quality triggered-sweep oscilloscope. A function generator or other source of test signals is also handy. An *electronic* bench vise, such as a Panavise 396, and a bench-mounted punch press, such as a Roper type XX and its accessory kit, are extremely useful. The latter saves a lot of drilling and is quicker, neater, and quieter.

Some test aids that you will probably want to pick up include a glomper clip or two, a state checker, and some sort of bounceless push button. In the absence of a scope, a good vom can sometimes be used.

"BAD" AND "BURNED-OUT" ICs

New CMOS integrated circuits from a reliable source are almost always good and, with a little common-sense handling, practically indestructible. Possibly you will get two to five bad circuits per hundred from a quality distributor, and maybe slightly more from secondary sources unless you are buying obvious garbage. In general, the ICs are the most reliable part of your circuit and the hardest part to damage.

If your CMOS circuit doesn't work, chances are it is your fault and not the ICs. Typical problems include: forgetting to tie down inputs; forgetting to debounce and sharpen input clocking signals; getting supplies connected wrong, totally unbypassed, or backwards; putting ICs in upside down; doing a pc layout topside and reversing all the connections; missing or loosening a pin on a socket or bending a pin over; misreading resistors (have you ever noticed the color-code similarity between a 15-ohm resistor and a one-megohm resistor?); and, of course, causing solder splashes and hairline pc opens and shorts.

The key rule is this: **ALWAYS BLAME YOURSELF FIRST AND THE ICs LAST.** *Always* assume that there is something incredibly wrong with your circuits when you first power them up. You'll be right almost every time. In fact, if things seem to work perfectly on the first try, this may mean that the real surprises are hiding, waiting to get you later on when it is more expensive to correct them. Anything that "has" to be correct is usually the mistake. And what seems like "impossible behavior" is really the poor IC trying its best to do a good job. With a little help and the right attitude you can help the ICs along.

SOME CONVENTIONS

In many of the circuit descriptions given in the rest of this book, we will not bother to state some obvious details. For instance, all

CMOS packages will need a source of power, usually +5 volts and ground in a TTL-compatible system, and +10 volts or so and ground in a CMOS-only system. We will usually call these terminals +V and ground, even though you are often free to run at +5 volts and −5 volts, ground and −10 volts, or whatever. In any case, what we usually call ground will equal your most-negative supply return.

Pinouts for most of the CMOS circuits appear in the next chapter. Remember that all inputs must go somewhere, even if they aren't needed. So, usually, we won't show enable pins, direct sets and resets, and so on, unless they are needed to show how the circuit works. But when you build the circuit, make sure all these pins go to proper +V or ground connections. Only unused CMOS *outputs* may remain unconnected. We will also see later that a positive-logic NOR circuit is the same thing as a negative-logic NAND circuit, but we will always be using the positive-logic symbol and part number, regardless of how the gate is actually used. Also, we will not keep track of fractional IC packages. By a "4001," we'll usually mean one-fourth of a 4001 quad NOR gate, and so on. Furthermore, when a 74C86 is used in a circuit, a 74HC86 or 74HCT86 can also be used, provided the proper power-supply limits are observed.

The address of many of the product sources mentioned in this chapter appear in the appendix.

Interface to Regular TTL

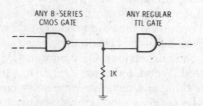

The gate and resistor combine to provide enough pulldown
current. Output high is above +3 volts.

Chapter 2

Some CMOS Integrated Circuits

This chapter is a catalog of the more important CMOS devices you're likely to need or use. It differs from the usual manufacturer's catalogs in several ways. First, it is an industry-wide sampling that favors nobody in particular, but tries to stick mostly to a hundred "mainstream" devices—those that are popular, easy to get, and easy to use. Second, we've tried to give you only the essential how-to-use-it information without burying key points in a lot of detail. And third, we've tried to point out the important usage hangups and restrictions on these ICs—things that can get you into trouble and are often creatively buried in the usual sources.

Most of the MSI and SSI devices are in the 4000, 4500, and 74C00 Series, along with a CMOS linear operational amplifier. The LSI selection includes a thousand-bit memory, a frequency counter on a chip, a stopwatch, a wristwatch, a tv numeric display, a modem, a bit-rate generator, a touch-tone dialer, a band-rate generator, and several a/d converters. This is a random selection of some of the more interesting LSI devices available. CMOS microprocessors have been saved for another book.

In addition, we've thrown in a few devices not yet available in CMOS logic that are very useful in CMOS systems. These include two transistor arrays, two regulators, two programmable read-only memories, a tv game chip, and a top-octave music generator.

Unless otherwise noted, our CMOS devices work over a 3- to 15-volt supply range. An input-low, or zero, state is less than one-fifth of the supply voltage. An input-high, or one, state is more than four-fifths of the supply voltage. Inputs above $+V$ or below ground must be current-limited to 10 milliamperes or less to protect the protection.

Most 4000/4500 B-series CMOS outputs provide 0.8 milliampere within half a volt of either supply rail, and 4 milliamperes of short-circuit current on a 5-volt supply. On a 10-volt supply, these values are 1.6 milliamperes and 11 milliamperes. Any "force-fed" currents above or below the supply rails must be limited to 10 milliamperes or less. A substantial speed penalty (3:1 reduction) is paid for 5-volt operation.

CMOS devices having clocked inputs need conditioning to eliminate noise and contact bounce (see Chapter 4). Rise and fall times to clocked logic should be no slower than 5 microseconds.

Remember that all CMOS inputs must go somewhere. If a particular feature isn't needed, it is disabled by connecting it to ground or (rarely—see data sheet) to the positive supply. Only unused CMOS *outputs* can remain unconnected. While CMOS will stand wide power-supply variations, a moderate amount of power-supply bypassing is needed. At least one 0.1-μF capacitor per six circuits is the minimum bypassing recommended.

The following pages list in numeric order the devices covered in this chapter.

Type Number	Description
4031	64-stage *Shift Register*, serial-in/serial-out
4032	Triple-adder, positive-logic *Arithmetic Unit*
4033	Decade *Counter* and 7-segment decoder with blanking
4034	8-bit bidirectional *Storage Register*
4035	4-stage *Shift Register*, parallel-in/parallel-out
4038	Triple-adder, negative-logic *Arithmetic Unit*
4040	12-stage binary ripple *Counter* (same as 74HC4040)
4041	Quad inverting/noninverting *Buffer*
4042	Quad latch *Storage Register*
4043	Quad *Flip-Flop*, R/S NOR logic
4044	Quad *Flip-Flop*, R/S NAND logic
4046	Phase-locked loop, *Special Device*
4047	Astable and monostable *Multivibrator*
4049	Hex *Inverter*/translator (same as 74HC4049)
4050	Hex *Buffer*/translator (same as 74HC4050)
4051	1-of-8 *Analog Switch* (same as 74HC4051)
4052	Dual 1-of-4 *Analog Switch* (same as 74HC4052)
4053	Triple 1-of-2 *Analog Switch* (same as 74HC4053)
4060	14-stage binary ripple *Counter* with oscillator (same as 74HC4060)
4063	4-bit magnitude comparator *Arithmetic Unit*
4066	Quad *Analog Switch*, low-impedance (same as 74HC4066)
4067	1-of-16 *Analog Switch*
4068	8-input NAND *Gate*
4069	Hex *Inverter*
4070	Quad EXCLUSIVE-OR *Gate*
4071	Quad 2-input OR *Gate*
4072	Dual 4-input OR *Gate*
4073	Triple 3-input AND *Gate*
4075	Triple 3-input OR *Gate* (same as 74HC4075)
4076	4-stage tri-state *Storage Register*
4077	Quad 2-input EXCLUSIVE-NOR *Gate*
4078	8-input NOR *Gate* (same as 74HC4078)
4081	Quad 2-input AND *Gate*
4082	Dual 4-input AND *Gate*
4089	Binary rate multiplier, *Special Device*
4093	Quad 2-input NAND *Schmitt Trigger*
4097	Dual 1-of-8 *Analog Switch*
4502	Hex *Inverter*/driver, three-state
4503	Hex *Buffer*/driver, three-state
4508	Dual 4-bit latch *Storage Register*, three-state

Type Number	Description
4510	Decimal up-down *Counter*
4511	Decimal *Counter* with 7-segment decoder/driver
4512	1-of-8 *Data Selector*, three-state
4514	1-of-16 *Decoder*, high output (same as 74HC4514)
4515	1-of-16 *Decoder*, low output
4516	Binary up-down *Counter*
4518	Dual decimal synchronous *Counter*
4520	Dual binary synchronous *Counter* (same as 74HC4520)
4522	Programmable decimal divide-by-n *Counter*
4526	Programmable binary divide-by-n *Counter*
4527	Decimal rate multiplier, *Special Device*
4528	Dual retriggerable *Monostable*
4531	12-input parity generator *Arithmetic Unit*
4532	8-input priority *Encoder*, special device
4539	Dual 4-input *Data Selector*
4543	See 74HC4543
4555	Dual 1-of-4 *Decoder*, noninverting
4556	Dual 1-of-4 *Decoder*, inverting
4584	Hex *Schmitt Trigger*
4702	Bit-rate generator, *Special Device*
5024	Top-octave music generator, *Special Device* (not CMOS)
5101	Static *Random-Access Memory*, 256 × 4
5369	17-stage oscillator/divider, *Special Device*
5841	Television numeric display, *Special Device*
7045	Stopwatch, *Special Device*
7106	3½-digit A/D converter, LCD drive, *Special Device*
7107	3½-digit A/D converter, LED drive, *Special Device*
7200	Wristwatch, *Special Device*
7207	Time Base, *Special Device*
7208	Frequency counter, *Special Device*
7603	*Programmable Read-Only Memory*, 32 × 8 (not CMOS)
7611	*Programmable Read-Only Memory*, 256 × 4
7805	5-volt positive voltage regulator, *Linear Device* (not CMOS)
7812	12-volt positive voltage regulator, *Linear Device* (not CMOS)
8500	TV game controller, *Special Device* (not CMOS)
40174	Hex-D *Storage Register*
40175	Quad-D *Storage Register*

Type Number	Description
40192	Decade up-down synchronous *Counter*
40193	Binary up-down synchronous *Counter*
74C00	Quad 2-input NAND *Gate*
74C02	Quad 2-input NOR *Gate*
74C04	Hex *Inverter*
74C08	Quad 2-input AND *Gate*
74C10	Triple 3-input NAND *Gate*
74C11	Triple 3-input AND *Gate*
74C14	Hex *Schmitt Triggers*
74C20	Dual 4-input NAND *Gate*
74C27	Triple 3-input NOR *Gate*
74C30	8-input NAND *Gate*
74C32	Quad 2-input OR *Gate*
74C42	1-of-10 *Decoder*
74C48	BCD-to-7-segment *Decoder/driver*
74C73	Dual JK *Flip-Flop* with preclear only
74C74	Dual Type-D *Flip-Flop* with preset and preclear
74C75	Quad latch *Storage Register*
74C76	Dual JK *Flip-Flop* with preset and preclear
74C83	4-bit binary full adder *Arithmetic Unit*
74C85	4-bit magnitude comparator *Arithmetic Unit*
74C86	Quad EXCLUSIVE-OR *Gate*
74C90	Decimal ripple *Counter*
74C93	Binary ripple *Counter*
74C107	Dual JK *Flip-Flop* with preclear
74C148	8-bit priority *Encoder*
74C150	1-of-16 *Data Selector*
74C151	1-of-8 *Data Selector*
74C154	1-of-16 *Data Distributor*
74C155	Dual 1-of-4 *Data Distributor*
74C157	Quad 1-of-2 *Data Selector*
74C160	Synchronous decimal *Counter* with asynchronous clear
74C161	Synchronous binary *Counter* with asynchronous clear
74C164	8-stage *Shift Register*, serial-in/parallel-out
74C165	8-stage *Shift Register*, parallel-in/serial-out
74C174	Hex-D *Storage Register*
74C175	Quad-D *Storage Register*
74C192	Decimal up-down synchronous *Counter*
74C193	Binary up-down synchronous *Counter*
74C915	7-segment-to-BCD *Converter*, three-state
74HC123	Dual retriggerable monostable *Multivibrator*
74HC125	Quad *Buffer*, three-state

Type Number	Description
74HC126	Quad *Buffer*, three-state
74HC153	Dual 1-of-4 *Data Selector*
74HC4002	See 4002
74HC4015	See 4015
74HC4020	See 4020
74HC4024	See 4024
74HC4040	See 4040
74HC4049	See 4049
74HC4050	See 4050
74HC4051	See 4051
74HC4052	See 4052
74HC4053	See 4053
74HC4060	See 4060
74HC4066	See 4066
74HC4075	See 4075
74HC4078	See 4078
74HC4514	See 4514
74HC4520	See 4520
74HC4543	BCD-to-decimal decoder *Driver* (same as 4543)
MC14410	Touch-Tone dialer, *Special Device*
MC14411	Baud-rate generator, *Special Device*
MC14412	Modem, *Special Device*
MC14433	3½-digit A/D converter, *Special Device*

The following pages list devices covered in this chapter by function.

FUNCTION INDEX

Analog Switches:

Quad	4016
Quad, low-impedance	4066, 74HC4066
1-of-8	4051, 74C151, 74HC4051
Dual 1-of-8	4097
Dual 1-of-4	4052, 74C153, 74HC4052
Triple 1-of-2	4053, 74HC4053
1-of-16	4067, 74C150

Arithmetic Units:

Triple-adder, + logic	4032
Triple-adder, − logic	4038
Magnitude comparator	4063, 74C85
Parity generator, 12-input	4531
Full 4-bit adder	4008, 74C83

Buffers and Inverters:

Counters:

Octal

Decimal

Binary (Hexadecimal)

Ripple

Programmable divide-by-n

Data Selectors:

Decoders:

Dual 1-of-4, noninverting	4555, 74C155
Dual 1-of-4, inverting	4556, 74C156
1-of-10	4028, 74C42
1-of-16 high	4514, 74C154, 74HC4514
1-of-16 low	4515
BCD-to-7-segment driver	4543, 74HC4543
BCD-to-7-segment driver with latch	4511
BCD-to-7-segment driver with latch	74C48
7-segment-to-BCD with latch	74C915

Encoders

Priority encoder, 8-to-3-line	4532, 74C148

Flip-Flops:

Dual Type-D	4013, 40175, 74C74
Dual JK with preclear	74C73, 74C107
Dual JK with preset and preclear	4027, 74C76
Quad R/S, NOR logic	4043
Quad R/S, NAND logic	4044

Gates:

Quad 2-input AND	4081, 74C08
Quad 2-input NAND	4011, 74C00
Quad 2-input OR	4071, 74C32
Quad 2-input NOR	4001, 74C02
Quad EXCLUSIVE-OR (obsolete)	4030
Quad EXCLUSIVE-OR	4070, 74C86
Quad EXCLUSIVE-NOR	4077
Dual 3-input NOR plus inverter	4000
Triple 3-input AND	4073, 74C11
Triple 3-input NAND	4023, 74C10
Triple 3-input OR	4075, 74HC4075
Triple 3-input NOR	4025, 74C27
Dual 4-input AND	4082
Dual 4-input NAND	4012, 74C20
Dual 4-input OR	4072
Dual 4-input NOR	4002, 74C25, 74HC4002
8-input NAND	4068, 74C30
8-input NOR	4078, 74HC4078

Linear Devices:

Transistor array, common-emitter (not CMOS) 3081
Transistor array, common-collector (not CMOS) 3082
CMOS operational amplifier 3130
5-volt positive regulator (not CMOS) 7805
12-volt positive regulator (not CMOS) 7812

Multivibrators:

Single, astable and monostable 4047
Dual, retriggerable monostable 4528, 74C123
Single CMOS timer L555
Dual CMOS timer L556

Programmable Read-Only Memories:

32 8-bit words (not CMOS) 7603
256 4-bit words (not CMOS) 7611

Random-Access Memories:

1024 1-bit words 2102
256 4-bit words 2112
256 4-bit words 5101

Schmitt Triggers:

Hex inverting 4584, 74C14
Quad 2-input NAND 4093

Shift Registers:

4-stage, parallel-in/parallel-out 4035, 40195
Dual 4-stage, serial-in/parallel-out . . . 4015, 4094, 74HC4015
8-stage, parallel-in/serial-out 4014
8-stage, parallel-in/serial-out, asynchronous . . . 4021, 74C165
18-stage, serial-in/serial-out 4006
64-stage 4031

Special Devices:

MSI
  Binary rate multiplier 4089
  Decimal rate multiplier 4527
  Phase-locked loop 4046
  Priority encoder, 8-input 4532
LSI
  A/D converter, 3½-digit MC14433
  A/D converter, 3½-digit, LCD 7106
  A/D converter, 3½-digit, LED 7107

Baud-rate generator MC14411
Bit-rate generator. 4702
Frequency counter 7208
Frequency synthesizer 0320
Modem MC14412
Oscillator/divider, 17-stage. 5369
Stopwatch. 7045
Television numeric display 5841
Time base 7207
Top-octave music generator (not CMOS) 5024
Touch-Tone dialer. MC14410
TV game controller 8500
Wristwatch. 7200

Storage Registers:

Quad latch. 4042, 74C75
Quad-D 40175, 74C175
Quad tri-state 4076
Dual quad latch 4508
Hex-D 40174; 74C174
8-stage bidirectional 4034

L555/ICM 7555

CMOS TIMER—ASTABLE OR MONOSTABLE

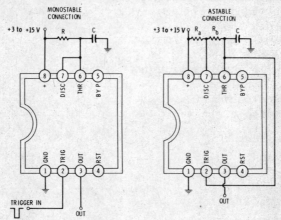

This circuit may be used for astable or monostable in timing applications from microseconds to hours. Complete details for its use appear in Chapter 4.

As a monostable, the circuit is triggered by bringing the Trigger input momentarily below 2 volts. The output pulse width is determined by R and C, and the curves are shown in Fig. 4-44. R can vary from 1K to 100 megohms. C can range from 100 pF up.

As an astable, the circuit is free running. The charging time is determined by R_a and R_b in series with C. The discharging time is determined by C and R_b. Design curves appear in Fig. 4-40. The minimum value of R_a is 1K; the maximum value of $R_a + R_b$ is 100 megohms.

The RST input (pin 4) will drive the output low if it is grounded. If unused, it should be tied to the positive supply voltage. The Bypass input should be bypassed to ground with a suitable capacitor (0.1 µF upward) in critical timing applications.

The output is high during the monostable on time and low otherwise. The output is high during the astable charging time and low during the discharge time.

Operating current is approximately 80 microamperes.

DUAL CMOS TIMER—ASTABLE OR MONOSTABLE

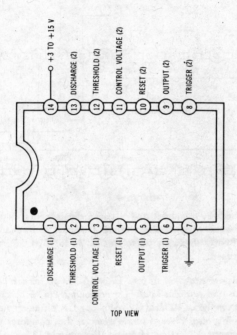

TOP VIEW

This device consists of two independent 555 timers in a single package with a common power-supply connection. Both devices may be used for astable or monostable operation in timing applications ranging from microseconds to hours. Like the 555 timer, complete details for use of the timer appears in Chapter 4.

Operating current is approximately 160 microamperes.

0320
(HUGHES)

FREQUENCY SYNTHESIZER

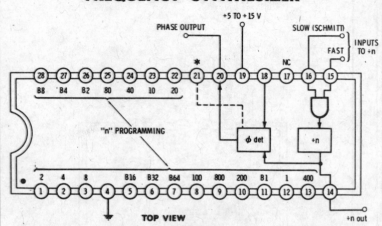

This is a divide-by-n counter and a phase detector in a single package. It is used in frequency synthesizers where a number of frequencies are to be generated from a single crystal reference. Up to 1020 different frequencies can be obtained.

Input signals are accepted at pins 15 or 16. Pin 16 accepts 5-kHz to 1-MHz clock signals and provides internal conditioning. Pin 15 accepts signals to 5 MHz (5-volt supply) or 10 MHz (12-volt supply). A divide-by-n output appears on pin 14. This also drives a phase detector that compares the divide-by-n output frequency against the frequency input to pin 18.

Polarity of the phase detector is set by pin 21. If pin 21 is grounded, the phase detector output is low if the pin-14 frequency is low with respect to the pin-18 frequency, and high if the pin-14 frequency is high with respect to the pin-18 frequency. Making pin 21 positive reverses this output sense. Use a grounded pin 21 if increasing the voltage increases the frequency in any vco used with this chip.

The number n can range from 3 to 1023; n is made of two parts, a binary number from 0 to 128 and a binary-coded decimal number from 0 to 1000. Thumbwheel switches are often used to enter the bcd portion of the division number, while the binary inputs are often hard-wired. The binary inputs make frequency offsets particularly easy, and they facilitate code conversions needed to get to channel numbers or channel frequencies.

Total package current at 5 megahertz is 1 milliampere on a +5-volt supply.

STATIC RANDOM-ACCESS MEMORY
(1024 X 1, 16-Pin, Separate I/O)

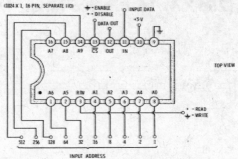

This is a static random-access memory organized as 1024 × 1 in a 16-pin package with separate input and output leads. Information may be rapidly read into, and nondestructively read out of, memory at system speeds. No clocking or refresh is needed. Storage is volatile with data being held only as long as supply power is applied.

To read, pin 13 is grounded and pin 3 is made positive. A binary address applied to the ten input address pins will select an internal storage cell and will output the data in that cell.

To write, pin 13 is grounded and pin 3 is made positive. A binary address is applied to the ten input address pins to select an internal storage cell. The write input, pin 3, is then brought low and returned high. *All address lines must be stable immediately before, during, and immediately after the low state on pin 3.*

All inputs and outputs are TTL and CMOS compatible. The output will drive one TTL load. Making pin 13 positive will float the outputs and ignore write commands. Any number of outputs from separate devices may be connected in parallel as long as only one circuit is enabled at a time.

Access time varies with the manufacturer and the grade of the device. A 800-ns read time and a 400-ns write pulse are typical for a nonpremium unit.

Supply power is 70 mA or less, again depending on the grade of the device and the manufacturer.

Note that input addresses may be redefined in any manner convenient for circuit layout.

2112

STATIC RANDOM-ACCESS MEMORY
(256 X 4, 16 Pin, Common I/O)

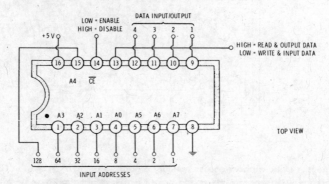

This is a static random-access memory organized as 256 words of four bits each in a 16-pin package with common input/output leads. Information may be rapidly read into, and nondestructively read out of, memory at system speeds. No clocking or refresh is needed. Storage is volatile, with data being held only as long as supply power is applied.

To read, pin 14 is grounded and pin 13 is made positive. A binary address is applied to the eight address pins to select four internal storage cells. The data selected appears on the four i/o lines. *Nothing else may be connected to the i/o lines that source data during the read time.*

To write, pin 14 is grounded and pin 13 is made positive. A binary address is similarly applied to the address pins. The write input, pin 13, is brought low and back high again, *at the same time enabling an external source of data for only the time the write input is low.* All address lines must be stable immediately before, during, and immediately after the low state on pin 13.

Inputs and outputs are TTL and CMOS compatible. Outputs drive one TTL load. Making pin 14 positive floats the outputs and ignores write commands. Note that very careful managing of the input/output lines is needed to make sure only one source exists at any particular time. This i/o management is eased in the 18-pin 2111.

Access times of 500 ns and less are typical. Consult specific data sheets for timing restrictions. Supply current is typically 60 mA.

Note that input addresses may be redefined in any manner convenient for circuit layout.

NPN TRANSISTOR ARRAY, COMMON-EMITTER

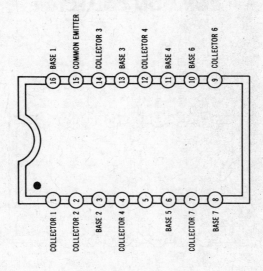

TOP VIEW

This package contains seven identical high-current NPN transistors connected in a common-emitter configuration. It is primarily intended for driving 7-segment, common-anode, LED displays from CMOS devices.

For each transistor, typical beta is 30, maximum collector current is 100 milliamperes, collector-to-emitter breakdown voltage is 16 volts, and the maximum package power dissipation is 750 milliwatts, while the maximum for any one transistor is 500 milliwatts.

3082
(RCA)

NPN TRANSISTOR ARRAY, COMMON-COLLECTOR

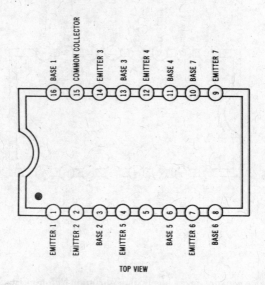

TOP VIEW

This package contains seven identical high-current NPN transistors connected in a common-collector configuration. It is primarily intended for driving 7-segment, common-cathode, LED displays from CMOS devices.

For each transistor, typical beta is 30, maximum collector current is 100 milliamperes, collector-to-emitter breakdown voltage is 16 volts, and the maximum package power dissipation is 750 milliwatts, while the maximum for any one transistor is 500 milliwatts.

CMOS OPERATIONAL AMPLIFIER

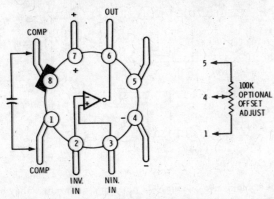

This is a CMOS linear operational amplifier with an extremely high input impedance, a 15-megahertz unity-gain bandwidth, and a 10-volt-per-microsecond slew rate.

Details on its use appear in Chapter 7. The high input impedance is very useful in sample/hold, pH meter, comparator, and precision rectifier applications.

Low-frequency open-loop gain is 100,000. A compensation capacitor of 30 pF or more is normally added between pins 1 and 8, as shown, for stable operation. Input offset is typically 8 millivolts and is trimmable with potentiometer, as shown.

The available output current is 20 milliamperes in either direction. The device is somewhat noisy and quite sensitive to output capacitive loading.

The optimum input reference is one-half the supply voltage. Inputs can be referenced as far negative as the pin-4 voltage, but input voltages more than 0.3 volt negative with respect to pin 4 will draw input current and can reverse the output sense in comparator circuits.

Total power-supply voltage can range from 5 to 15 volts (±2.5 to ±7.5). A current of 10 milliamperes is typical with a 15-volt supply. The 3140 is a similar device with bipolar output and higher supply voltages.

4000

DUAL 3-INPUT NOR GATE PLUS INVERTER

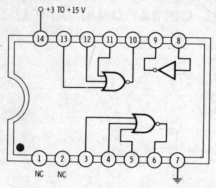

+3 TO +15 V

TOP VIEW

The package contains two 3-input NOR gates and an inverter. They may be used separately.

An input *low* drives the inverter output *high,* and vice versa.

On the NOR gates, any input *high* drives the output *low.* All inputs *low* drives the output *high.*

The gates may be combined. A NOR gate followed by an inverter gives a 3-input OR gate. This output routed to the remaining gate gives a 5-input NOR function.

Propagation delay is 25 nanoseconds at 10 volts and 60 nanoseconds at 5 volts. Total package current is 0.3 mA at 5 volts and 0.6 mA at 10 volts at a 1-megahertz data rate.

QUAD 2-INPUT NOR GATE

+3 TO +15V

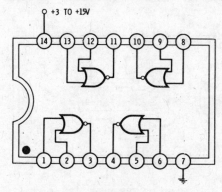

TOP VIEW

All four positive-logic NOR gates may be used independently. On any one gate, with *either* or both inputs *high*, the output is *low*; with *both* inputs *low*, the output is *high*. Functionally equivalent to the 7402 (TTL) and the 74C02 (CMOS).

Propagation delay is 25 nanoseconds at 10 volts and 60 nanoseconds at 5 volts. Total package current at 1 megahertz is 0.4 milliampere at 5 volts and 0.8 milliampere at 10 volts.

4002

DUAL 4-INPUT NOR GATE

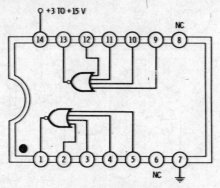

TOP VIEW

Both positive-logic NOR gates may be used independently.

On either gate, one or more inputs *high* provides a *low* output. If all four inputs are *low*, the output will go *high*.

The A-series devices have a very poor and very lopsided response. Use B-series devices for all but totally noncritical applications. Functionally equivalent to the 7425 (TTL) and the 74C25 (CMOS).

Propagation delay is 25 nanoseconds at 10 volts and 60 nanoseconds at 5 volts. Total package current at 1 megahertz is 0.4 milliampere at 5 volts and 0.8 milliampere at 10 volts.

VARIABLE-LENGTH (To 18 Stages) SHIFT REGISTER (Serial in, Serial Out)

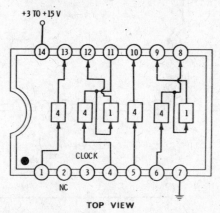

TOP VIEW

This package contains four separate shift registers. Two of them are four stages long. Two of them may be used as either a four-stage or a five-stage register.

On any single register, data entered on the input is stored internally on the negative edge of the common clock and appears at the output, in serial-in, serial-out form, four or five clock cycles later, depending on the register length. There is no inversion of data; a positive level at an input appears as a positive output level four or five clock cycles later.

Registers may be connected in series for total lengths of 4, 5, 8, 9, 10, 12, 13, 14, 16, 17, or 18 stages. With longer register lengths, some, but not all, intermediate outputs are available.

Do not attempt to enter data on pin 11 or pin 8.

The clock must be conditioned to be noiseless and have only one abrupt negative transition per desired shift. Clock rise and fall times should be 5 microseconds or faster.

Maximum clock frequency is 5 megahertz at 10 volts and 2.5 megahertz at 5 volts. Total package current is 0.8 milliampere at 5 volts and 1.6 milliamperes at 10 volts.

4007

DUAL CMOS PAIR PLUS INVERTER

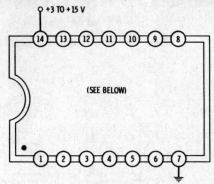

+3 TO +15 V

(SEE BELOW)

TOP VIEW

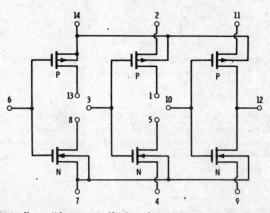

This is basically a "do-it-yourself" kit of CMOS transistors. You can use it for simple logic, transmission gates, buffers, drivers, CMOS variable resistors, discharge transistors for capacitors, analog-to-digital converters, input translators, oscillators, etc. Do not allow any pin voltage to exceed the pin-14 voltage. Do not allow any pin voltage to go below the pin-7 voltage.

Propagation time is 20 nanoseconds at 10 volts and 35 nanoseconds at 5 volts. Total package current is 0.7 milliampere at 5 volts and 1.4 milliamperes at 10 volts at a 1-megahertz clock rate.

4-BIT FULL ADDER

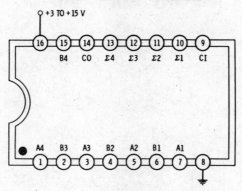

TOP VIEW

This is an arithmetic package that gives the positive-logic sum of two 4-bit binary numbers. Packages may be cascaded for more bits.

Functionally equivalent to the 7483 (TTL) and the 74C83 (CMOS).

Input word A is applied to the A inputs with weights $A1 = 1$, $A2 = 2$, $A3 = 4$, and $A4 = 8$. Input word B is applied to the B inputs with weights $B1 = 1$, $B2 = 2$, $B3 = 4$, and $B4 = 8$. The positive-logic binary sum of the two input words appears on the Σ outputs, with $\Sigma1 = 1$, $\Sigma2 = 2$, $\Sigma3 = 4$, and $\Sigma4 = 8$ weightings. Should a carry result, it will appear on the CO terminal with a weighting of 16.

The carry input should be *grounded* on the package working on the four least-significant bits of a binary sum. The CO of this package should go to the CI input of the package working on the next four most-significant bits, and so on. Internal look-ahead carry is done to increase operating speed.

Addition time is 900 nanoseconds at 5 volts and 325 nanoseconds at 10 volts, with newer devices being significantly faster. Total package current is 1.6 milliamperes at 5 volts and 3.2 milliamperes at 10 volts at a 1-megahertz word rate.

4009

HEX INVERTING BUFFER

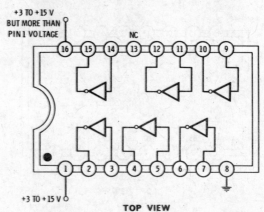

+3 TO +15 V
BUT MORE THAN
PIN 1 VOLTAGE

NC

+3 TO +15 V

TOP VIEW

This device should not be used for new designs. Use the 4049 instead.

The 4009 will self-destruct if supply voltages are applied in the wrong sequence. The voltage on pin 1 must always be equal to or less than the voltage on pin 16.

Device is functionally equivalent to the 7404 (TTL) and 74C04 (CMOS) devices.

HEX NONINVERTING BUFFER

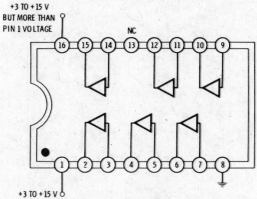

+3 TO +15 V
BUT MORE THAN
PIN 1 VOLTAGE

NC

+3 TO +15 V

TOP VIEW

This device should not be used for new designs. Use the 4050 instead.

The 4010 will self-destruct if supply voltages are applied in the wrong sequence. The voltage on pin 1 must always be equal to or less than the voltage on pin 16.

4011

QUAD 2-INPUT NAND GATE

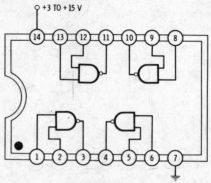

TOP VIEW

All four positive-logic NAND gates may be used independently.

On any one gate, with *either* or both inputs *low,* the output will be *high;* with both inputs *high,* the output will be *low.*

Propagation delay is 25 nanoseconds at 10 volts and 60 nanoseconds at 5 volts. Total package current at 1 megahertz is 0.4 milliampere at 5 volts and 0.8 milliampere at 10 volts.

Device is functionally equivalent to 7400 (TTL) and 74C00 (CMOS).

DUAL 4-INPUT NAND GATE

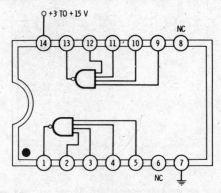

TOP VIEW

Both positive-logic NAND gates may be used independently.

On either gate, one or more inputs *low* provides a *high* output. If all four inputs are *high,* the output will go *low.*

Functionally equivalent to the 7440 (TTL) device.

The A-series devices have a very poor and very lopsided response. Use B-series devices for all but totally noncritical applications.

Propagation delay is 25 nanoseconds at 10 volts and 60 nanoseconds at 5 volts. Total package current at 1 megahertz is 0.4 milliampere at 5 volts and 0.8 milliampere at 10 volts.

4013

DUAL D FLIP-FLOP

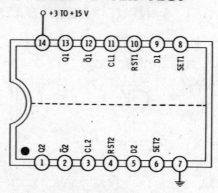

+3 TO +15 V

| 14 | 13 | 12 | 11 | 10 | 9 | 8 |
| | Q1 | Q̄1 | CL1 | RST1 | D1 | SET1 |

| Q2 | Q̄2 | CL2 | RST2 | D2 | SET2 | |
| 1 | 2 | 3 | 4 | 5 | 6 | 7 |

TOP VIEW

Each flip-flop may be used independently. There are two modes, *clocked* and *direct*.

In the *clocked* mode, the direct set and reset inputs must remain at ground. The input to the D line decides what the flip-flop is going to do. The actual operation doesn't happen until the positive edge (ground-to-positive transition) of the clock.

If D is positive, clocking makes the Q output positive and Q̄ grounded. If D is grounded, clocking makes the Q output grounded and Q̄ positive.

In the *direct* mode, a positive set input forces Q positive and Q̄ to ground. A positive reset input forces Q to ground and Q̄ positive. Should both set and reset be simultaneously positive, both Q and Q̄ will also go positive. This is usually a disallowed state. The last direct input to go to ground will determine the final state of the Q and Q̄ outputs. The direct inputs override the clocked inputs.

Each flip-flop may be made to binary-divide by cross coupling the Q̄ output to the D input.

The clock input must be noiseless and have only a single ground-to-positive edge transition per desired clocking. Clock rise and fall times should be 5 microseconds or faster.

Maximum clock frequency is 10 megahertz at 10 volts and 4 megahertz at 5 volts. Total package current at a 1-MHz clock frequency is 0.8 milliampere at 5 volts and 1.6 milliamperes at 10 volts.

Functionally equivalent to the 7474 (TTL) and 74C74 (CMOS) devices.

8-STAGE SHIFT REGISTER
(Parallel-in/Serial-Out; Clocked Load)

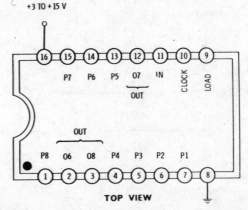

+3 TO +15 V

TOP VIEW

This package may be used as a 6-, 7-, or 8-stage shift-right register, either as serial-in/serial-out or as parallel-in/serial-out. Stages may be cascaded for longer lengths.

As a serial-in/serial-out register, the load input should be *grounded*. Data presented to the IN terminal gets shifted into the first stage on the ground-to-positive transition (positive edge) of the clock input. In six successive clockings, this data appears at output O6. Another clocking transfers to output O7 and yet another to O8. Additional clockings will lose this bit of data unless stages are cascaded or data is recirculated.

To parallel-load data, apply an 8-bit word to the P1 through P8 terminals, having the P1 bit nearest the input of the register and the P8 bit nearest the output. The load terminal is made *positive*. The high state of the load terminal must overlap the ground-to-positive clock transition. The parallel word gets synchronously loaded into the register on the positive clock edge. The load terminal must then return and stay grounded for normal register operation.

The clock must be noiseless and have only a single ground-to-positive transition per desired shifting. Clock rise and fall times should preferably be faster than 5 microseconds.

Maximum clock frequency is 5 megahertz at 10 volts and 2.5 megahertz at 5 volts. Total package current at a 1-megahertz clock rate is 2 milliamperes at 5 volts and 4 milliamperes at 10 volts.

Functionally equivalent to the 74165 (TTL) and 74C165 (CMOS). The 4021 is a similar register with immediate load.

4015

DUAL 4-STAGE SHIFT REGISTER
(Serial-in/Parallel-Out)

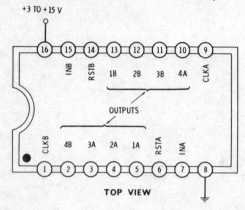

TOP VIEW

This package contains two completely separate 4-stage shift registers that may be used in serial-in/serial-out or serial-in/parallel-out modes. Each register has its own clock and reset pins. Registers may be cascaded for longer lengths.

On either register, data entered on the IN pin appears at the 1 output on the positive edge (ground-to-positive transition) of the clock. A second clocking transfers the data to the 2 output, a third to the 3 output, and a fourth to the 4 output. The reset input must be *grounded* during normal operation. Making the reset input *positive* forces all outputs to ground and keeps them there until the reset input returns to ground.

The clock must be noise free and have only one ground-to-positive transition per desired register clocking. The clock rise and fall times should be reasonably fast, preferably faster than 5 microseconds.

Maximum clock frequency is 5 megahertz at 10 volts and 2.5 megahertz at 5 volts. Total current per package at a 1-megahertz clocking rate is 2 milliamperes at 5 volts and 5 milliamperes at 10 volts.

Functionally equivalent to the 7491 and 74164 (TTL) and the 74HC4015 and 74C164 (CMOS) devices.

QUAD DIGITAL OR ANALOG
BILATERAL SWITCH

+3 TO +15 V (DIGITAL MODE)
+5 V (ANALOG MODE)

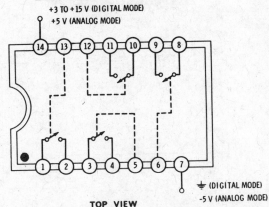

TOP VIEW

(DIGITAL MODE)
-5 V (ANALOG MODE)

All four switches may be used separately or in combination.

On any single switch, when the control voltage equals the pin-7 voltage, the switch remains OFF and behaves as a very high impedance. When the control voltage equals the pin-14 voltage, the switch turns ON and behaves as a nearly linear, bilateral, 300-ohm resistor.

Signals routed through the switch may be digital or analog, but they must never exceed the pin-14 voltage nor go below the pin-7 voltage.

Switches may be shorted together in any pattern, and there is no difference between the input and output terminal of any switch.

For instance, if all four switches are connected with one common terminal, the package may be used as a 1-of-4 data selector, a 1-to-4 data distributor, a 1-of-4 analog commutator, or a 1-of-4 analog multiplexer.

If more than one switch is connected to a common point, external logic must normally guarantee that only one switch is turned on at a time.

Maximum switch frequency is 10 megahertz at 10 volts and 5 megahertz at 5 volts. Package dissipation depends on the loading. Dissipation should be kept under 100 milliwatts total.

The 4066 is an improved version with lower ON resistance.

See Chapter 7 for more details.

4017

DIVIDE-BY-10 COUNTER WITH 1-OF-10 OUTPUTS (Synchronous)

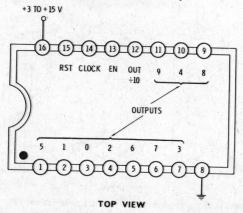

TOP VIEW

This is a fully synchronous decade, or divide-by-10, counter. It may be used to obtain a 1-of-10 decoded output or a square-wave output one-tenth the frequency of the input.

For normal operation, the clock enable and the reset should be at *ground*. The counter advances one count on the positive edge (ground-to-positive transition) of the clock. On any count, the decoded output goes *positive;* the others remain at ground. The OUT terminal is high for counts 0 through 4 and low for counts 5 through 9.

Making the reset input positive returns the counter to count zero. In this state, the "0" output and the OUT terminal are positive; the other outputs are at ground. The reset must be returned to ground to allow counting to continue. A positive voltage on the clock enable will inhibit (prevent count advance) clock operation.

The clock must be noiseless and have only one ground-to-positive transition per desired count. The clock rise time should be faster than 5 microseconds. An external gate will allow division by 1 through 10. Maximum clock frequency is 5 megahertz at 10 volts and 2.5 megahertz at 5 volts. Total package current at 1 megahertz is 0.4 milliampere at 5 volts and 0.8 milliampere at 10 volts, unloaded.

DIVIDE-BY-2-THRU-10 COUNTER
(Synchronous)

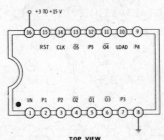

TOP VIEW

This is a specialized walking-ring synchronous counter. It can be made to divide an input clock by 2 through 10. The output is a square wave for even divisions and nearly a square wave for odd divisions. Because of the walking-ring code, decoding of intermediate states is not easily done.

Feedback must be provided to program this counter. In normal operation, Reset and Load are held at ground, and the counter advances one count for each ground-to-positive transition (positive edge) of the clock. Feedback is as follows:

05 to IN divides by 10	04 AND 05 to IN divides by 9*
04 to IN divides by 8	03 AND 04 to IN divides by 7*
03 to IN divides by 6	02 AND 03 to IN divides by 5*
02 to IN divides by 4	01 AND 02 to IN divides by 3*
01 to IN divides by 2	

* external gate needed

The feedback line is also the output. The counter is reset to zero by bringing the Reset positive. It is also possible to parallel-load the counter with the P1 through P5 inputs by bringing the Load input positive. Note that this is a specialized code.

The clock must be noiseless and have only one ground-to-positive transition per desired count. Clock rise and fall times should be faster than 5 microseconds.

Maximum clock frequency is 5 megahertz at 10 volts and 2.5 megahertz at 5 volts. Propagation delay is 200 nanoseconds at 10 volts and 500 nanoseconds at 5 volts. Total package current at 1 megahertz is 0.4 milliampere at 5 volts and 0.8 milliampere at 10 volts.

This package is ideal for digital sine-wave generator use (see Chapter 6). On the A-series devices, the available drive current on outputs 01, 02, 03, and 04 is very limited. Use B-series devices if these outputs are used.

Note: Do not confuse this device with the MC4018 (74418) MTTL binary ripple down counter.

4019

4-POLE, DOUBLE-THROW DATA SELECTOR

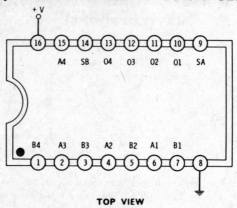

TOP VIEW

This package contains four AND/OR select gates with common select logic. It is usually used as a 4-pole, double-throw data selector. The four outputs are controlled by the Select A (SA) and Select B (SB) inputs.

If SA is grounded and SB is grounded, all four outputs remain grounded. If SA is positive and SB is grounded, all four outputs follow the A inputs. If SA is grounded and SB is positive, all four outputs follow the B inputs. If SA and SB are both positive, the four outputs provide the OR logic function of the inputs. In this mode, if either A1 OR B1 is positive, output 01 will be positive. If both A1 and B1 are grounded, output 01 will be grounded.

Propagation delay is 85 nanoseconds at 10 volts and 200 nanoseconds at 5 volts. Total package current is 1.6 milliamperes at 5 volts and 3.2 milliamperes at 10 volts.

Note that this is a data selector only. It cannot be used as a data distributor.

Device is functionally equivalent to the 74157 (TTL) and 74C157 (CMOS) devices. The 4519 is a similar device, except that the EXCLUSIVE-OR of the inputs is provided for both SA and SB positive.

14-STAGE (÷16,384) BINARY RIPPLE COUNTER

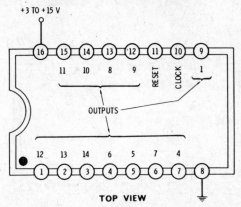

TOP VIEW

This is a binary ripple counter that counts in the up direction using positive logic.

The reset input is normally held at ground. Every time the clock changes from positive to ground, the counter advances one count. The 1 output divides the input clock by $2^1 = 2$. The 4 output divides the input clock by $2^4 = 16$, up to the 14 output which divides by $2^{14} = 16,384$. *There are no outputs for the second and third stages.*

Making the reset input positive forces all outputs to ground and holds them there until the reset returns to ground.

The clock input must be conditioned to be noiseless and must fall only once per desired count. Clock rise and fall times should be faster than 5 microseconds.

Since this is a ripple counter, the outputs change in sequential order. Incorrect counts will briefly result during the settling time. Device is functionally and pin-for-pin equivalent to the 74HC4020 for LS TTL levels.

Maximum input frequency is 7 megahertz at 10 volts and 2.5 megahertz at 5 volts. Total package current is 0.2 milliampere at 5 volts and 0.4 milliampere at 10 volts at a 1-megahertz clocking rate. Consult specific manufacturer's data sheet for propagation delay times.

The 4060 is a slower device with internal oscillator and clock conditioning.

4021

8-STAGE SHIFT REGISTER
(Parallel-in/Serial-Out; Load-Anytime)

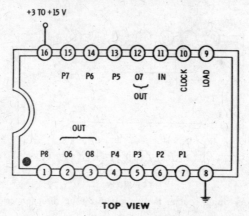

+3 TO +15 V

16 15 14 13 12 11 10 9

P7 P6 P5 07 IN CLOCK LOAD

OUT

OUT

P8 06 08 P4 P3 P2 P1

1 2 3 4 5 6 7 8

TOP VIEW

This package may be used as a 6-, 7-, or 8-stage shift-right register, either as a serial-in/serial-out or as a parallel-in/serial-out device. Stages may be cascaded for longer lengths.

As a serial-in/serial-out register, the load input should be *grounded*. Data presented to the IN terminal gets shifted into the first stage on the ground-to-positive transition (positive edge) of the clock input. In six successive clockings, this data appears at output 06. Another clocking transfers to output 07 and yet another to 08. Additional clockings will lose this bit of data unless stages are cascaded or data is recirculated.

To parallel-load data, apply an 8-bit word to the P1 through P8 terminals, having the P1 bit nearest the input of the register and the P8 bit nearest the output. The Load terminal is made high, and the data immediately loads into the register. The Load terminal must return and stay grounded for normal register operation.

The clock must be noiseless and have only a single ground-to-positive transition per desired shifting. Clock rise and fall times should be faster than 5 microseconds.

Maximum clock frequency is 5 megahertz at 10 volts and 2.5 megahertz at 5 volts. Total package current at a 1-megahertz clock rate is 2 milliamperes at 5 volts and 4 milliamperes at 10 volts.

Device is functionally equivalent to the 74165 (TTL) and 74C165 (CMOS) devices. The 4014 is a similar register with synchronous load.

DIVIDE-BY-8 COUNTER WITH
1-OF-8 OUTPUTS
(Synchronous)

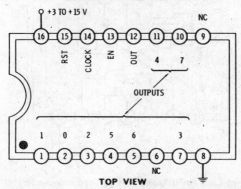

TOP VIEW

This is a fully synchronous octal, or divide-by-8, counter. It may be used to obtain a 1-of-8 decoded output or a square-wave output one-eighth the frequency of the input.

For normal operation, the clock enable and the reset should be at *ground*. The counter advances one count on the positive edge (ground-to-positive transition) of the clock. On any count, the decoded output goes *positive;* the others remain at ground. The OUT terminal is high for counts 0 through 3 and low for counts 4 through 7.

Making the reset input positive returns the counter to count zero. In this state, the "0" output and the OUT terminal are positive; the other outputs are at ground. The reset must be returned to ground to allow counting to continue. A positive voltage on the clock enable, "EN," will inhibit clock operation and prevent count advance.

The clock must be noiseless and have only one ground-to-positive transition per desired count. The clock rise time should be faster than 5 microseconds.

Maximum clock frequency is 5 megahertz at 10 volts and 2.5 megahertz at 5 volts. Total package current at 1 megahertz is 0.4 milliampere at 5 volts and 0.8 milliampere at 10 volts, unloaded.

An external gate will allow division by 1 through 8. The 4017 is a similar decade counter.

4023

TRIPLE 3-INPUT NAND GATE

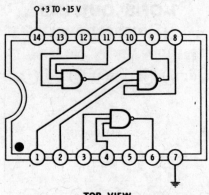

TOP VIEW

All three positive-logic NAND gates may be used independently. On any one gate, with one or more inputs *low*, the output will be *high*. With all inputs *high*, the output will be *low*.

The A-series response is lopsided. Use B-series for new designs. Device is functionally equivalent to the 7410 (TTL) and 74C10 (CMOS) devices.

Propagation delay is 25 nanoseconds at 10 volts and 60 nanoseconds at 5 volts. Total package current at 1 megahertz is 0.6 milliampere at 5 volts and 1.2 milliamperes at 10 volts.

7-STAGE (÷128) BINARY RIPPLE COUNTER

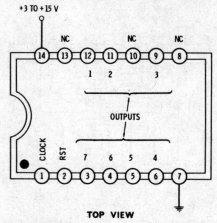

TOP VIEW

This is a binary ripple counter that counts in the up direction using positive logic.

The reset input is normally held at ground. Every time the clock changes from positive to ground, the counter advances one count. The 1 output divides the input clock by $2^1 = 2$. The 2 output divides the input clock by $2^2 = 4$, up to the 7 output which divides by $2^7 = 128$. Making the reset input positive forces all outputs to ground and holds them there until the reset returns to ground.

The clock input must be conditioned to be noiseless and fall only once per desired count. Clock rise and fall times should be faster than 5 microseconds.

Since this is a ripple counter, the outputs change in sequential order. Incorrect counts will briefly result during the settling time. Device is functionally and pin-for-pin equivalent to the 74HC4024 for LS TTL levels.

Maximum input freqency is 7 megahertz at 10 volts and 2.5 megahertz at 5 volts. Total package current is 0.2 milliampere at 5 volts and 0.4 milliampere at 10 volts for a 1-megahertz clocking rate. Consult the specific manufacturer's data sheet for propagation delay times.

Note: Device should not be confused with the MC4024 (74424) MTTL dual voltage-controlled astable IC.

4025

TRIPLE 3-INPUT NOR GATE

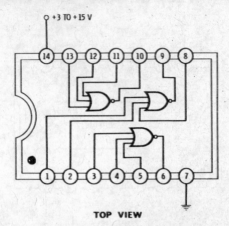

TOP VIEW

All three positive-logic NOR gates may be used independently.

On any one gate, with one or more inputs *high*, the output is *low*. With all three inputs *low*, the output is *high*.

The A-series response is lopsided. Use B-series for new designs. Device is functionally equivalent to the 7427 (TTL) and the 74C27 (CMOS).

Propagation delay is 25 nanoseconds at 10 volts and 60 nanoseconds at 5 volts. Typical package current at 1 megahertz is 1.2 milliamperes at 5 volts and 2.4 milliamperes at 10 volts.

4026

DECADE (÷10) COUNTER WITH 7-SEGMENT DECODED OUTPUT (Synchronous)

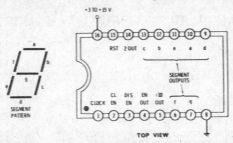

SEGMENT PATTERN

TOP VIEW

This synchronous decade, or divide-by-10, counter provides internal decoding to drive a 7-segment display. It does *not* have internal count storage, nor does it provide enough output current to directly drive higher-current displays. A divide-by-10 square-wave output is also available.

In normal operation, reset and clock enable are held at ground and the display enable is held positive. The counter advances one count on each ground-to-positive (positive edge) transition of the clock input.

There are two types of outputs. At the ÷10 output, a square wave that is high for counts 0 through 4 and low for counts 5 through 9 results. At the *a* through *g* outputs, a *high* state is produced if a display segment is to be *lit*. Segments *b* and *c* are used for the "1" output. Note that the "6" output includes segment *a* and the "9" output includes segment *d*. Available drive current is 1.2 milliamperes at 5 volts and 5 milliamperes at 10 volts. The "2 OUT" goes low only on count 2.

Fluorescent displays and newer LED displays may be directly driven by the outputs. High-current light-emitting diode and neon displays require external drivers. Liquid-crystal displays require external "ac" drive. Note that the outputs go *high* if the segment is to be *lit*.

The counter is reset to zero by bringing the RST terminal high. An a-b-c-d-e-f high output results, along with a high on the ÷10 output line. The reset must be returned to ground when counting is to continue. A high on the CL EN inhibits the clock operation and ignores counts. A low on the DIS EN turns *off* the display by making the *a* through *g* outputs low. This may be used to conserve display power or to provide a variable duty cycle brightness control. A slightly delayed EN OUT also follows the DIS EN input.

The clock must be noiseless and have only one ground-to-positive transition per desired count. Clock rise and fall times should be faster than 5 μs.

Maximum clock frequency is 5 megahertz at 10 volts and 2.5 megahertz at 5 volts. Total package current at 1-megahertz clock rate with unloaded outputs is 0.4 milliampere at 5 volts and 0.8 milliampere at 10 volts.

4027

DUAL JK FLIP-FLOP

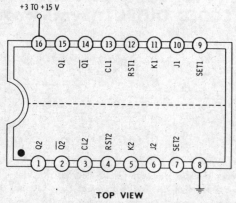

TOP VIEW

Each flip-flop may be used independently. There are two modes, *clocked* and *direct.*

In the *clocked* mode, the direct set and clear inputs must remain at ground. Inputs to the J and K lines decide what the flip-flop is going to do. The actual operation doesn't happen until the positive edge (ground-to-positive transition) of the clock.

If J and K remain grounded, clocking does nothing.
If J is positive and K grounded, clocking forces Q positive and $\overline{Q}$ to ground.
If J is grounded and K positive, clocking forces Q to ground and $\overline{Q}$ positive.
If both J and K are positive, clocking alternates the Q and $\overline{Q}$ states.

In the *direct* mode, a positive set input forces Q positive and $\overline{Q}$ to ground. A positive reset input forces Q to ground and $\overline{Q}$ positive. Should both set and reset simultaneously go positive, both Q and $\overline{Q}$ will also go positive. This is usually a disallowed state. The last direct input to go to ground will determine the final state of the Q and $\overline{Q}$ outputs. The direct inputs override the clocked inputs.

The clock input must be noiseless and have only a single ground-to-positive edge transition per desired clocking. Clock rise and fall times should be 5 microseconds or faster.

Maximum clock frequency is 8 megahertz at 10 volts and 3 megahertz at 5 volts. Total package current is 0.4 milliampere at 5 volts and 0.8 milliampere at 10 volts for a 1-megahertz clock frequency.

Device is functionally equivalent to the 7473, 7476, 74107, 74111, and 74167 (all TTL), and the 74C73, 74C76, and 74C107 (all CMOS).

BCD TO DECIMAL (1-of-10) DECODER

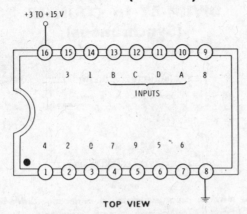

+3 TO +15 V

16	15	14	13	12	11	10	9
3	1		B.	C	D	A	8

INPUTS

4	2	0	7	9	5	6
1	2	3	4	5	6	7	8

TOP VIEW

This package will decode a standard binary-coded-decimal 4-bit code into 1-of-10 outputs. It can also convert any 3-bit code into 1-of-8 outputs.

The bcd code is input on terminals 10 through 13, with the least-significant or $2^0 = 1$ bit on "A," the $2^1 = 2$ bit on "B," the $2^2 = 4$ bit on "C," and the $2^3 = 8$ bit on "D." Positive logic with a "1" positive and a zero grounded is used.

One output goes positive for a given input; the other nine remain at ground. Available output current is 1 milliampere at 5 volts and 2 milliamperes at 10 volts.

For instance, with $A = 1$, $B = 1$, $C = 1$, and $D = 0$, output 7 goes positive and the rest remain low. All outputs remain low if an invalid bcd code (greater than 1001) is applied. Note that early RCA versions of this circuit produce an "8" output on invalid states "10," "12," and "14;" and a "9" output on invalid states "11," "13," and "15."

As a 1-of-8 decoder, the D input is *grounded*. Any 3-bit code applied to the A, B, and C inputs will provide one of the outputs high with the rest low. However, the output numbering will be correct only for a "straight binary" input code. Outputs are simply relabeled for any other code.

Propagation time is 100 nanoseconds at 10 volts and 250 nanoseconds at 5 volts. Total package current at a 1-megahertz word rate with unloaded outputs is 0.25 milliampere at 5 volts and 0.5 milliampere at 10 volts. Device is functionally equivalent to the 7442 (TTL) and 74C42 (CMOS) devices.

4029

UP-DOWN DIVIDE-BY-10 OR
DIVIDE-BY-16 COUNTER
(Synchronous)

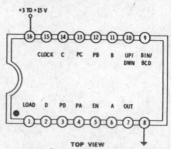

TOP VIEW

This package may be programmed to be a decade (divide-by-10) or a hexadecimal (divide-by-16) up-down, presettable counter.

In normal operation, LOAD and EN are held low. If pin 9 is grounded, the package counts by tens. If pin 9 is positive, the package counts by sixteens. If pin 10 is positive, the package counts "frontwards," or up-counts. If pin 10 is grounded, the package counts "backwards," or down-counts.

The package advances or retards one count on the ground-to-positive transition (positive edge) of the clock input. Outputs are weighted $A = 1$, $B = 2$, $C = 4$, and $D = 8$ in either the binary or bcd positive-logic code. The OUT terminal provides a *grounded* output on count 9 (bcd) or count 15 (binary mode). This is useful in cascading fully synchronous counters.

The EN input allows the clock to operate when grounded. It prevents clocking when positive. By connecting the OUT terminal of the first decade to the EN terminal of the second decade and driving the counters from a common clock, fully synchronous counting will result.

The counter can be parallel-loaded by presenting a word weighted $PA = 1$, $PB = 2$, $PC = 4$, and $PD = 8$ to the preset inputs and bringing the LOAD terminal positive. For instance, if $A = B = C = D = $ ground, a 0000 state will get entered, resetting the counter. Note that there is no separate reset pin; reset is done by parallel-loading a 0000.

The up-down control can be changed only when the clock is positive.

The clock must be noise free and have only one ground-to-positive transition per desired count. A rise time of faster than 5 microseconds on the clock is recommended.

Maximum clock frequency is 5 megahertz at 10 volts and 2.5 megahertz at 5 volts. Total package current at a 1-megahertz clock rate with unloaded outputs is 0.4 milliampere at 5 volts and 0.8 milliampere at 10 volts.

QUAD EXCLUSIVE-OR GATE

+3 TO +15 V

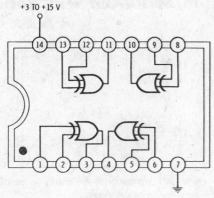

TOP VIEW

Early versions of this device have a very *low* input impedance and behave erratically in pulse circuits. Use the 4070 or the equivalent 4507 instead.

Device is functionally equivalent to the 7486 (TTL) and 74C86 (CMOS) devices.

4031

64-STAGE STATIC SHIFT REGISTER
(Serial-In/Serial-Out With Recirculate)

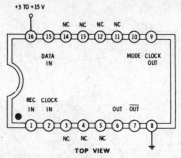

This is a fully static shift register with the ability to recirculate data. It is arranged as 64 serial-in/serial-out stages.

In normal operation, data to be stored is routed to the Data In terminal and the Mode input is grounded. Data enters the register on the ground-to-positive transition (positive edge) of the clock.

In 64 successive clock pulses, entered data will appear as an output on pin 6 and as its complement on pin 7. The normal output (pin 6) may be routed as an input to a following register, cascading stages in multiples of 64. Pin 6 and pin 7 can each drive one TTL load.

If the Mode input is made positive, data at the REC IN terminal gets entered on the next positive clock. For recirculation, the REC IN terminal can be connected to pin 6 of this register or any cascaded register. Alternately, the Mode control can be used to select the data stream on pin 1 (Mode positive) or on pin 15 (Mode grounded).

The clock must be noise free and have only one ground-to-positive transition per desired clocking. Unlike most CMOS circuits, the clock has a very high capacitance of 60 picofarads that must be driven with a rise time of one microsecond or less. Whatever is driving the clock must have a minimum source and sink current of 1 milliampere to drive this capacitance. As packages are cascaded with a common clock, the clock capacitance multiplies accordingly and the input drive current needed goes up proportionately.

A slightly delayed replica of the clock appears at pin 9 which can be used to drive one additional register. "Daisy chain" clocking like this introduces some recirculation problems—see manufacturer's data sheet for recommendations.

Maximum clock frequency is 4 megahertz at 10 volts and 2 megahertz at 5 volts. Total package current unloaded at a 1-megahertz clock rate is 1.2 milliamperes at 5 volts and 2.4 milliamperes at 10 volts.

TRIPLE SERIAL ADDER
(Positive Logic)

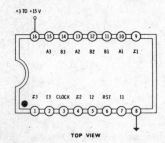

TOP VIEW

This package contains three separate adders that perform binary addition of *sequential* words of any length. All three adders must be used in a common system as they share a common clock and a common carry reset. Positive logic is used (high = 1; ground = 0).

In normal operation, the Invert "I" inputs are held low. Then the CARRY RST input is briefly made high, during which time the clock provides a ground-to-positive transition. Both clock and reset must then return to ground. This operation clears any carry information remaining from a previous addition.

Data words A and B, obtained from a shift register or sequentially accessed memory, are routed to the adder one bit at a time. The bit-by-bit sum of A and B appears at the Σ output. One clock must go with each bit-by-bit sum.

The words presented to the A and B inputs *must arrive least-significant bit first, with a sign bit following the most-significant bit.*

Suppose the first adder is to handle two 4-bit words. The carry is reset from any previous additions. Then the least-significant bit of A1 and B1 is routed to the adder. The sum of these two bits is computed on the positive clock edge; any carry is internally stored. The next two significant bits are then combined with any carry, the sum is computed and provided at the Σ output, and any internal carry is stored. The process continues for the third, fourth, and sign bits.

The words may be of any length, but both words must be of equal length. The longer the word, the more clock cycles needed for a complete addition. *Input data must change immediately after clocking; otherwise, glitches in the output will result.* The 1 (invert) inputs may be used to independently invert the output data where a complementary output is needed.

Total add time per bit is 250 nanoseconds at 10 volts and 800 nanoseconds at 5 volts. Maximum clock rate is 3 megahertz at 10 volts and 1.5 megahertz at 5 volts. Total package current at a 1-megahertz clock rate is 0.8 milliampere at 5 volts and 1.6 milliamperes at 10 volts.

4033

DECADE (÷10) COUNTER WITH 7-SEGMENT DECODED OUTPUT (Synchronous)

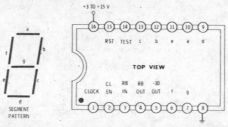

This synchronous decade, or divide-by-10, counter provides internal decoding to drive a 7-segment display. It does *not* have internal count storage, nor does it provide enough output current to directly drive high-current display types. A divide-by-10 square-wave output is also available.

In normal operation, reset and clock enable are held at ground and the ripple blanking input is connected to ground or a more-significant count stage. The counter advances one count on each ground-to-positive (positive edge) transition of the clock input.

There are two types of outputs. At the ÷10 output, a square wave that is high for counts 0 through 4 and low for counts 5 through 9 results. At the *a* through *g* outputs, a *high* state is produced if a display segment is to be *lit*. Segments *b* and *c* are used for the "1" output. Note that the "6" output includes segment *a* and the "9" output includes segment *d*.

Fluorescent displays and newer common-cathode LED displays may be directly driven by the outputs. High-current light-emitting diode and neon displays require external drivers. Liquid-crystal displays require external "ac" drive. Note that the outputs go *high* if the segment is to be lit.

The counter is reset to zero by bringing the RST terminal high. This results in an a-b-c-d-e-f low, along with a high on the ÷10 output. The RST input must be returned to ground when counting is to continue. A high on the Test input puts all outputs high for lamp or display test.

To automatically extinguish all right-hand zeros, ground the RB IN terminal of the most-significant stage and connect its RB OUT to the RB IN of the next most-significant stage, and so on down the line. This zero blanking is defeated by making all RB IN terminals positive.

The clock must be noiseless and have only one ground-to-positive transition per desired count. Rise and fall times should be 5 microseconds or faster.

Maximum clock frequency is 5 megahertz at 10 volts and 2.5 megahertz at 5 volts. Total package current at a 1-megahertz clock rate with unloaded outputs is 0.4 milliampere at 5 volts and 0.8 milliampere at 10 volts.

BIDIRECTIONAL BUS REGISTER

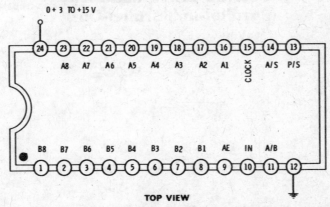

0 + 3 TO +15 V

TOP VIEW

This is a very versatile interface that lets you transfer 8 bits worth of data between two bus systems. It is also a universal 8-bit shift register. Note the 24-pin package.

There are four control lines that decide what the package is to do.

If the A/B line is high, the eight A-bus lines are *inputs* and the eight B-bus lines are *outputs*. If the A/B line is low, the eight B-bus lines are *inputs* and the eight A-bus lines are *outputs*.

The AE line floats the A outputs tri-state style if the AE line is low; it allows the A outputs to appear if the AE line is high. For instance, if the AE and A/B lines are both low, data would be accepted from the B inputs and held internally to be output on the A lines at a later time.

The asynchronous/synchronous (A/S) line lets transfers take place *immediately* if it is high. If the A/S line is low, transfers take place on the ground-to-positive transition (positive edge) of a clock input.

The parallel/serial (P/S) line picks parallel operation if it is high, selecting data from the A or B buses and asynchronously or synchronously transferring the data. If the P/S line is high, data from the IN terminal is serially entered, transferring one stage on each positive clock edge. *The A/S line must be low if the P/S line is high.* Asynchronous serial operation is not allowed.

When used, the clock has to be noise free and have only one ground-to-positive transition per desired clocking. Clock rise and fall times should be faster than ten microseconds.

Maximum clock frequency is 3 megahertz at 10 volts and 1.5 megahertz at 5 volts. Total package current is 2 milliamperes at 5 volts and 4 milliamperes at 10 volts at a 1-megahertz clocking rate.

4035

4-STAGE SHIFT REGISTER
(Parallel-In/Parallel-Out)

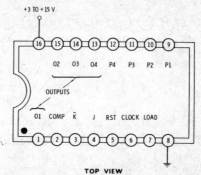

TOP VIEW

This is a very flexible 4-stage shift register. It can be used for any shift-right application ranging from serial-in/serial-out through parallel-in/parallel-out. Parallel loading is clocked.

To build a serial-in/serial-out or serial-in/parallel-out register, connect J and $\overline{K}$ together and ground COMP, RST, and LOAD. Input data routed to the J and $\overline{K}$ inputs will get entered into the first stage and appear at output 01 after a ground-to-positive (positive edge) clock transition. A second clocking transfers this data to the 02 output, a third to 03, and a fourth to 04.

To parallel-load data, information is presented to the P1 through P4 inputs with P1 nearest the input and P4 at the output. The Load input is next brought high. On the next ground-to-positive (positive edge) clock transition, the parallel information gets loaded into the register. The Load must remain high till after clocking, then drop.

If the RST input is made positive, all stages immediately go to their output-low state. The RST input must return low before clocking can continue. The COMP input will complement (change) all outputs when it is high, giving the complement of the data in the register. If the J input is made high and the $\overline{K}$ input is made low, clocking will change or complement only the first bit of the register while passing on information in the usual way. This unusual connection is sometimes useful in sequence generators.

The clock must be noise free and have only one ground-to-positive transition per desired shift. Clock rise and fall times should be faster than 5 μs.

Propagation delay time is 100 ns at 10 volts and 250 ns at 5 volts. Maximum clock frequency is 3 MHz at 10 volts and 1.5 MHz at 5 volts. Total package current at a 1-MHz clock rate is 2 mA at 5 volts and 4 mA at 10 volts. Functionally equivalent to the TTL 74178 and 74195 types and the CMOS 40195 and 74C195 types.

4038

TRIPLE SERIAL ADDER
(Negative Logic)

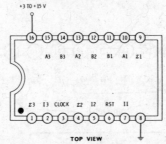

TOP VIEW

This package contains three separate adders that perform binary addition of *sequential* words of any length. All three adders must be used in a common system as they share a common clock and a common carry reset. *Negative logic* is used (ground = 1; high = 0).

In normal operation, the I (invert) inputs are held low. The CARRY RST input is briefly brought low, during which time the clock provides a positive-to-ground transition. Both clock and reset must then return positive. This operation clears any carry information remaining from a previous addition.

Data words A and B, obtained from a shift register or sequentially accessed memory, are routed to the adder one bit at a time. The bit-by-bit sum of A and B appears at the Σ output. One clock must go with each bit-by-bit sum.

The words presented to the A and B inputs must arrive least-significant bit first, with a sign bit following the most-significant bit.

Suppose the first adder is to handle two 4-bit words. The carry is reset from any previous additions. Then the least-significant bit of A1 and B1 is routed to the adder. The sum of these two bits is computed on the negative clock edge; any carry is internally stored. The next two significant bits are then combined with any carry, the sum is computed and provided at the output, and any new carry is internally stored. The process continues for the third, fourth, and sign bits.

The words may be of any length, but both words must be of *equal* length. The longer the word, the more clock cycles needed for a complete addition. *Input data must change immediately after clocking; otherwise, glitches in the output will result.* The I (invert) inputs may be used to independently invert the output data where a complementary output is needed.

Total add time per bit is 250 nanoseconds at 10 volts and 800 nanoseconds at 5 volts. Maximum clock rate is 3 megahertz at 10 volts and 1.5 megahertz at 5 volts. Total package current at a 1-megahertz clock rate is 0.8 milliampere at 5 volts and 1.6 milliamperes at 10 volts.

4040

12-STAGE (÷4096) BINARY RIPPLE COUNTER

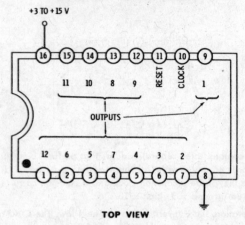

TOP VIEW

This is a binary ripple counter that counts in the up direction using positive logic.

The reset input is normally held at ground. Every time the clock changes from positive to ground, the counter advances one count. The 1 output divides the input clock by $2^1 = 2$. The 2 output divides the input clock by $2^2 = 4$. The 3 output divides the input clock by $2^3 = 8$, up to the 12 output which divides by $2^{12} = 4096$.

Making the reset input positive forces all outputs to ground and holds them there until the reset returns to ground.

The clock input must be conditioned to be noiseless and fall only once per desired count. Clock rise and fall times should be faster than 5 microseconds.

Since this is a ripple counter, the outputs change in sequential order and incorrect counts will briefly result during the settling time. Device is functionally and pin-for-pin equivalent to 74HC4040 for LS TTL levels.

Maximum input frequency is 6 megahertz at 10 volts and 2 megahertz at 5 volts. Total package current at a 1-megahertz clock rate is 0.4 milliampere at 5 volts and 0.8-milliampere at 10 volts. Consult the manufacturer's data sheet for propagation times.

QUAD TTL BUFFER

+3 TO +15 V
(+5 V FOR TTL COMPATABILITY)

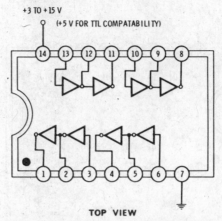

TOP VIEW

This package contains four buffers having both inverting and noninverting outputs. All four buffers may be used independently.

A high level on pin 3 appears as a low level on pin-2 output and as a high level on pin-1 output, and vice versa.

When used with a 5-volt supply, the outputs are TTL compatible, with the complementary outputs sinking 1.6 mA or one TTL load, and the true outputs sinking 3.2 mA or two TTL loads.

The package is also useful for complement generation and as rise-time improvers in pulse-shaper, astable, and monostable circuits, and in digital-to-analog converters.

Note that this package must be used with a 5-volt supply if TTL output compatibility is needed. If you need a buffer between TTL and a higher supply voltage, use the 4049 or 4050 instead.

Propagation time is 45 nanoseconds at 10 volts and 75 nanoseconds at 5 volts. Total package current with unloaded outputs at 1 megahertz is 1.6 milliamperes at 5 volts and 3.2 milliamperes at 10 volts. Dissipation increases with slow rise and fall input signals. Consult the manufacturer's data sheet for curves.

4042

QUAD LATCH

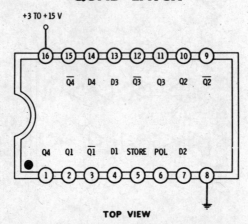

+3 TO +15 V

| 16 | 15 | 14 | 13 | 12 | 11 | 10 | 9 |
| | $\overline{Q4}$ | D4 | D3 | $\overline{Q3}$ | Q3 | Q2 | $\overline{Q2}$ |

| Q4 | Q1 | $\overline{Q1}$ | D1 | STORE | POL | D2 | |
| 1 | 2 | 3 | 4 | 5 | 6 | 7 | 8 |

TOP VIEW

This package contains four latches having a common *level-controlled* storage-command input. A choice of control polarity is available.

If the POL input is low and the STORE input is low, data sent to the D input appears at its respective true and complement outputs. When the Store input is brought high, the data on the input during the positive transition gets stored internally and appears in true form at output Q and its complement at $\overline{Q}$.

If the POL input is high and the STORE input is high, data sent to the D input appears at its respective true and complement outputs. When the Store input is brought low, the data on the input during the negative transition gets stored internally and appears in true form at output Q and its complement at $\overline{Q}$.

Note that the output *follows* the input for one state of the Store control. This is not a true edge-clocked device and *stages may NOT be cascaded.*

The Store input must be noise free. Recommended rise and fall times must be faster than 5 microseconds.

Functionally equivalent to the TTL 7475 and 7477 devices, and the CMOS 74C75 and 74C77 devices.

Propagation delay times are 75 nanoseconds at 10 volts and 150 nanoseconds at 5 volts. Typical package current at a 1-megahertz store rate is 1.2 milliamperes at 5 volts and 2.4 milliamperes at 10 volts.

QUAD R/S FLIP-FLOP
(NOR Logic)

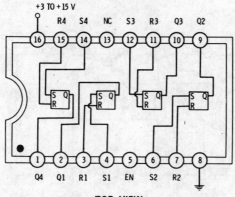

TOP VIEW

This package contains four independent set-reset flip-flops sharing a common-output tri-state enable control.

On any one flip-flop, Set and Reset should normally be *low*. If Set is made *high*, the output goes and stays *high*. If Reset is made *high*, the output goes and stays *low*.

If both Set and Reset go *high*, the output goes *high*. This is a normally disallowed state, and the last input to go low determines the final flip-flop state.

The outputs are tri-state. They float if the EN (enable) control is *low*, and they are connected to the flip-flops if the EN control is *high*.

Note that these are unclocked, simple flip-flops. They should not be cascaded, nor are they suitable for counting or shift-register use.

Propagation delay time is 75 nanoseconds at 10 volts and 175 nanoseconds at 5 volts. At a 1-megahertz clock rate, the total package current is 1 milliampere at 5 volts and 2 milliamperes at 10 volts.

The 4044 is a similar device with NAND (normally high) inputs.

4044

QUAD R/S FLIP-FLOP
(NAND Logic)

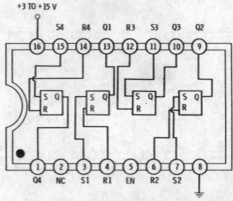

+3 TO +15 V

S4 R4 Q1 R3 S3 Q3 Q2

16 15 14 13 12 11 10 9

S Q R   S Q R   S Q R   S Q R

1 2 3 4 5 6 7 8

Q4 NC S1 R1 EN R2 S2

TOP VIEW

This package contains four independent set-reset flip-flops sharing a common-output tri-state enable control.

On any one flip-flop, Set and Reset should normally be *high*. If Set is made *low*, the Q output goes and stays *high*. If Reset is made *low*, the Q output goes and stays *low*.

If both Set and Reset go *low*, the output goes *low*. This is a normally disallowed state, and the last input to go high determines the final state of the flip-flop.

The outputs are tri-state. They float if the EN (enable) control is *low*, and they are connected to the flip-flops if the EN control is *high*.

Note that these are unclocked, simple flip-flops. They should not be cascaded, nor are they suitable for counting or shift-register use.

Propagation delay time is 75 nanoseconds at 10 volts and 175 nanoseconds at 5 volts. The total package current at a 1-megahertz clock rate is 1 milliampere at 5 volts and 2 milliamperes at 10 volts.

The 4043 is a similar device with NOR (normally low) inputs.

PHASE-LOCKED LOOP

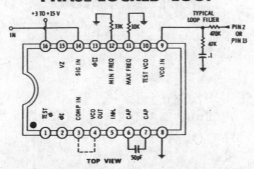

This is a very versatile phase-locked-loop circuit. More information on it appears in Chapter 7. In normal operation, no connections are made to pins 1, 10, and 15, and the INH input is made *low*.

The voltage-controlled oscillator frequency is determined by the voltage on pin 9, the capacitor between pins 6 and 7 (50-pF min), the maximum frequency resistor on pin 11 (10K to 1 megohm), and the minimum frequency resistor on pin 12 (10K to infinity—larger than pin-11 resistor). The output appears at pin 4, and is usually routed to the COMP input, either directly or via a divide-by-n counter.

There are two possible phase detectors. The $\phi1$ system is an EXCLUSIVE-OR system that offers good noise performance, but is harmonic sensitive and MUST have square waves on both pins 3 and 14. It is limited to a narrow frequency range.

The $\phi2$ system is a logic frequency/phase detector that operates over a wide frequency range (to 1000:1 and beyond), accepts any input duty cycle, and is not harmonic sensitive. It has relatively poor noise rejection.

The selected phase detector is routed to the vco input via a loop filter. The series resistor and capacitor set the time constant of the loop, while the shunt resistor sets the damping. Normally, the damping resistor is much smaller than the series resistor. Some typical values using a high frequency and the $\phi2$ detector are shown in the circuit above.

When properly connected, the frequency of the vco will track an input frequency applied to pin 14. If a divide-by-n counter is placed between pins 3 and 4, the vco will run at n times the input frequency. *If the input is a low frequency, a noise free, sharp rise and fall input must be provided.* At higher frequencies, pin-14 inputs may be capacitor coupled or directly driven from logic.

Total supply current varies with frequency and connections. Consult the manufacturer's data sheet and Chapter 7 for more information.

4047

MULTIVIBRATOR

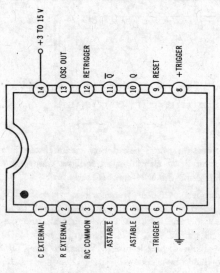

TOP VIEW

This circuit may be used for astable or monostable applications. Complete details for its use appear in Chapter 4.

As a monostable, the circuit may be triggered with either a positive or negative trigger pulse, depending upon which input is used. The output pulse width is determined by an external R and C (the curves are shown in Fig. 4-46). R can vary from 10K to 1 megohm. C can range from 0.001 microfarad up. There are complementary outputs.

As an astable, the circuit is free running, which depends on the external R and C. Design curves are given in Fig. 4-48. The output frequency at the OSC output (pin 4) is twice that of the complementary outputs (pins 10 and 11), whose duty cycle are both 50%. The duty cycle of the waveform at the OSC output is not guaranteed to be 50%.

Typical quiescent current is 5 microamperes at 5 volts, 10 microamperes at 10 volts, and 20 microamperes at 15 volts.

4049

HEX INVERTING BUFFER & TTL DRIVER

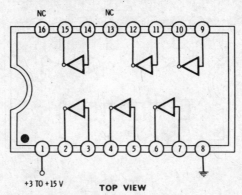

TOP VIEW

+3 TO +15 V

All six buffers may be used independently. The buffers may be used as simple inverters, as voltage translators, or as current drivers for interfacing TTL or other logic.

On any buffer, the input *low* drives the output *high*, and vice versa.

The voltage on pin 1 sets the voltage swing at the *output* only. Input voltages up to +15 volts are safely accepted regardless of the selected output voltage.

With a +5-volt supply on pin 1, the output is TTL compatible. It provides 3.2 milliamperes or a fan-out to *two* regular TTL gates, or *four* LS TTL gates.

Note the unusual supply connections. This package should not normally be used with slow-rise-time inputs such as pulse shapers, monostable and astable circuits, etc. The internal power dissipation can become too great in these and other linear applications, particularly at high frequencies and high supply voltages.

Functionally equivalent to the 7404 (TTL) and the CMOS 74C04 and 74HC4049 devices.

Propagation delay is 25 nanoseconds at 10 volts and 35 nanoseconds at 5 volts. Total package current (unloaded) at 1 megahertz is 0.8 milliampere at 5 volts and 1.6 milliamperes at 10 volts.

4050

HEX NONINVERTING BUFFER & TTL DRIVER

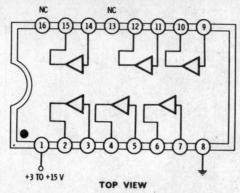

TOP VIEW

+3 TO +15 V

All six buffers may be used independently. The buffers may be used as rise-time improvers, as voltage translators, or as current drivers for interfacing TTL and other logic.

On any buffer, the input *low* provides a *low* output, and vice versa.

The voltage on pin 1 sets the voltage swing at the *output* only. Input voltages up to +15 volts are safely accepted regardless of the selected output voltage.

With a +5-volt supply on pin 1, the output is TTL compatible. It provides 3.2 milliamperes or a fan-out to two regular TTL gates, or four LS TTL gates.

Note the unusual supply connections. This package should not normally be used with slow-rise-time inputs such as pulse shapers, monostable and astable circuits, etc. The internal power dissipation can become too great in these and other linear applications, particularly at high frequencies and high supply voltages.

Propagation delay is 30 nanoseconds at 10 volts and 60 nanoseconds at 5 volts. Total package current (unloaded) at 1 megahertz is 0.8 milliampere at 5 volts and 1.6 milliamperes at 10 volts.

4051

1-OF-8 SWITCH

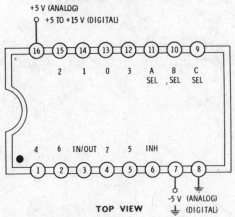

TOP VIEW

This package may be used as a 1-of-8 analog data multiplexer or demultiplexer, or as a 1-of-8 digital selector or distributor.

In the analog mode, −5 volts is applied to pin 7, and digital-control signals of low = ground and high = +5 are applied to the A, B, C, and INH inputs. If INH is high, no channel is selected. If INH is low, the channel selected is determined by the binary word input to A = 1, B = 2, and C = 4. Analog signals may be any value between +5 and −5 volts.

In the digital mode, pin 7 is grounded, and digital-control signals of low = ground and high = pin-16 voltage are applied to the A, B, C, and INH inputs. If INH is high, no channel is selected. If INH is low, the channel selected is determined by the binary word input to the A = 1, B = 2, and C = 4 inputs. Digital signals controlled may be any value between the pin-16 voltage and ground.

In either mode, the OFF state is an open circuit and the ON state is a 120-ohm resistor. Pin 3 may be used as an input or an output, depending on whether information is to be gathered from eight possible sources or distributed to eight possible locations. For digital signals only, device is functionally equivalent to the 74151 (TTL) and the 74C151 (CMOS) devices.

The minimum permissible load resistance is 100 ohms, and not more than 25 milliamperes can be routed through the circuit.

Total package current at a 1-megahertz clock rate is 0.5 milliampere at 5 volts and 1 milliampere at 10 volts, open circuited. Propagation delay is 200 nanoseconds for the supply connections shown. See Chapter 7 for more information.

4052

DUAL 1-OF-4 SWITCH

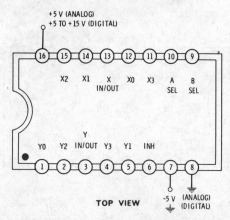

+5 V (ANALOG)
+5 TO +15 V (DIGITAL)

| 16 | 15 | 14 | 13 | 12 | 11 | 10 | 9 |

X2 X1 X X0 X3 A B
IN/OUT SEL SEL

Y
Y0 Y2 IN/OUT Y3 Y1 INH

| 1 | 2 | 3 | 4 | 5 | 6 | 7 | 8 |

TOP VIEW

-5 V (ANALOG)
(DIGITAL)

This package may be used as a dual 1-of-4 analog data multiplexer or demultiplexer, or as a dual 1-of-4 digital selector or distributor.

In the analog mode, —5 volts is applied to pin 7, and digital-control signals of low = ground and high = +5 are applied to the A, B, and INH terminals. If INH is high, no channel is selected. If INH is low, the channels selected are determined by the binary word input to the A = 1 and B = 2 inputs. Analog signals may be any value between +5 and —5 volts.

In the digital mode, pin 7 is grounded, and digital-control signals of low = ground and high = pin-16 voltage are applied to the A, B, and INH terminals. If INH is high, no channels are selected. If INH is low, the channels selected are determined by the binary word input to the A = 1 and B = 2 inputs. Digital signals controlled may be any value between the pin-16 voltage and ground.

In either mode, the OFF state is an open circuit, and the ON state is a 120-ohm resistor. Pins 3 and 13 may be used as either an input or an output, depending on whether information is to be gathered from four possible sources or distributed to four possible locations. For digital signals only, device is functionally equivalent to the 74153 (TTL) and the 74C153 (CMOS) devices.

The minimum permissible load resistance is 100 ohms, and not more than 25 milliamperes can be routed through the circuit.

Total package current at a 1-megahertz clock rate is 1 milliampere at 5 volts and 2 milliamperes at 10 volts, open circuited. Propagation delay is 200 nanoseconds for the supply connections shown. See Chapter 7 for more information.

TRIPLE 1-OF-2 SWITCH

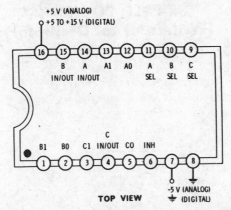

+5 V (ANALOG)
+5 TO +15 V (DIGITAL)

TOP VIEW

-5 V (ANALOG)
(DIGITAL)

This package may be used as a triple 1-of-2 analog data multiplexer or as a triple 1-of-2 digital selector or distributor. The three sections may be independently controlled, but they share a common inhibit terminal.

In the analog mode, −5 volts is applied to pin 7, and digital-control signals of low = ground and high = +5 are applied to the INH and the A, B, and C terminals. If INH is high, no channels are selected. If INH is low, the channels selected depend on the A, B, and C SEL inputs. For instance, if A SEL is low, A0 is connected to A IN/OUT; if A SEL is high, A1 is connected to A IN/OUT, and similarly for the B and C channels.

In the digital mode, pin 7 is grounded, and digital-control signals of low = ground and high = pin-16 voltage are applied to the A, B, C, and INH terminals. If INH is high, no channels are selected. If INH is low, the channels selected depend on the A, B, and C SEL inputs. For instance, if A SEL is low, A0 is connected to A IN/OUT; if A SEL is high, A1 is connected to A IN/OUT, and similarly for the B and C channels.

In either mode, the OFF state is an open circuit and the ON state is a 120-ohm resistor. Pins 4, 14, and 15 may be used as either an input or an output, depending on whether information is to be gathered from two possible sources or distributed to two possible locations.

The minimum permissible load resistance is 100 ohms, and not more than 25 milliamperes can be routed through the switches.

Total package current at a 1-megahertz clock rate is 1.2 milliamperes at 5 volts and 2.4 milliamperes at 10 volts, open circuited. Propagation delay is 200 nanoseconds for the supply connections shown.

The 4051 is a similar 1-of-8 device, while the 4052 is a dual 1-of-4 device. See Chapter 7 for more information.

4060

14-STAGE (÷16,384) BINARY RIPPLE COUNTER
(With Internal Oscillator)

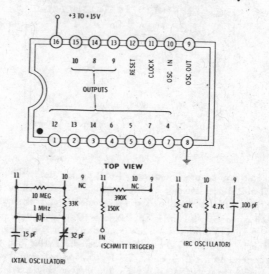

This is a binary ripple counter that counts in the up direction using positive logic. Feedback available at the clock input allows input conditioning or use of a self-contained crystal or resistor-capacitor oscillator.

The reset input is normally held at ground. Each time the clock changes from positive to ground, the counter advances one count. The 4 output divides the input clock by $2^4 = 16$, up through the 14 output which divides by $2^{14} = 16,384$. No outputs are available for divisions of 2, 4, 8, and 2048.

Since this is a ripple counter, the outputs change in sequential order. Incorrect counts will briefly result during settling times.

The crystal oscillator, Schmitt trigger, and RC oscillator connections are shown in the circuit above. When used as a divider-only, the clock input must be bounceless and noise free, and have rise and fall times faster than 5 microseconds. Bringing the reset input positive for a minimum of half a microsecond resets all counter stages to zero. For LS TTL levels, device is functionally and pin-for-pin equivalent to 74HC4060.

Power-supply current is 0.4 milliampere at 5 volts and 0.8 milliampere at 10 volts for a 1-megahertz clocking rate. An extra 2 milliamperes is needed if the clock is used in an astable or crystal mode. Maximum clock frequency is 1.75 megahertz at 5 volts and 4 megahertz at 10 volts.

4-BIT MAGNITUDE COMPARATOR

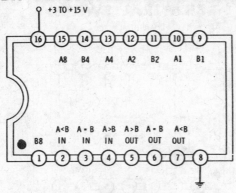

TOP VIEW

This package compares two 4-bit binary words and provides outputs to determine whether they are equal or which is greater.

In normal operation, pins 2 and 4 are grounded and pin 3 is made high. The A word is weighted $A1 = 1$, $A2 = 2$, $A4 = 4$, and $A8 = 8$. The B word is weighted $B1 = 1$, $B2 = 2$, $B4 = 4$, and $B8 = 8$, using positive logic. If $A = B$, pin 6 goes high and pins 4 and 7 remain low.

To cascade more bits, the three outputs of the first package are connected to the three respective inputs of the next package, and so on. The first package then becomes the least-significant bits of the word.

Compare time is 625 nanoseconds at 5 volts and 250 nanoseconds at 10 volts. Power-supply current at a 1-megahertz word rate is 0.4 milliampere at 5 volts and 0.8 milliampere at 10 volts.

Device is functionally equivalent to the 7485 (TTL) and 74C85 (CMOS) devices.

The 4085 is a similarly functioning device with different pinouts. The 4585 is a somewhat similar device with different pinouts and a reversed input sense on the A>B line.

4066

QUAD DIGITAL OR ANALOG BILATERAL SWITCH

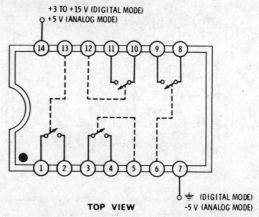

+3 TO +15 V (DIGITAL MODE)
+5 V (ANALOG MODE)

(DIGITAL MODE)
-5 V (ANALOG MODE)

TOP VIEW

All four switches may be used separately or in combination.

On any single switch, when the control voltage equals the pin-7 voltage, the switch remains OFF and behaves as a very high impedance. When the control voltage equals the pin-14 voltage, the switch turns ON and behaves as a nearly linear, bilateral, 90-ohm resistor.

Signals routed through the switch may be digital or analog, but they must never exceed the pin-14 voltage nor go below the pin-7 voltage.

Switches may be shorted together in any pattern, and there is no difference between the input and output terminals of any switch.

For instance, if all four switches are connected with one common terminal, the package may be used as a 1-of-4 data selector, a 1-of-4 data distributor, a 1-of-4 analog commutator, or a 1-of-4 analog multiplexer.

If more than one switch is connected to a common point, external logic usually must guarantee that only one switch is turned on at a time.

Maximum switching frequency is 10 megahertz at 10 volts and 5 megahertz at 5 volts. Package dissipation depends on the loading. Dissipation should be kept under 100 milliwatts total.

This is an improved version of the 4016, having a lower ON resistance. However, the 4016 remains a better choice for ultralow-leakage applications, such as sample-hold circuits.

See Chapter 7 for more information.

1-OF-16 ANALOG SWITCH

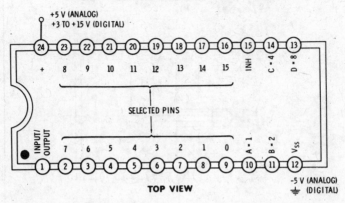

This package may be used as a 1-of-16 analog data multiplexer or demultiplexer, or as a 1-of-16 digital selector or distributor. In the analog mode, −5 volts is applied to pins 12 and 15 and +5 volts is applied to pin 24.

The channel selected is determined by the A, B, C, and D inputs weighted A=1, B=2, C=4, and D=8, with a zero defined as −5 volts and a 1 defined as +5 volts. For instance, with A=1, B=0, C=1, and D=1, channel 13 (pin 18) is connected to the common input/output terminal.

In the digital mode, ground is applied to pins 12 and 15, and +3 to +15 volts to pin 24. The channel selected is similarly determined, with a zero being ground and a 1 being defined as the pin-24 voltage. Device is functionally equivalent to the 74150 (TTL) and the 74C150 (CMOS), for digital signals only.

The ON resistance is 200 ohms. On-channel frequency response extends to 40 megahertz. Off-channel cross talk is −40 decibels (1/100 amplitude) for frequencies less than 1 megahertz. The inhibit pin will turn all channels off if brought to the pin-24 voltage. More than one channel can be on during settling times. This can be eliminated by using the Inhibit before, during, and immediately after channel selection.

All inputs and outputs must be less than the pin-24 voltage and more than the pin-12 voltage.

Select and inhibit propagation times are 200 nanoseconds at 10 volts and 400 nanoseconds at 5 volts. Power dissipation depends on the loading and the frequency of operation. See Chapter 7 for more information.

4068

8-INPUT NAND GATE

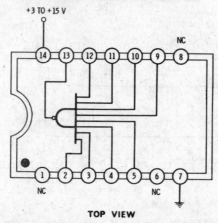

+3 TO +15 V

NC

14 13 12 11 10 9 8

1 2 3 4 5 6 7

NC   NC

TOP VIEW

This package contains a single positive-logic, 8-input NAND gate.

If one or more inputs are *low*, the output will be *high*. If all eight inputs are *high*, the output will be *low*.

Device is functionally equivalent to the 7430 (TTL) and 74C30 (CMOS) devices.

Propagation delay is 130 nanoseconds at 10 volts and 325 nanoseconds at 5 volts. Total package current is 0.5 milliampere at 5 volts and 1 milliampere at 10 volts.

Note that this is a very slow device. It should not be used in high-speed applications, particularly on a 5-volt or lower supply.

HEX INVERTER

+3 TO +15 V

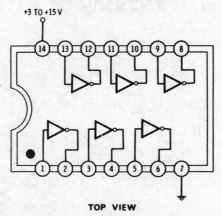

TOP VIEW

All six inverters may be used independently. On any inverter, a *low* input provides a *high* output, and vice versa.

Device is functionally equivalent to the 7404 (TTL) and 74C04 (CMOS) devices.

This is a "low-power" version of the 4049. It will not directly drive regular TTL, nor can it be used for voltage translation.

In addition, this device is only singly buffered, which means *the 4069B will perform no better in astable and pulse circuits than ordinary A-series devices.* Thus, while the 4069B has the output drive typical of other B-series devices, it has far less internal gain.

Propagation delay is 25 nanoseconds at 10 volts and 50 nanoseconds at 5 volts. Total package current at 1 megahertz is 0.5 milliampere at 5 volts and 1 milliampere at 10 volts.

4070

QUAD EXCLUSIVE-OR GATE

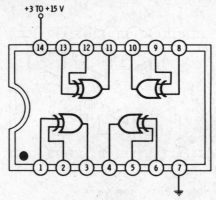

TOP VIEW

All four EXCLUSIVE-OR gates may be used independently.

On any one gate, if one input is high but not both, the output will be *high*. If both inputs are *high* or both inputs are *low*, the output will be *low*.

The gate can be used as a comparator by noting that identical inputs give a low output, and different inputs, a high output. It can also be used as a controllable inverter by noting that a low on one input passes on whatever is on the other input; while a high complements whatever is on the other input.

Propagation delay is 70 nanoseconds at 10 volts and 175 nanoseconds at 5 volts. Supply current at a 1-megahertz data rate is 0.2 milliampere at 5 volts and 0.4 milliampere at 10 volts.

This device is identical to the 4508 and replaces the obsolete 4030. It is functionally equivalent to the 7486 (TTL) and the 74C86 (CMOS).

QUAD 2-INPUT OR GATE

+3 TO +15 V

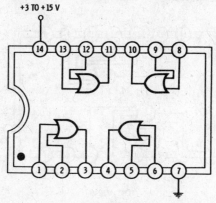

TOP VIEW

All four positive-logic OR gates may be used independently.

On any one gate, with either or both inputs *high*, the output will be *high*. With both inputs *low*, the output will be *low*.

Device is functionally equivalent to the 7432 (TTL) and the 74C32 (CMOS) devices.

Propagation delay is 80 nanoseconds at 10 volts and 190 nanoseconds at 5 volts. Total package current at 1 megahertz is 0.5 milliampere at 5 volts and 1 milliampere at 10 volts.

4072

DUAL 4-INPUT OR GATE

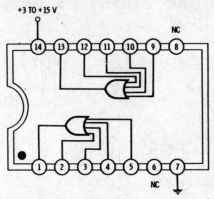

TOP VIEW

Both positive-logic OR gates may be used independently.

On any one gate, one or more inputs *high* provides a *high* output. With all four inputs *low*, the output will be *low*.

Propagation delay is 80 nanoseconds at 10 volts and 190 nanoseconds at 5 volts. Total package current at 1 megahertz is 0.5 milliampere at 5 volts and 1 milliampere at 10 volts.

TRIPLE 3-INPUT AND GATE

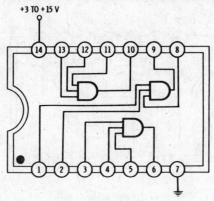

TOP VIEW

All three positive-logic AND gates may be used independently.

On any one gate, with one or more inputs *low*, the output will be *low*. With all three inputs *high*, the output will be *high*.

Propagation delay is 70 nanoseconds at 10 volts and 150 nanoseconds at 5 volts. Total package current at 1 megahertz is 0.5 milliampere at 5 volts and 1 milliampere at 10 volts.

Device is functionally equivalent to the 7411 (TTL) and the 74C11 (CMOS) devices.

4075

TRIPLE 3-INPUT OR GATE

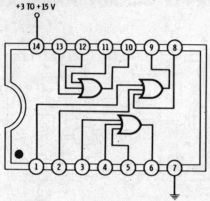

+3 TO +15 V

TOP VIEW

All three positive-logic OR gates may be used independently.

On any one gate, one or more inputs *high* provides a *high* output. With all three inputs *low*, the output will be *low*.

Propagation delay is 80 nanoseconds at 10 volts and 190 nanoseconds at 5 volts. Total package current at 1 megahertz is 0.5 milliampere at 5 volts and 1 milliampere at 10 volts.

Device is functionally and pin-for-pin equivalent to the 74HC4075 at LS TTL levels.

QUAD D REGISTER, TRI-STATE

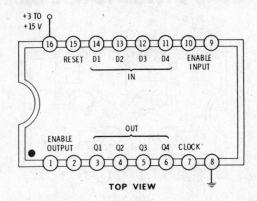

TOP VIEW

This is a quad storage register. In normal operation, pins 1, 2, 9, 10, and 15 are grounded. Data presented to the D inputs will get entered and stored on the ground-to-positive transition (positive edge) of the clock.

If either enable-input pin (9 or 10) is made positive, input data will be ignored and the outputs will stay the same even if repeatedly clocked.

If either enable-output pin (1 or 2) is made positive, the outputs will go to a high-impedance floating state. Data can still be entered and loaded on the clock positive edge.

If the Reset input (15) is made positive, all outputs go and stay in the low state.

The clock input must have one, and only one, positive transition per desired entry. This waveform must be debounced and noise free, and have rise and fall times faster than 5 microseconds. Maximum clock frequency is 9 megahertz at 10 volts and 3.6 megahertz at 5 volts.

Supply current is 0.4 milliampere at 5 volts and 0.8 milliampere at 10 volts for a 1-megahertz clock rate. Propagation delay is 120 nanoseconds at 10 volts and 260 nanoseconds at 5 volts.

Device is functionally equivalent to the 74173 (TTL) and the 74C173 (CMOS) devices.

4077

QUAD EXCLUSIVE-NOR GATE

+3 TO +15 V

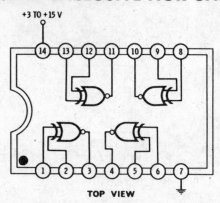

TOP VIEW

All four EXCLUSIVE-NOR gates may be used independently. On any one gate, if one input is high, but not both, the output will be low. If both inputs are high or both inputs are low, the output will be high.

The gate can be used as a comparator by noting that identical inputs give a high output and different inputs give a low output. It can also be used as a controllable inverter by noting that a high on one input will pass on whatever is on the other input, while a low complements it.

Propagation delay is 70 nanoseconds at 10 volts and 175 nanoseconds at 5 volts. Supply current at a 1-megahertz data rate is 0.2 milliampere at 5 volts and 0.4 milliampere at 10 volts.

8-INPUT NOR GATE

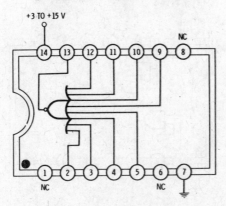

TOP VIEW

This package contains a single positive-logic, 8-input NOR gate.

If one or more inputs are *high*, the output will be *low*. If all eight inputs are *low*, the output will be *high*.

Propagation delay is 170 nanoseconds at 10 volts and 425 nanoseconds at 5 volts. Total package current is 0.4 milliampere at 5 volts and 0.8 milliampere at 10 volts.

Note that this is a very slow device. It should not be used in high-speed applications, particularly at a 5-volt or lower supply voltage. Device is functionally and pin-for-pin equivalent to a 74HC4078 for LS TTL levels.

4081

QUAD 2-INPUT AND GATE

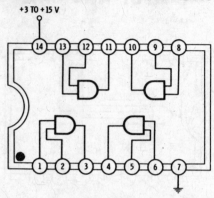

+3 TO +15 V

TOP VIEW

All four positive-logic AND gates may be used independently.

On any one gate, with either or both inputs *low*, the output will be *low*. With both inputs *high*, the output will be *high*.

Propagation delay is 70 nanoseconds at 10 volts and 150 nanoseconds at 5 volts. Total package current at 1 megahertz is 0.5 milliampere at 5 volts and 1 milliampere at 10 volts.

Device is functionally equivalent to the 7408 (TTL) and the 74C08 (CMOS) devices.

DUAL 4-INPUT AND GATE

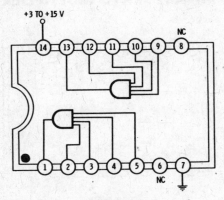

+3 TO +15 V

NC

NC

TOP VIEW

Both positive-logic AND gates may be used independently.

On either gate, with one or more inputs *low*, the output will be *low*. With all four outputs *high*, the output will be *high*.

Propagation delay is 70 nanoseconds at 10 volts and 150 nanoseconds at 5 volts. Total package current at 1 megahertz is 0.5 milliampere at 5 volts and 1 milliampere at 10 volts.

4089

BINARY RATE MULTIPLIER

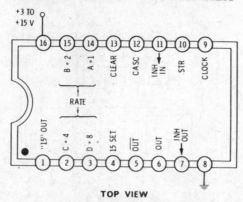

TOP VIEW

This is a special-purpose logic block that may be used to multiply an output pulse rate by a selected amount.

In normal use, pins 4, 10, 11, 12, and 13 are grounded. A clock is routed to pin 9. An input "rate" word, weighted A=1, B=2, C=4, and D=8, is applied to the rate inputs.

An output is provided at pin 6 and its complement is at pin 5. This output will be one-*sixteenth* the clock input multiplied by the input rate word. For instance, with a 16-kHz clock input, a rate word of 0000 produces zero output. A rate word of 0001 produces 1 kHz, 0010 produces 2 kHz, 1011 produces 11 kHz, and so on.

The output pulse rate is an average, and the pulses are usually unevenly spaced. Jitter is inherent in any rate-multiplier circuit. Rate multipliers can be used only where a certain total or a long-term average is all that is needed.

The 15-Set and Clear inputs (pins 4 and 13, respectively) are used to synchronize the start of operation to the zero or maximum count. The Inhibit input stops output pulses if it is high.

Rate multipliers are cascaded by connecting the Inhibit output of the first package to the Inhibit input and strobe of the second package, and the Output of the first package to the Cascade input of the second package. Clock inputs are connected together.

Operating current at a 1-megahertz clock rate is 0.5 milliampere at 5 volts and 1 milliampere at 10 volts. Maximum clock frequency is 4.5 megahertz at 10 volts and 2 megahertz at 5 volts.

The 4527 is a similar decimal unit.

QUAD 2-INPUT NAND SCHMITT TRIGGER

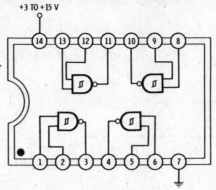

TOP VIEW

All four positive-logic NAND gates may be used independently.

While this package can be used as an ordinary NAND gate, internal hysteresis on the inputs makes this device ideal for noisy or slowly changing input levels. It is also ideal in astable and monostable applications.

If either or both inputs are *low*, the output will be *high*. If both inputs are *high*, the output will be *low*.

On a positive-going waveform, the output will change at 2.9 volts with a 5-volt supply and at 5.9 volts with a 10-volt supply.

On a negative-going waveform, the output will change at 2.3 volts with a 5-volt supply and at 3.9 volts with a 10-volt supply. Thus the hysteresis, dead band, or noise immunity is 0.6 volt with a 5-volt supply and 2 volts with a 10-volt supply.

Propagation delay is 300 nanoseconds at 5 volts and 150 nanoseconds at 10 volts. Total package current is 0.4 milliampere at 5 volts and 0.8 milliampere at 10 volts for a 1-megahertz clock frequency.

Device is functionally equivalent to the 74132 (TTL) and 74C132 (CMOS) devices.

4097

DUAL 1-OF-8 ANALOG SWITCH

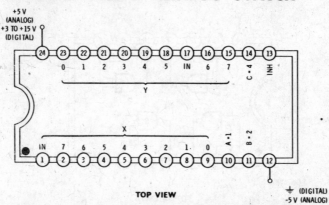

TOP VIEW

This package may be used as a dual 1-of-8 analog data multiplexer or demultiplexer, or as a 1-of-8 digital selector or distributor.

In the analog mode, −5 volts is applied to pins 12 and 13, and +5 volts to pin 24. The channel selected is determined by the A, B, and C inputs weighted A = 1, B = 2, and C = 4, with a zero defined as −5 volts and a one defined as +5 volts. For instance, with A = 1, B = 1, and C = 0, Channel three is selected on both the X and Y sides, connecting pin 1 to pin 6 and separately connecting pin 20 to pin 17.

In the digital mode, ground is applied to pins 12 and 13, and +3 to +15 volts is applied to pin 24. Channels are similarly selected with a zero defined as ground and a one being the pin-24 voltage.

The ON resistance is 200 ohms. On-channel frequency response extends to 40 megahertz. Off-channel cross talk is −40 decibels (1/100 amplitude) for frequencies less than 1 megahertz. The Inhibit pin turns both sides off if brought to the pin-24 voltage. More than one channel can be on during settling times. This can be eliminated by using the Inhibit before, during, and immediately after channel selection.

If any inputs are present in the absence of supply voltage, they must be current-limited to 10 milliamperes or less. All inputs and outputs must be less than the pin-24 voltage and more than the pin-12 voltage.

Select and inhibit propagation delay is 200 nanoseconds at 10 volts and 400 nanoseconds at 5 volts. Power dissipation depends on the loading, supply voltage, and the frequency of operation.

The 4067 is a similar 1-of-16 device, while the 4051 is a single 1-of-8 with built-in logic translation for analog use.

See Chapter 7 for more information.

TRI-STATE HEX INVERTER

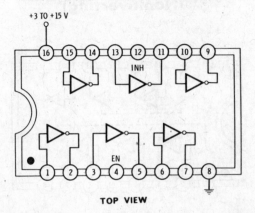

+3 TO +15 V

TOP VIEW

All six inverters may be used separately, but they share common inhibit and output enable functions.

If the EN terminal (pin 4) is grounded and the INH (pin 12) is grounded, a *high* on any inverter's input produces a *low* output, and vice versa.

If the INH input (pin 12) is made positive, all inverter outputs simultaneously go to *ground*.

If the EN input (pin 4) is made positive, all inverter outputs float and behave as an open circuit.

The output can drive two regular TTL loads or six LS TTL loads. In any bus system, only one source should be tri-state enabled at any particular time.

Propagation delay is 30 nanoseconds at 10 volts and 40 nanoseconds at 5 volts. Total package current at a 1-megahertz input rate is 3 milliamperes at 5 volts and 6 milliamperes at 10 volts.

The 4503 is a noninverting hex tri-state driver with different pinouts and a different enable arrangement.

4503

TRI-STATE HEX BUFFER
(Noninverting)

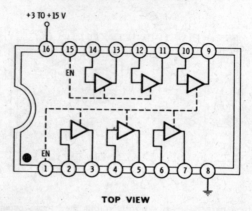

TOP VIEW

The six noninverting buffers are arranged into one group of four and a separately enabled group of two.

If either EN terminal is grounded, a *high* on any buffer's input results in a *high* on that buffer's output, and an input *low* provides a *low* output. Pin 15 controls the buffers that are outputting on pins 11 and 13, while pin 1 controls the remaining four.

If an EN input is made positive, all the buffer outputs that are controlled by it float and behave as open circuits.

The outputs can each drive two regular TTL loads or six LS TTL loads. In any bus system, only one source should be tri-state enabled at any particular time.

Propagation delay is 30 nanoseconds at 10 volts and 40 nanoseconds at 5 volts. Total package current at a 1-megahertz input rate is 3 milliamperes at 5 volts and 6 milliamperes at 10 volts.

Device is functionally equivalent to the 74126 (TTL) and 74C126 (CMOS) devices. The 4502 is a tri-state hex inverter with different pinouts and a different enable arrangement.

DUAL 4-BIT LATCH (Tri-State)

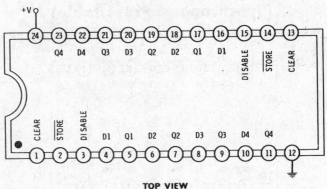

TOP VIEW

This package contains two separate 4-bit, set-reset latches, each with independent Store, Clear, and Output disable controls. The two latches may be combined to form a single 8-bit latch.

In normal operation, Clear and Disable are grounded, and information to be stored is presented to the D inputs. If the $\overline{\text{Store}}$ is high, the Q outputs will follow the inputs. If the $\overline{\text{Store}}$ input is brought low, the last information present when it was high is stored internally and appears at the Q outputs.

Bringing the Clear high resets all latches to zero. Bringing the Disable high floats the tri-state outputs but does not change internal storage or loading.

Note that this is NOT a clocked-logic block. The $\overline{\text{Store}}$ acts as a hold-follow or set-reset logic block. Stages cannot be cascaded.

Propagation delay is 350 nanoseconds at 5 volts and 175 nanoseconds at 10 volts. Supply current for a 1-megahertz data rate is 0.4 milliampere at 5 volts and 0.8 milliampere at 10 volts.

4510

DIVIDE-BY-10 BCD UP-DOWN COUNTER
(Synchronous, Presettable)

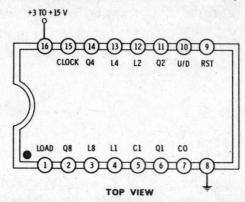

TOP VIEW

This decade up-down counter is presettable and may be cascaded by using internal carry/borrow logic. In normal operation, Carry In (CI), Reset (RST), and Load are held low. The count will advance by one on the ground-to-positive (positive edge) transition of the clock if the Up-Down (U/D) control is low. The output appears as a bcd code, weighted Q1 = 1, Q2 = 2, Q4 = 4, and Q8 = 8.

Stages are cascaded by connecting the Carry Out (CO) of the first stage to the CI of the second, and driving both from a common clock. The CI of the first stage should be grounded in this mode.

Data may be parallel-loaded by placing information on the Load 1 (L1) through Load 8 (L8) lines and bringing the Load input momentarily high. The counter may be reset by momentarily bringing the RST terminal high. Optionally, the counter may also be reset by parallel-loading a 0000. *Note that the clock must be in its low state during loading or resetting.*

The CI input may also be used as an enable. Bringing CI low allows counting; bringing CI high inhibits it.

The clock must be bounceless and have only one ground-to-positive transition per desired count. The clock rise and fall times should be faster than 5 microseconds.

Maximum clock frequency is 3 megahertz at 10 volts and 1.5 megahertz at 5 volts. Total supply current at a 1-megahertz clock rate is 0.4 milliampere at 5 volts and 0.8 milliampere at 10 volts.

Device is functionally equivalent to the TTL 7490 and 74190, and to the CMOS 74C90 and 74C190 devices. The 4516 is a similar binary (divide-by-16) counter.

7-SEGMENT LATCH & DRIVER

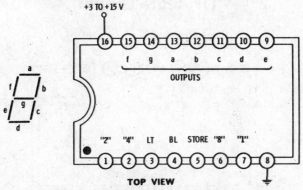

TOP VIEW

This package accepts a bcd input code, stores it, and converts it to a 7-segment, high-current, positive-logic readout drive signal.

In normal operation, LT and BL are made high and Store is held low. A bcd input code on the 1, 2, 4, and 8 lines appears as a 7-segment, positive-logic output code. For instance, 0110, or bcd six, on the inputs will make outputs c, d, e, f, and g high, and outputs a and b low. There are no tails on the six and nine, and the one is right-justified.

The outputs are designed to source current, and up to 25 milliamperes to ground may be provided. Output-current sinking capability is much less.

Current limiting must be provided if LED displays are driven with this package. For a 5-volt supply, 150 ohms is a typical value.

Output shorts will damage the device.

If the Store input is made high, the value of the bcd input at the instant Store goes high is held internally. With Store high, the last value is held for display. With Store low, the input variations are followed.

If the Blanking input (BL) is made low, all outputs go low, extinguishing the display. The Blanking input may also be used as a brightness control by duty-cycle-modulating it with a high-frequency waveform. If the Lamp Test (LT) input is grounded, all of the lamps will light, regardless of the input code or the BL input state.

There is no provision for automatic internal, leading-edge blanking, although any invalid (bcd 10 through 15) input code will also extinguish the display.

Turn-On delay time is 700 nanoseconds at 5 volts and 350 nanoseconds at 10 volts. Current consumption is dominated entirely by the output loading and must be limited to 25 milliamperes per segment.

4512

8-CHANNEL DATA SELECTOR
(Tri-State Output)

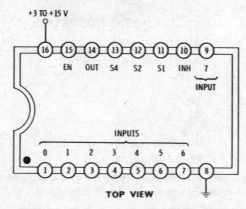

This package selects one of eight inputs and provides that input as an output. It may be used as a data selector or to generate any four-variable logic function (see Chapter 3). Inhibit and disable controls are also available.

In normal operation, INH and EN are grounded, and a select code weighted $S1 = 1$, $S2 = 2$, and $S4 = 4$ is applied to select the proper input. That input appears as an output. For instance, with select code 101, input 5 is selected and routed to the output.

If INH is made high, the output goes low independent of the input selected.

If EN is made high, the output goes to a floating, high-impedance state, independent of all other inputs.

As a four-variable function generator, three of the variables are routed to the S1, S2, and S4 inputs. The fourth variable, its complement, a permanent one, or a permanent zero, is routed as needed to the inputs to generate the truth table.

Note that this is a selector only. It cannot be reversed to act as a data distributor.

Total select time is 225 nanoseconds at 5 volts and 75 nanoseconds at 10 volts. Package current at a 1-megahertz select rate is 0.6 milliampere at 5 volts and 1.2 milliamperes at 10 volts.

1-OF-16 DECODER, HIGH OUTPUT

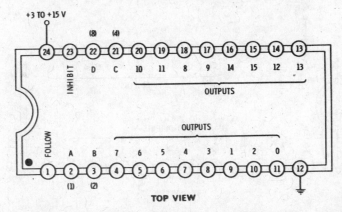

TOP VIEW

This package contains a latch followed by a 1-of-16 decoder. It may be used to generate one output out of sixteen high as a decoder or can be used to distribute positive-logic data to one of sixteen outputs.

In normal operation, the Follow input is made high and the Inhibit input is grounded. An input code weighted A = 1, B = 2, C = 4, and D = 8 selects the desired output. For instance, code 1101 will select output 13. The selected output goes high; the rest remain low.

Making the Inhibit pin positive forces all outputs to ground. The Inhibit may also be used as a data input whose complement will appear at the output.

If the Follow input is grounded, the state of the A, B, C, and D input lines is internally stored. Note that this is a hold-follow logic and NOT a D-type input. Address changes immediately appear at the output when Follow is high. When Follow is low, the last address is internally stored.

Propagation delay times are 800 nanoseconds at 5 volts and 300 nanoseconds at 10 volts.

Supply current is 1 milliampere at 5 volts and 2 milliamperes at 10 volts for a 1-megahertz clocking frequency.

For LS TTL levels, the device is functionally and pin-for-pin equivalent to the 74HC4514.

4515

1-OF-16 DECODER, LOW OUTPUT

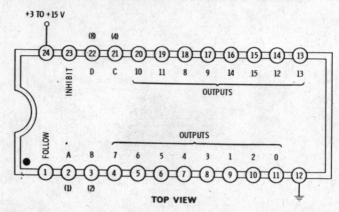

This package contains a latch followed by a 1-of-16 decoder. It may be used to generate one output out of sixteen low as a decoder or can be used to distribute negative-logic data to one of sixteen outputs.

In normal operation, the Follow input is made high and the Inhibit input is grounded. An input code weighted $A = 1$, $B = 2$, $C = 4$, and $D = 8$ selects the desired output. For instance, code 1101 will select output 13. The selected output goes low; all others remain high.

Making the Inhibit pin positive forces all outputs high. The Inhibit may also be used as a data input.

If the Follow input is grounded, the state of the A, B, C, and D input lines is internally stored. Note that this is a hold-follow logic and NOT a D-type input. Address changes immediately appear at the output when follow is high. When Follow is low, the last address is internally stored.

Propagation delay times are 800 nanoseconds at 5 volts and 300 nanoseconds at 10 volts.

Supply current is 1 milliampere at 5 volts and 2 milliamperes at 10 volts for a 1-megahertz clocking frequency.

Device is functionally equivalent to the 74154 (TTL) and the 74C154 (CMOS).

DIVIDE-BY-16 BINARY UP-DOWN COUNTER
(Synchronous, Presettable)

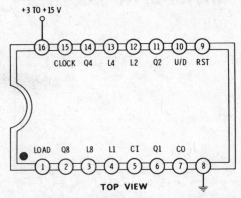

+3 TO +15 V

16	15	14	13	12	11	10	9
	CLOCK	Q4	L4	L2	Q2	U/D	RST

LOAD	Q8	L8	L1	CI	Q1	CO	
1	2	3	4	5	6	7	8

TOP VIEW

This base-16 up-down counter is presettable and may be cascaded by using internal carry/borrow logic. In normal operation, Carry In (CI), Reset (RST), and Load are held low. The count will advance one count on the ground-to-positive (positive edge) transition of the clock if the Up-Down (U/D) control is high, and will subtract one count on the positive clock edge if the U/D control is low. The output appears as a binary code, weighted $Q1=1$, $Q2=2$, $Q4=4$, and $Q8=8$.

Stages are cascaded by connecting the Carry Out (CO) of the first stage to the CI of the second, and driving both from a common clock. The CI of the first stage should be grounded in this mode.

Data may be parallel-loaded by placing information on the Load 1 (L1) through Load 8 (L8) lines and bringing the Load input momentarily high. The counter may be reset by momentarily bringing the RST terminal high. Optionally, the counter may also be reset by parallel-loading a 0000. *Note that the clock must be in its low state during loading or resetting.*

The CI input may also be used as an enable. Bringing CI low allows counting; bringing CI high inhibits it.

The clock must be bounceless and have only one ground-to-positive transition per desired count. The clock rise and fall times should be faster than 5 microseconds.

Maximum clock frequency is 3 megahertz at 10 volts and 1.5 megahertz at 5 volts. Total supply current at a 1-megahertz clock rate is 0.4 milliampere at 5 volts and 0.8 milliampere at 10 volts.

Device is functionally equivalent to the 74191 (TTL) and the 74C191 (CMOS) devices. The 4510 is a similar bcd (divide-by-10) counter.

4518

DUAL SYNCHRONOUS DIVIDE-BY-10 COUNTER

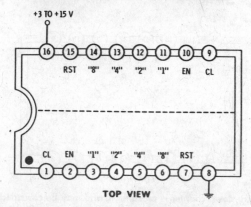

+3 TO +15 V

| 16 | 15 | 14 | 13 | 12 | 11 | 10 | 9 |
| RST | "8" | "4" | "2" | "1" | EN | CL |

CL | EN | "1" | "2" | "4" | "8" | RST

| 1 | 2 | 3 | 4 | 5 | 6 | 7 | 8 |

TOP VIEW

This package contains two separate synchronous divide-by-10 counters using the bcd 1-2-4-8 output code. They count in the up direction only and are not presettable. Each counter may be used separately.

Normally, RST is grounded and EN is made high. With these connections, the counter advances one count on each ground-to-positive (positive edge) clock transition. Outputs follow the 1-2-4-8 bcd code, and the outputs all change state synchronously without significant ripple delays.

As an option, RST and CL may both be grounded. In this condition, the positive-to-ground (negative edge) transition of the EN input will advance the counter one count. This is useful for negative-edge triggering and for ripple-cascading decades.

If RST is made high, the counter resets to the 0000 state and remains there after RST once again returns to a low state.

Counters are synchronously cascaded by common clocking and detecting a positive-logic state 9 (8 AND 1) on the first counter and routing this to the EN input of the next counter.

Note that the EN input should not be changed to low while the clock is low, nor should the CL input be changed to high while EN is high unless a count advance is specifically wanted.

Clock and Enable signals should be noise and bounce free and have only one desired transition in the right direction per wanted count. Clock and Enable rise and fall times should be faster than 10 microseconds.

Maximum clock frequency is 6 megahertz at 10 volts and 2.5 megahertz at 5 volts. Total package current at a 1-megahertz clock rate is 0.8 milliampere at 5 volts and 1.6 milliamperes at 10 volts.

DUAL SYNCHRONOUS DIVIDE-BY-16 COUNTER

+3 TO +15 V

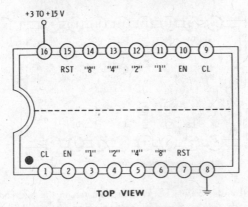

TOP VIEW

This package contains two separate divide-by-16 counters using the binary 1-2-4-8 output code. They count in the up direction only and are not presettable. Each counter may be used separately.

Normally, RST is grounded and EN is made high. With these connections, the counter advances one count on each ground-to-positive (positive edge) clock transition. Outputs follow the 1-2-4-8 binary code, and the outputs all change state synchronously without significant ripple delays.

As an option, RST and CL may both be grounded. In this condition, the positive-to-ground (negative edge) transition of the EN input will advance the counter one count. This is useful for negative-edge triggering and for ripple-cascading decades. If RST is made high, the counter resets to the 0000 state and remains there after RST once again returns to a low state.

Counters are synchronously cascaded by common clocking and detecting a positive-logic state 15 (1 AND 2 AND 4 AND 8) on the first counter and then routing this to the EN input of the next counter. Note that the EN input should not be changed to low while the clock is low, nor should the CL input be changed to high while EN is high unless a count advance is specifically wanted.

Clock and Enable signals should be noise and bounce free and have only one desired transition in the right direction per wanted count. Clock and enable rise and fall times should be faster than 10 microseconds.

Maximum clock frequency is 6 megahertz at 10 volts and 2.5 megahertz at 5 volts. Total package current at a 1-megahertz clock rate is 0.8 milliampere at 5 volts and 1.6 milliamperes at 10 volts.

For LS TTL levels, device is functionally and pin-for-pin equivalent to the 74HC4520.

4522

DECIMAL DIVIDE-BY-N COUNTER

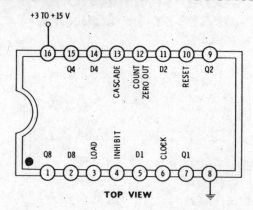

+3 TO +15 V

TOP VIEW

This base-ten counter counts backwards from ten in the bcd code. Stages may be cascaded for "by-decades" setting of any count. It is used as a divide-by-n counter, as a frequency synthesizer, or for any "by-decades" programmable counter use. In normal operation, Inhibit, Load, and Reset are grounded and Cascade is made positive. The counter backs up one count on each ground-to-positive transition of the clock.

Making the Preset input high will load a count into the package determined by the bcd code presented to the D1, D2, D4, and D8 inputs. Making the Reset input high resets the counter to 0000. A Count Zero output is provided. If this is connected back to the Load input, the counter will continue to divide by the word input on the D inputs, automatically reloading the maximum count every time the counter reaches zero.

To cascade stages as a "by-decades" frequency divider, connect the Zero Output of the first stage both to its own and the following Load inputs. Connect the Zero Output of the second stage back to the Cascade input of the first stage. Make the Cascade input of the second stage high. Make both Inhibits and both Resets low. Connect the Q4 output of the first stage to the Q2 input of the second stage. The output will appear on the Zero Output of the first stage as a narrow pulse.

Propagation delay time is 200 nanoseconds at 5 volts and 100 nanoseconds at 10 volts. Maximum clock frequency is 1 megahertz at 5 volts and 2.5 megahertz at 10 volts. The clock must be bounceless and conditioned to have only one positive transition per desired count. Its rise and fall times must be better than 5 microseconds.

Supply current for a 1-megahertz clock rate is 1.5 milliamperes at 5 volts and 3 milliamperes at 10 volts. Device is functionally equivalent to the 74160 (TTL) and 74C160 (CMOS) devices.

BINARY DIVIDE-BY-N COUNTER

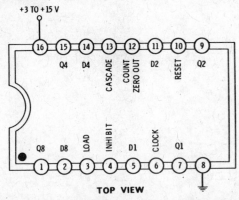

TOP VIEW

This base-sixteen counter counts backwards from sixteen in the binary code. Stages may be cascaded for hexadecimal setting of any count. It is used as a divide-by-n counter, as a frequency synthesizer, or for any hexadecimal programmable counter use. In normal operation, Inhibit, Load, and Reset are grounded and Cascade is made positive. The counter backs up one count on each ground-to-positive transition of the clock.

Making the Preset input high will load a count into the package, as determined by the binary code presented to the D1, D2, D4, and D8 inputs. Making the Reset input high resets the counter to 0000. A Count Zero output is provided. If this is connected back to the Load input, the counter will continue to divide by the word input on the D inputs, automatically reloading the maximum count every time the counter reaches zero.

To cascade stages as a hexadecimal frequency divider, connect the Zero Output of the first stage both to its own and the following Load inputs. Connect the Zero Output of the second stage back to the Cascade input of the first stage. Make the Cascade input of the second stage high. Make both Inhibits and both Resets low. Connect the Q4 output of the first stage to the Q2 input of the second stage. The output will appear on the Zero Output of the first stage as a narrow pulse.

Propagation delay time is 200 nanoseconds at 5 volts and 100 nanoseconds at 10 volts. Maximum clock frequency is 1 megahertz at 5 volts and 2.5 megahertz at 10 volts. The clock must be bounceless and conditioned to have only one positive transition per desired count. Its rise and fall times must be better than 5 microseconds.

Supply current for a 1-megahertz clock rate is 1.5 milliamperes at 5 volts and 3 milliamperes at 10 volts. Device is functionally equivalent to the 74161 (TTL) and 74C161 (CMOS) devices.

4527

DECIMAL RATE MULTIPLIER

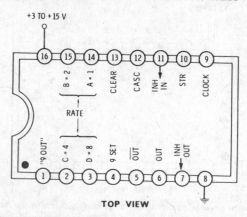

+3 TO +15 V

TOP VIEW

This is a special-purpose logic block that may be used to multiply an output pulse rate by a selected amount.

In normal use, pins 4, 10, 11, 12, and 13 are grounded. A clock is routed to pin 9. An input "rate" word, weighted A = 1, B = 2, C = 4, and D = 8, is applied to the rate inputs.

An output is provided at pin 6 and its complement at pin 5. This output will be one-*tenth* the clock input multiplied by the input rate word. For instance, with a 10-kHz clock input, a rate word of 0000 produces zero output; 0001 produces 1 kHz; 0010 produces 2 kHz; up through 1001 which produces 9 kHz and so on.

The output pulse rate is an average, and the pulses are usually unevenly spaced. Jitter is inherent in any rate-multiplier circuit. Rate multipliers can be used only where a certain total or a long-term average is all that is needed. Rate multipliers are cascaded by connecting the output of the first package to the Cascade input of the second package. Clock inputs are connected together. The Inhibit output of the first package goes to the Inhibit input and Strobe of the second package.

The 9-Set and Clear inputs are used to synchronize the start of operation to the zero or maximum count. The Inhibit input stops output pulses if it is high.

Operating current at a 1-megahertz clock rate is 0.5 milliampere at 5 volts and 1 milliampere at 10 volts. Maximum clock frequency is 4.5 megahertz at 10 volts and 2 megahertz at 5 volts.

Device is functionally equivalent to the 74C167 (CMOS) device. The 4089 is a similar binary unit.

4528

DUAL RETRIGGERABLE MONOSTABLE

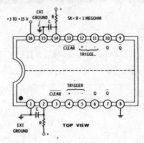

Both retriggerable monostables may be used separately. In normal operation, the Clear inputs are held high and pins 1 and 15 are grounded. For positive-edge triggering, use the + input and tie the − input high. For negative-edge triggering, use the − input and tie the + input to ground.

Triggering will drive the Q output high and the $\overline{Q}$ output low for a length of time set by the resistor and capacitor. The ON time can be shortened by bringing the Clear input low. A low Clear input also prevents output pulses during power-up times.

The timing resistor can range from 10K to 10 megohms, while the timing capacitor can range from 20 picofarads upwards. The on time is determined by the product of the resistance and the capacitance. Thus, a 100-pF capacitor and a 1-megohm resistor will have an on time of 100 microseconds. The resistor is usually varied to change the on time.

The circuit is retriggerable; if more than one triggering edge arrives during the on time, the RC product determines the delay *after* the last triggering edge arrives.

The timing capacitor is discharged every time *after* triggering. *This causes a capacitor-dependent delay between the time you trigger and the time you actually get any output.* This delay is almost a microsecond with a 1000-pF timing capacitor and a 5-volt supply; several *milli*seconds can elapse before you get an output after triggering with microfarad-sized timing capacitors.

Do not use this chip if you need both long time delays and an output immediately after triggering. Always use the largest possible timing resistor for a given ON time to minimize start-up delay.

Note that *external* grounds must be provided on pins 1 and 15.

Minimum input pulse width is 70 nanoseconds at 5 volts and 30 nanoseconds at 10 volts. Minimum output pulse width is 550 nanoseconds at 5 volts and 350 nanoseconds at 10 volts.

This device is identical to the 4098 in function and pinout.

4531

12-INPUT PARITY GENERATOR

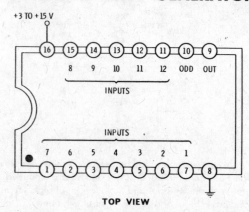

TOP VIEW

This package tests up to twelve inputs for parity. It is used for error testing in serial data communications.

In normal operation, pin 10 is grounded. If an even number of inputs are at a zero state, a zero appears at the output. If an odd number of inputs are at a zero state, a one appears at the output. Making pin 10 positive reverses the output sense.

All unused inputs may be grounded. For an even = low output, make pin 10 positive for an odd number of active inputs, and ground pin 10 for an even number of active inputs.

Response time is 450 nanoseconds at 5 volts and 175 nanoseconds at 10 volts. Supply current is 0.2 milliampere at 5 volts and 0.4 milliampere at 10 volts for a 1-megahertz input data rate.

PRIORITY ENCODER (8-Level)

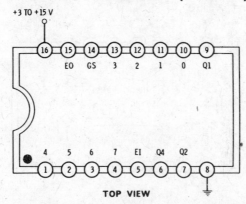

+3 TO +15 V

TOP VIEW

This is a specialized package that may be used to rank eight inputs in order of importance. It also serves as a keyboard encoder or other 1-of-8 binary encoder. It is functionally equivalent to the 74148 (TTL) and 74C148 (CMOS).

There are eight inputs (0 through 7) and three binary-weighted outputs (Q1 = 1, Q2 = 2, and Q4 = 4). For normal operation, the Enable Input (EI) is made high.

With no input, all outputs remain low. If only one input is made high, the outputs assume the binary code for that input. For instance, an input on line 6 (pin 3) will make the Q1 output low, Q2 high, and Q4 high, since six is a binary 110.

If two inputs are simultaneously made high, the one with the highest number (the highest priority) gets encoded as an output, and other inputs are ignored. Inputs 4 and 6 together output a 110, while 4 and 7 output a 111, and so on. As higher-priority inputs are returned low, the output code drops back to the next highest-priority input until all inputs finally go low.

If the EI is made low, all outputs go and stay low. There are two outputs. The "GS" output is essentially a "keypressed" detector; it goes high if any input is present along with a high EI input. This is useful in keyboard systems. It also tells the difference between no inputs and a zero priority input. The Enable Output (EO) is useful for expansion. It goes high only if EI is high and no inputs are made to this package.

Operation is unclocked, and there is no internal memory. At any time, the highest ranked input appears as its binary equivalent on the outputs.

Propagation delay time is 120 nanoseconds at 10 volts and 250 nanoseconds at 5 volts. Total package current at a 1-megahertz input rate is 1 milliampere at 5 volts and 2 milliamperes at 10 volts.

4539

DUAL 4-INPUT DATA SELECTOR

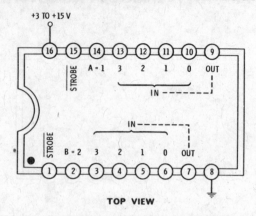

+3 TO +15 V

TOP VIEW

This package contains two 4-input data selectors. Each half of the package may be selectively disabled, but both halves share common select logic.

In normal operation, the $\overline{\text{Strobe}}$ inputs are grounded and a code, weighted A = 1 and B = 2, is applied to the select inputs. For instance, a 11 code will connect the 3 input to its respective output. The logic state on pin 3 will appear on pin 7, and the logic state on pin 13 will separately appear on pin 9.

Bringing the $\overline{\text{Strobe}}$ input positive grounds the output; pin 15 high grounds pin 9 and pin 1 high grounds pin 7.

Note that this is a data selector only. It is not bidirectional. Note further that there is common select logic, limiting somewhat the independent use of both circuit halves, particularly in data-selector logic applications.

Propagation time is 215 nanoseconds at 5 volts and 95 nanoseconds at 10 volts. Supply current is 0.5 milliampere at 5 volts and 1.0 milliampere at 10 volts.

DUAL 1-OF-4 DECODER, NONINVERTING

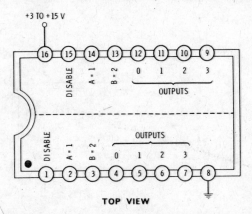

TOP VIEW

Both 1-of-4 decoders may be used separately, either as decoders or data distributors. They may also be combined with an external inverter for a 1-of-8 decoder or distributor.

In normal operation, pins 1 and 14 are grounded. A select code, weighted $A = 1$ and $B = 2$, is applied, and the selected output goes high. For instance, with a 1 on pin 2 and a 0 on pin 3, output "1" (pin 5) goes high. With a 0 on pin 14 and a 1 on pin 13, output "2" (pin 10) goes high. Note that both halves of the circuit have separate select and disable inputs.

Making the Disable input positive drives all outputs low. The Disable input may also be used as a data input for distributor use. A "1" on the Disable provides a "0" at the selected output, and vice versa. The Disable input can also be used to form a 1-of-8 distributor or decoder by driving one side from a new $C = 4$ input and the other from its complement.

Select time is 225 nanoseconds at 5 volts and 90 nanoseconds at 10 volts. Power-supply current for a 1-megahertz select rate is 0.5 milliampere at 5 volts and 1 milliampere at 10 volts.

Device is functionally equivalent to the 74155 (TTL) and the 74C155 (CMOS) devices.

4556

DUAL 1-OF-4 DECODER, INVERTING

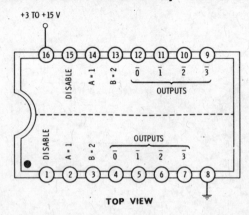

TOP VIEW

Both 1-of-4 decoders may be used separately, either as decoders or data distributors. They may also be combined with an external inverter for a 1-of-8 decoder or distributor.

In normal operation, pins 1 and 14 are grounded. A select code, weighted A=1 and B=2, is applied, and the selected output goes low. For instance, with a 1 on pin 2 and a 0 on pin 3, output $\overline{1}$ (pin 5) goes high. With a 0 on pin 14 and a 1 on pin 13, output $\overline{2}$ (pin 10) goes high. Note that both halves of the circuit have separate select and disable inputs.

Making the Disable input positive drives all outputs high. The Disable input may also be used as a data input for distributor use. A 1 on the Disable provides a 1 at the selected output, and vice versa. The Disable input can also be used to form a 1-of-8 distributor or decoder by driving one side from a new C=4 input and the other from its complement.

Select time is 225 nanoseconds at 5 volts and 90 nanoseconds at 10 volts. Power-supply current for a 1-megahertz select rate is 0.5 milliampere at 5 volts and 1 milliampere at 10 volts.

HEX SCHMITT TRIGGER

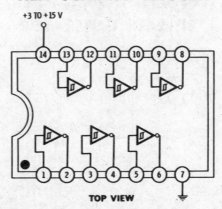

+3 TO +15 V

TOP VIEW

All six inverters may be used independently. While this package can be used as ordinary inverters, internal hysteresis on the inputs makes this device ideal for noisy or slowly changing input levels. It is also well suited to debouncing, contact conditioning, and astable and monostable circuits.

If an input is *low,* the output will be *high,* and vice versa.

On a positive-going waveform, the output will change at 2.9 volts with a 5-volt supply and at 5.9 volts with a 10-volt supply.

On a negative-going input waveform, the output will change at 2.3 volts with a 5-volt supply and at 3.9 volts with a 10-volt supply. Thus, the hysteresis, dead band, or noise immunity is 0.6 volt with a 5-volt supply and 2 volts with a 10-volt supply.

Propagation delay is 200 nanoseconds at 5 volts and 90 nanoseconds at 10 volts. Total package current is 0.3 milliampere at 5 volts and 0.6 milliampere at 10 volts for a 1-megahertz clocking rate.

Device is functionally equivalent to the TTL 7414, and the CMOS 40106 and 74C14 devices.

4702
(FAIRCHILD)

BIT-RATE GENERATOR

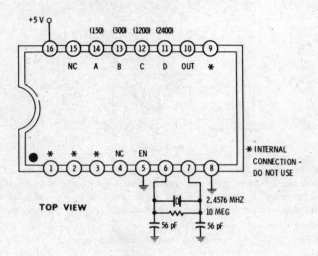

This special-purpose package synthesizes frequencies often used in serial data communications, particularly with UARTs.

In normal operation, pin 5 is grounded. If pins 11 through 14 are *not* grounded, an output frequency of 1760 hertz results. This is sixteen times the 110-baud rate used on serial teletypewriters.

Grounding only pin A generates sixteen times 150 baud. Grounding only pin B produces 16 × 300 baud. Pin C generates 16 × 1200 baud, and pin D generates 16 × 2400 baud. The output can drive one regular TTL load.

Operating current varies with loading. Five milliamperes from a 5-volt supply is typical.

TOP-OCTAVE MUSIC GENERATOR
(Not CMOS)

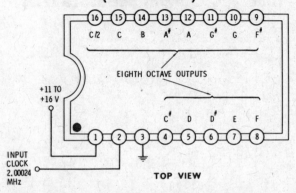

This package generates twelve equally tempered musical notes.

In normal operation, a 2.000240-MHz clock is applied to pin 2. This should be a square wave swinging between the pin-1 and pin-3 voltage. The twelve notes of the uppermost musical octave (octave No. 8) are produced at their respective pins, along with an optional C7 output.

The outputs can source or sink 0.7 milliampere. Lower-octave notes are usually obtained by CMOS binary dividers such as the 4024.

Two output duty cycles are available. The —0 has a 50/50 duty cycle, while the —1 has a 30/70 duty cycle. The —1 has odd and even harmonics available for formant tone coloring, while the —0 has only odd harmonics available.

The input-clock line should be carefully shielded from the outputs. The outputs are high level and should also be properly isolated and shielded from any related low-level audio circuitry.

Translation by octaves may be obtained by putting a CMOS binary divider between the clock source and the clock input.

Typical supply current is 24 milliamperes.

5101
(INTEL)

STATIC RANDOM-ACCESS MEMORY, CMOS
(256 × 4, 22-Pin, Separate I/O)

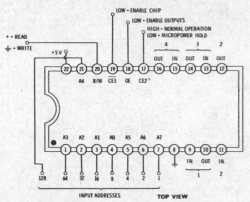

This is a static random-access memory organized as 256 × 4 in a 22-pin *narrow* package with separate input and output leads. Information may be rapidly read into and nondestructively read out of memory at system speeds. No clocking or refresh is needed. Nonvolatile, permanent storage can be simulated by retaining power supply in a micropower holding mode.

To read, pins 18 and 19 are grounded and pins 17 and 20 are made positive. A binary address applied to the eight input address pins will select four internal storage cells and will output the data in those cells on the four output lines.

To write, pins 18 and 19 are grounded and pins 17 and 20 are made positive. A binary address is then applied to the eight input address pins. The Write input (pin 20) is then brought low and returned high. *All address lines must be stable immediately before, during, and immediately after the low state on pin 20.*

All inputs and outputs are TTL and CMOS compatible. Making pin 19 positive will float the outputs and ignore write commands. Making pin 18 positive will float the outputs but still allow writing into memory. Making pin 17 low drops the chip into a micropower-hold mode in which only 15 *micro*amperes of supply power are needed but all other commands are ignored.

Access times of 650 nanoseconds for read and write are typical. Operating supply current is 13 milliamperes in the normal mode and 15 microamperes in the storage mode. This latter current is easily provided by a small battery in the system. Supply-voltage changeover must be transient-free for data holding.

17-STAGE OSCILLATOR/DIVIDER

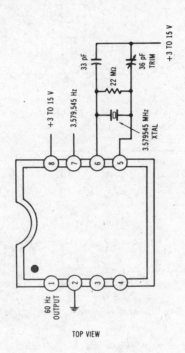

TOP VIEW

This is a package containing a binary counter that is primarily used to generate a precise 60-Hz reference from a commonly available 3.579545-megahertz TV-color-reference crystal frequency.

There are two buffered outputs. The crystal-frequency output at pin 7 is available for setting the exact frequency against a frequency counter or standard. The buffered divider output at pin 1 is a 60-Hz reference frequency.

Typical operating current is 1.2 milliamperes at 10 volts.

5841
(NATIONAL)

TELEVISION NUMERIC DISPLAY

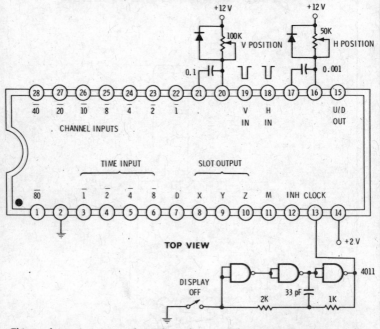

This package puts time, channel number, or video-game scores on a television display.

Grounding the M input displays only the channel; making it positive displays channel and time. Grounding the D input provides a five-slot time display; making it positive provides an eight-slot time display. The channel inputs are continuously applied in negative-logic form to the channel inputs. The time inputs must be multiplexed externally. This is done using the X, Y, and Z slot outputs to select the proper time in the proper slot. A 0000 time input, corresponding to a negative-logic binary 15, is read as a colon.

To interface the tv, properly conditioned H and V pulses must be applied to pins 18 and 19, respectively. Output video appears on pin 15, and is buffered and then summed into existing video inside the television set. Display position is controlled by the H and V controls. Horizontal display size is determined by the clock frequency. Disabling the clock stops the display. Supply current is 800 microamperes.

The 5318 is a non-CMOS clock chip that is compatible with the 5841.

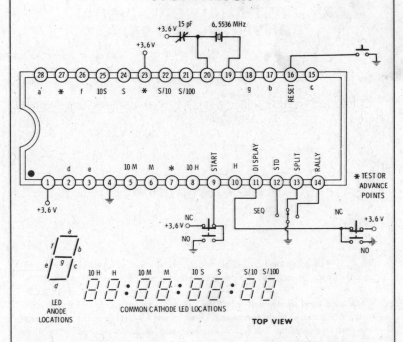

STOPWATCH

7045
(INTERSIL)

This is a complete stopwatch on a chip, capable of reading out in hundredths of a second, through hours, in any of four operating modes. It can also be used as a 24-hour clock.

The circuit is normally powered from three nickel-cadmium rechargeable batteries connected in series, having a nominal supply voltage of +3.6 volts. The selection of sequential, standard, split, or rally modes is made as shown. Bringing Start to ground and then returning it high starts the timing action. Repeated pressings of this button will activate the various operating modes, when selected.

Grounding the Reset input clears the stopwatch to all zeros.

The circuit may be directly connected to the LED display without any drivers or resistors. Approximately 15 milliamperes of digit drive current are provided. With the digits turned off, supply current is 180 microamperes.

7106

(INTERSIL)

3½-DIGIT ANALOG-TO-DIGITAL CONVERTER, LCD DRIVE

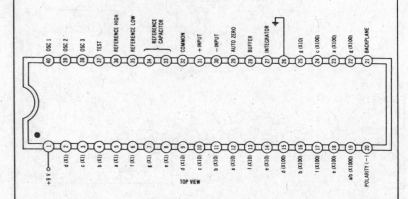

TOP VIEW

This is a self-contained 3½-digit a/d converter that is used with a liquid-crystal display. It includes the necessary backplane drive signal at pin 21.

When used as a simple voltmeter, the full-scale voltage is ±199.9 millivolts. A ±1.999-volt full-scale range can be obtained by changing the values of a reference resistor, an integrator resistor, and an auto-zero capacitor. See Chapter 10 for more information.

3½-DIGIT ANALOG-TO-DIGITAL CONVERTER, LED DRIVE

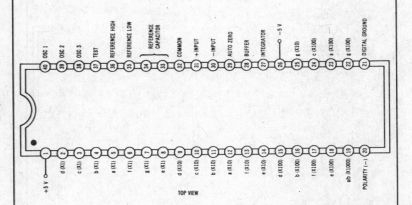

TOP VIEW

This is a self-contained 3½-digit a/d converter that is used with a 0.3- or 0.43-inch common-anode LED display without any additional drivers.

When used as a simple voltmeter, the full-scale voltage is ± 199.9 millivolts. A ± 1.999-volt full-scale range can be obtained by changing the values of a reference resistor, an integrator resistor, and an auto-zero capacitor. See Chapter 10 for more information.

The 7107 is designed to be used with ± 5-volt supplies.

7200
(INTERSIL)

WRISTWATCH

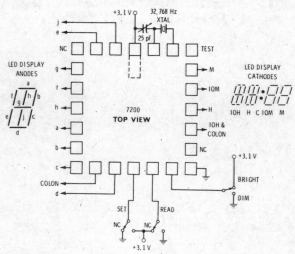

This package contains an entire wristwatch circuit, less only display, crystal, trimmer, control push buttons, and two 1.55-volt silver-oxide batteries. It is connected as shown.

Pressing Read once gives the time in hours and minutes. Pressing it again gives the day and date. Pressing it again gives a continuous readout of seconds for one minute.

Pressing Set once lets Read set the date. Pressing it again lets Read set the hour. The colon is off for a.m. and on for p.m. Pressing Set again lets Read set the day. Pressing Set again lets Read set the minutes. Pressing Set once more lets Read *hold* the seconds at 00 until one second after Read is released. Usually Read is released on a time tone.

A 10K resistor may temporarily be connected from Test to a 1.55-volt source (usually the center tap of the two cells). The 1-hertz square wave appearing across this resistor is used to adjust the trimmer capacitor to get the correct long-term accuracy. This is done with the display off.

Operating current with the display off is 4.0 microamperes. Display current varies from 20 milliamperes with 2 segments lit to 42 milliamperes with 7 segments lit. Oscillator stability is 1.3 parts per million, equal to 24 seconds per year.

Note that 9-segment readouts are needed for the hours positions if the day is to be read out.

TIME BASE FOR FREQUENCY COUNTER

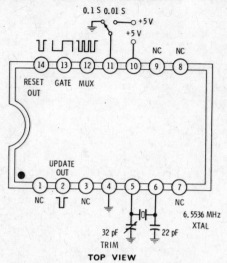

TOP VIEW

This package provides the time base and "housekeeping" for a frequency counter.

Pins 5 and 6 form a crystal oscillator as shown. The Gate output will be high for 0.1 second and low for 0.1 second if pin 11 is grounded, and will be high for 0.01 second and low for 0.01 second if pin 11 is high.

The MUX output is a 1.6-kHz square wave useful for multiplexing displays. The Update output is a brief negative-going pulse coincident with the rising edge of the gate output. It is used to transfer a count to display latches. The Reset output is a brief negative-going pulse that follows the Update output. It is used to reset a counter.

Operating current is 0.25 milliampere from a 5-volt supply.

7208
(INTERSIL)

FREQUENCY COUNTER

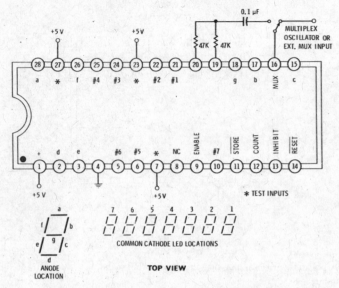

a
f g b
e g c
d

ANODE LOCATION

7 6 5 4 3 2 1

COMMON CATHODE LED LOCATIONS

TOP VIEW

This is a complete frequency counter on a chip. It will directly drive a seven-digit, seven-segment, common-cathode LED display.

In normal use, Enable is made +5 volts, Store is grounded, Inhibit is grounded, and Reset is grounded. A 1.6-kHz multiplex scanning signal is applied to pin 16, either from an external source or by using the self-contained oscillator shown. The display will advance one count on each ground-to-positive transition of the count input.

The count input has some internal conditioning equal to one-third the supply voltage. *Count input amplitude must be limited to less than the positive supply voltage and more than ground.*

Grounding Enable will turn off the display. Making Inhibit positive will prevent counts from being entered and may be used as a count gate. Grounding Reset makes the count 0000000. Making Store positive will hold the last count in the counter at the time this input rises.

The circuit may be directly connected to the LED display without any drivers or resistors. Twenty-five milliamperes of digit drive is provided. Maximum count frequency is 2 megahertz.

With the digits turned off, supply current is 210 microamperes at 5 volts.

The 7207 is a companion time-base chip.

PROGRAMMABLE READ-ONLY MEMORY, 32 × 8 (Not CMOS)

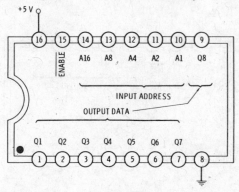

TOP VIEW

This is a nonvolatile memory that may be field-programmed to output any of 32 words of 8 bits each.

In normal operation, pin 15 is grounded, and the binary address selecting 1 of 32 words is applied to the address inputs. The preprogrammed data outputs for that word are then activated.

Input-address currents are typically 100 microamperes. Outputs can source 2 milliamperes and sink 16 milliamperes, handling up to 10 regular TTL loads. Making pin 15 positive floats the tri-state outputs.

Programming consists of applying excess output voltage to open internal fuses. A blown fuse corresponds to a "1" output. Consult the manufacturer's data sheet for programming information. The use of a programming machine or service is strongly recommended.

Typical access time is 40 nanoseconds. Operating current is 90 milliamperes from a +5-volt supply.

The 7603 is a tri-state output device. The 7602 is a similar open-collector device.

7611
(HARRIS)

PROGRAMMABLE READ-ONLY MEMORY, 256 × 4 (Not CMOS)

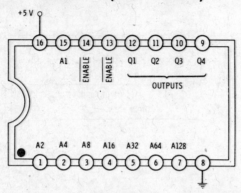

TOP VIEW

This is a nonvolatile memory that may be field-programmed to output any of 256 words of 4 bits each.

In normal operation, pins 13 and 14 are grounded, and the binary address selecting 1 of 256 words is applied to the address inputs. The preprogrammed data outputs for that word are then activated.

Input-address currents are typically 100 microamperes. Outputs can source 2-milliamperes and sink 16 milliamperes, handling up to 10 regular TTL loads. Making either, or both, pin 13 and 14 positive will float the tri-state outputs.

Programming consists of applying an excess output voltage in order to blow internal fuses. A blown fuse corresponds to a "1" output. Consult the manufacturer's data sheet for programming information. The use of a programming machine or service is strongly recommended.

Typical access time is 40 nanoseconds. Operating current is 90 milliamperes from a +5-volt supply.

The 7611 is a tri-state output device. The 7610 is a similar open-collector device.

REGULATOR, +5 VOLTS
(Not CMOS)

+IN   GND   +5 VOLTS OUT

FRONT VIEW

This +5-volt voltage regulator is recommended for supplies powering the 74HC/74HCT family — up to 750 milliamperes.

It must have a heat sink for high currents. The minimum applied supply voltage at a ripple trough and low line voltage must be more than 7 volts. The maximum applied supply voltage at a ripple peak and high line voltage must be less than 12 volts.

A 1-microfarad, high-quality, tantalum capacitor should be placed from output to ground for stability.

Standby current drain is 5 milliamperes.

7812

REGULATOR, +12 VOLTS
(Not CMOS)

+ IN   + 12 V OUT

FRONT VIEW

This +12-volt regulator is useful in power supplies up to 500 milliamperes.

It must have a heat sink for high currents. The minimum applied supply voltage at a ripple trough and low line voltage must be more than 14.5 volts. The maximum applied supply voltage at a ripple peak and high line voltage must be less than 20 volts.

A 1-microfarad, high-quality, tantalum capacitor should be placed from output to ground for stability.

Standby current drain is 5 milliamperes.

8500
(GENERAL INSTRUMENTS)

TV GAME CONTROLLER
(Not CMOS)

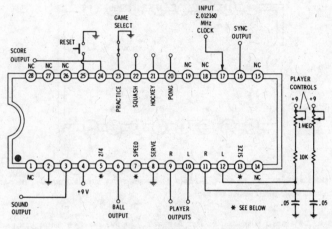

TOP VIEW

This IC contains most of the electronics needed for a video game. Pong, Hockey, Squash, or Practice is selected as shown by grounding pins 20 through 23.

To use the chip, provide a 9-volt supply and a 2.012160-MHz crystal clock reference. Connect the Ball, Player, and Score outputs to a 4-input OR circuit to generate composite video. Combine this composite video with the sync output to end up with video that is grounded for sync, +0.5 volt for black or normal, and +2 volts for white.

This final output is connected directly to the video circuitry of a television set or is routed to a suitable rf modulator. The sound output is optionally connected through a transistor amplifier to a speaker for full sound effects.

The pins marked * give several skill options. No connection on pin 5 gives two rebound angles. Grounding pin 5 gives four rebound angles. An open pin 7 gives a slow speed, while grounding pin 7 gives fast speed. A grounded pin 13 gives large bats, while an open pin 13 gives small bats.

Two rifle games are also possible using a photocell circuit on pins 18, 19, 26, and 27—see original data sheet for more details.

The circuit may be connected as shown in Chapter 10. The system is intended for an NTSC standard 525-line tv set with 60-hertz vertical scan rate.

Supply current is typically 24 milliamperes at 9 volts.

40174

HEX-D STORAGE REGISTER

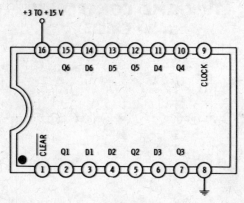

TOP VIEW

This package is used to store six bits of information simultaneously. In normal operation, pin 1 must be *high*.

Data to be stored is routed to the D inputs. On the positive edge (ground-to-positive transition) of the clock, information on the D inputs is internally stored and appears at the respective Q outputs.

Bringing pin 1 to ground forces all outputs to the low or zero state. The clock must be conditioned to be noise and bounce free, and have rise and fall times faster than 5 microseconds.

Package current at a 1-megahertz clock rate is 0.4 milliampere at 5 volts and 0.8 milliampere at 10 volts. Maximum clock frequency is 6 megahertz at 10 volts and 9 megahertz at 5 volts.

Device is functionally equivalent to the 74174 (TTL) and 74C174 (CMOS) devices.

40175

QUAD-D STORAGE REGISTER

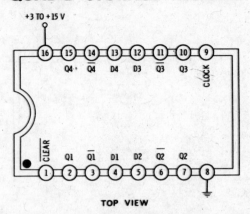

+3 TO +15 V

| 16 | 15 | 14 | 13 | 12 | 11 | 10 | 9 |
| Q4 | Q̄4 | D4 | D3 | Q̄3 | Q3 | | CLOCK |

| CLEAR | Q1 | Q̄1 | D1 | D2 | Q̄2 | Q2 | |
| 1 | 2 | 3 | 4 | 5 | 6 | 7 | 8 |

TOP VIEW

This package is used to store four bits of information and their complements. In normal operation, pin 1 must be *high*.

Data to be stored is routed to the D inputs. On the positive edge (ground-to-positive transition) of the clock, information on the D inputs is internally stored and appears at the respective Q outputs, with the complement appearing at the $\bar{Q}$ outputs.

Bringing pin 1 to ground forces all outputs to the low, or zero, state. The clock must be conditioned to be noise and bounce free, and have rise and fall times faster than 5 microseconds.

Package current at a 1-megahertz clock rate is 0.4 milliampere at 5 volts and 0.8 milliampere at 10 volts. Maximum clock frequency is 16 megahertz at 10 volts and 9 megahertz at 5 volts.

Device is functionally equivalent to the 74175 (TTL) and 74C175 (CMOS) devices.

40192

DECADE UP-DOWN COUNTER

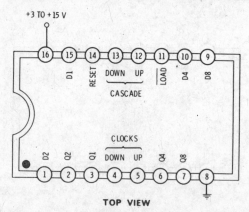

TOP VIEW

This is a synchronous up-down, base-ten counter having separate up and down clocks. Stages may be cascaded using internal carry-borrow logic.

In normal operation, Reset is held low and Load is held high. Both the Down clock and the Up clock are held high. If a clock is brought low and then high again, the count will advance or retard on the positive clock edge. *Input logic must guarantee that only one clock is allowed low at any particular time.*

The counter can be preset to any count by presenting the desired count on the D inputs and bringing the Load input low. The counter can be reset by bringing Reset high. This resets the Q1, Q2, Q4, and Q8 outputs to 0000.

To cascade stages, connect the Down output of the first stage to the Down clock of the second, and the Up output of the first stage to the Up clock of the second stage.

Propagation delay time is 225 nanoseconds at 5 volts and 95 nanoseconds at 10 volts. Maximum clock frequency is 4 megahertz at 5 volts and 8 megahertz at 10 volts. The clock must be conditioned to be noiseless and bounce free, with rise and fall times faster than 5 microseconds.

Supply current at a 1-megahertz clock rate is 0.8 milliampere at 5 volts and 1.6 milliamperes at 10 volts.

Device is functionally equivalent to the 74192 (TTL) and 74C192 (CMOS) devices.

40193

BINARY UP-DOWN COUNTER

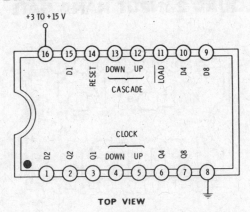

TOP VIEW

This is a synchronous up-down hexadecimal counter having separate up and down clocks. Stages may be cascaded using internal carry-borrow logic.

In normal operation, Reset is held low and Load is held high. Both the Down clock and the Up clock are held high. If a clock is brought low and then high again, the count will advance or retard on the positive clock edge. *Input logic must guarantee that only one clock is allowed low at any particular time.*

The counter can be preset to any count by presenting the desired count on the D inputs and bringing the Load input low. The counter can be reset by bringing the Reset high. This resets the Q1, Q2, Q4, and Q8 outputs to 0000.

To cascade stages, connect the Down output of the first stage to the Down clock of the second, and the Up output of the first stage to the Up clock of the second stage.

Propagation delay time is 225 nanoseconds at 5 volts and 95 nanoseconds at 10 volts. Maximum clock frequency is 4 megahertz at 5 volts and 8 megahertz at 10 volts. The clock must be conditioned to be noiseless and bounce free with rise and fall times faster than 5 microseconds.

Supply current at a 1-megahertz clock rate is 0.8 milliampere at 5 volts and 1.6 milliamperes at 10 volts.

Device is functionally equivalent to the 74193 (TTL) and 74C193 (CMOS) devices.

74C00

QUAD 2-INPUT NAND GATE

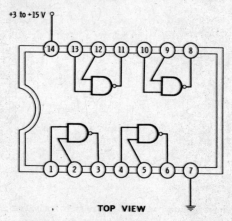

+3 to +15 V

TOP VIEW

All four positive-logic NAND gates may be used independently. On any one gate, when *either* input is *low*, the output is driven *high*. If *both* inputs are *high*, the output is *low*.

Propagation delay is 30 nanoseconds at 10 volts and 50 nanoseconds at 5 volts.

Device is functionally equivalent to a 4011 (CMOS) device.

QUAD 2-INPUT NOR GATE

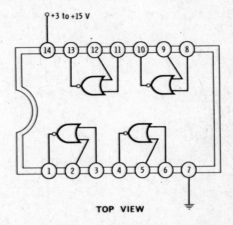

+3 to +15 V

TOP VIEW

All four positive-logic NOR gates may be used independently. On any one gate, with *either* input *high*, the output is *low*. When *both* inputs are *low*, the output is *high*.

Propagation delay is 30 nanoseconds at 10 volts and 50 nanoseconds at 5 volts.

Device is functionally equivalent to the 4001 (CMOS) device.

74C04

HEX INVERTER

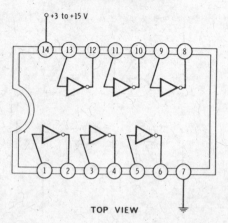

+3 to +15 V

TOP VIEW

All six inverters may be used independently. On any one inverter, the *low*-input condition drives the output *high*. The *high*-input condition drives the output *low*.

Propagation delay is 30 nanoseconds at 10 volts and 50 nanoseconds at 5 volts.

Device is functionally equivalent to the 4009, 4049, and 4069 (all CMOS) devices.

QUAD 2-INPUT AND GATE

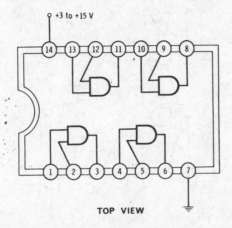

+3 to +15 V

TOP VIEW

All four positive-logic AND gates may be used independently. On any one gate, when *either* input is *low*, the output is *low*. When *both* inputs are *high*, the output is *high*.

Propagation delay is 15 nanoseconds average.

Device is functionally equivalent to the 4081 (CMOS) device.

74C10

TRIPLE 3-INPUT NAND GATE

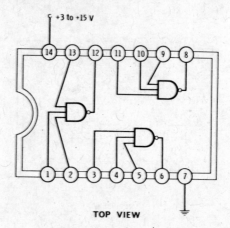

TOP VIEW

All three positive-logic NAND gates may be used independently. On any one gate, when *any* input *low*, the output is driven to a *high* state. When *all* three inputs are *high*, the output is driven to a *low* state.

Propagation delay is 9 nanoseconds average.

Device is functionally equivalent to the 4023 (CMOS) device.

TRIPLE 3-INPUT AND GATE

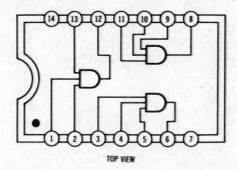

TOP VIEW

All three positive-logic AND gates may be used independently. On any one gate, with one or more inputs *low*, the output will be *low*. With all three inputs *high*, the output will be *high*.

Propagation delays are 70 nanoseconds at 10 volts and 140 nanoseconds at 5 volts.

Device is functionally equivalent to the 4073 (CMOS) device.

74C14

HEX SCHMITT TRIGGERS (Inverting)

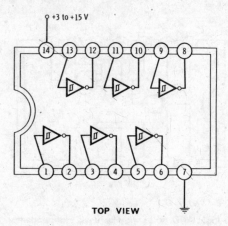

TOP VIEW

All six Schmitt triggers may be used independently. On any single trigger, a *low* input produces a *high* output. A *high* input produces a *low* output.

Unlike a normal CMOS gate, the inputs possess a snap action used to condition slowly changing or noisy inputs. The trip point for a *positive-going* signal is 6.8 volts. The trip point for a *negative-going* signal is 3.2 volt for a 10-volt supply. The minimum snap action or *hysteresis* range is 0.2 V_{CC}.

Propagation delay is 80 nanoseconds at 10 volts, and 220 nanoseconds at 5 volts.

Device is functionally equivalent to the 4584 and 40106 CMOS devices.

DUAL 4-INPUT NAND GATE

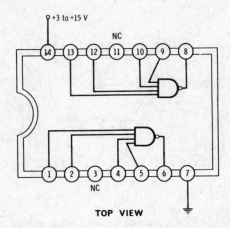

TOP VIEW

Both 4-input gates may be used independently. On either gate, any input-low condition drives the output *high*. When *all* inputs are *high*, the output is *low*.

Propagation delay is 10 nanoseconds typical.

Device is functionally equivalent to the 4012 (CMOS) device.

74C27

TRIPLE 3-INPUT NOR GATE

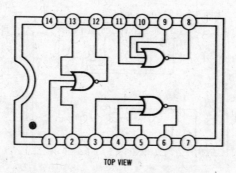

TOP VIEW

All three positive-logic NOR gates may be used independently. On any one gate, with one or more inputs *high*, the output will be *low*. With all three inputs *low*, the output will be *high*.

Propagation delays are 70 nanoseconds at 10 volts and 140 nanoseconds at 5 volts.

Device is functionally equivalent to the 4025 (CMOS) device.

8-INPUT NAND GATE

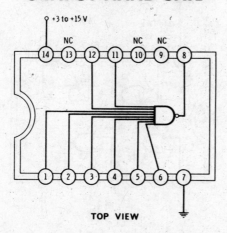

TOP VIEW

There is only a single gate per package. Any input-*low* condition drives the output *high*. When *all* inputs are *high*, the output is *low*.

Propagation delay is 55 nanoseconds at 10 volts and 125 nanoseconds at 5 volts.

Device is functionally equivalent to the 4068 (CMOS) device.

74C32

QUAD 2-INPUT OR GATE

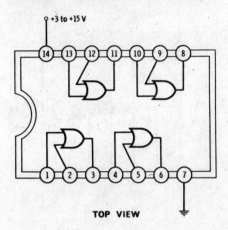

+3 to +15 V

TOP VIEW

All four positive-logic OR gates may be used independently. On any one gate, when *either* input is *high*, the output is driven *high*. When *both* inputs are *low*, the output is *low*.

Propagation delay is 12 nanoseconds average.

Device is functionally equivalent to the 4071 (CMOS) device.

BCD TO 1-OF-10 DECODER

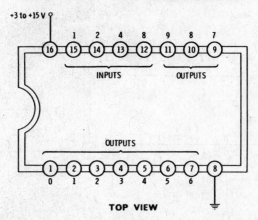

TOP VIEW

This package accepts a 1-2-4-8 Binary Coded Decimal (BCD) input code and provides a *grounded* output for the selected state. For instance, a 0111 input or "1"=1, "2"=1, "4"=1, and "8"=0 gives output line No. 7 a *low* state; all others remain high.

Outputs are CMOS compatible and can sink 2 milliamperes.

Note that the package can serve as a binary to 1-of-8 decoder simply by grounding pin 12.

Slight settling glitches and overlaps during address (input) changes are possible. Any input code over 1001 sends all outputs *high*.

Propagation delay is 90 nanoseconds at 10 volts and 200 nanoseconds at 5 volts.

Device is functionally equivalent to the 4028 (CMOS) device.

74C48

BCD TO 7-SEGMENT DECODER-DRIVER

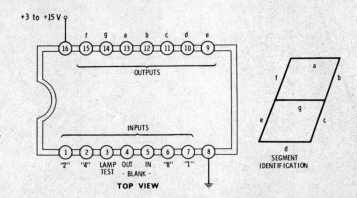

This package accepts a 1-2-4-8 positive-logic Binary Coded Decimal input and converts it to the proper pattern to light a 7-segment display. A *high* output is intended to light the segment. At 10 volts, the outputs can sink 16 milliamperes in the high state.

Current-limiting resistors must be used when driving a light-emitting diode display with this package. Incandescent and fluorescent readouts, as well as liquid-crystal displays, can be driven.

The Lamp Test input should remain high. Bringing the Lamp Test to ground simultaneously brings all the outputs to ground.

A low on the Blanking *input* will extinguish only character "0." A low on the Blanking output is provided to extinguish the character "0" of the next stage if leading-edge blanking is desired. A low on the Blanking *output* will extinguish the display. It is permissible to short this output to ground.

Propagation delay is 160 nanoseconds at 10 volts and 450 nanoseconds at 5 volts.

Device is functionally equivalent to the 4511 (CMOS) device.

DUAL JK LEVEL-TRIGGERED FLIP-FLOP
(With Preclear Only)

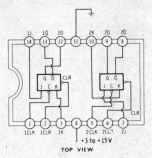

TOP VIEW

Contains two independent level-clocked JK flip-flops. Note the unusual supply connections. The same circuit with more normal supply pinouts is the 74C107.

This is a clocked logic block. There are two outputs: Q, and its complement Q̄. Under certain input conditions, Q and Q̄ can change whenever the Clock input goes to a low level. The Q and Q̄ outputs do not change for a change in the J and K inputs; the only time they can change is as the input clock goes to a low level.

If J and K are grounded, the clock does *nothing*. If J and K are made positive, the clock changes the output states on Q and Q̄, or *binarily divides*. If J is high and K is low, clocking makes Q high and Q̄ low. If J is low and K is high, clocking makes Q low and Q̄ high.

Information on the J and K inputs can be changed only once, immediately after clocking. Further changes can bring about an invalid operation. The clock must be conditioned to drop only once and then very rapidly.

The Clear input should be left or tied positive for normal operation. If the Clear input is grounded, the flip-flop immediately goes or stays in the state with the Q output low and the Q̄ output high.

Maximum toggle frequency is 11 megahertz at 10 volts and 4 megahertz at 5 volts.

Device is functionally equivalent to the 4027 (CMOS) device.

74C74

DUAL D EDGE-TRIGGERED FLIP-FLOP
(With Preset and Preclear)

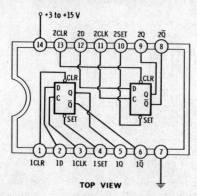

TOP VIEW

Contains two independent positive-edged-clocked D flip-flops. This is a clocked logic block. There are two outputs: Q, and its complement $\overline{Q}$.

The information presented to the D input goes on to the Q output whenever the clock input changes from a low to a high level. The only time the output can change is when the clock goes positive; changes on the D input are not passed on if the circuit is not clocked. If D is high, on clocking, Q goes high and $\overline{Q}$ goes low. If D is low, on clocking, Q goes low and $\overline{Q}$ goes high.

Information on the D input can be changed at any time. It is only its value at the instant of the positive clock edge that matters; this is what is entered into the flip-flop.

The Clear and Set inputs should be left or tied positive for normal operation. If the Clear input is grounded, the flip-flop *immediately* goes into the state with Q low and $\overline{Q}$ high. If the Set input is grounded, the flip-flop *immediately* goes into the state with Q high and $\overline{Q}$ low. Set and Clear should never be simultaneously grounded or a disallowed state will result.

Maximum toggle frequency is 25 megahertz. Current per package is 17 milliamperes. Device is functionally equivalent to the 4013 (CMOS).

QUAD LATCH (Level-Sensitive)

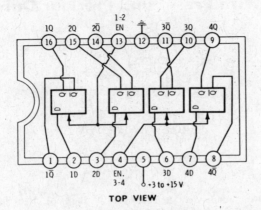

TOP VIEW

This package contains four memory elements. Note the unusual supply connections.

The memories are controlled in pairs with an *Enable* control. If the Enable control is high, the memories *follow* the input, thereby providing the input signal at Q and the complement of the input at $\overline{Q}$. A low at the D input appears as a low at Q and a high at $\overline{Q}$.

For use as a quad storage latch, both Enables are paralleled. Enable-high follows the input. Enable-low holds the previous value.

Note that this is *not* a clocked system and cannot be used as a shift-register element. Stages cannot be cascaded.

Propagation delay is 24 nanoseconds typical.

Device is functionally equivalent to the 4042 (CMOS).

74C76

DUAL JK LEVEL-TRIGGERED FLIP-FLOP
(With Preset and Preclear Only)

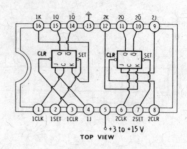

TOP VIEW

Contains two independent level-clocked JK flip-flops. Note the unusual supply connections. This is a clocked logic block. There are two outputs: Q, and its complement $\overline{Q}$. The same circuit with more normal supply pinouts is the 74HC112.

Under certain input conditions, Q and $\overline{Q}$ can change whenever the Clock input goes to a low level. The Q and $\overline{Q}$ outputs do not change for a change in the J and K inputs; the only time they can change is as the input clock goes to a low level.

If J and K are grounded, the clock does *nothing*. If J and K are made *positive*, the clock changes the output states on Q and $\overline{Q}$, or *binarily divides*. If J is high and K is low, clocking makes Q high and $\overline{Q}$ low. If J is low and K is high, clocking makes Q low and $\overline{Q}$ high.

Information on the J and K inputs can be changed only once, immediately after clocking. Further changes can bring about invalid operation. The clock must be conditioned to drop very rapidly per desired operation.

The Clear and Set inputs should be left, or tied, positive for normal operation. If the Clear input is grounded, the flip-flop *immediately* goes into the state with Q low and $\overline{Q}$ high. If the Set input is grounded, the flip-flop *immediately* goes into the state with Q high and $\overline{Q}$ low. Set and Clear should *never* be simultaneously grounded, or a disallowed state will result. Device is functionally equivalent to the 4027 (CMOS).

Maximum toggle frequency is 11 megahertz at 10 volts and 4 megahertz at 5 volts.

4-BIT FULL ADDER

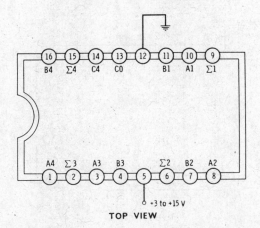

TOP VIEW

This is an arithmetic unit that provides the sum of two 4-bit binary numbers. Note the unusual supply connections.

The A number is weighted $A1=1$, $A2=2$, $A3=4$, $A4=8$ and is used as one input.

The B number is weighted $B1=1$, $B2=2$, $B3=4$, $B4=8$ and is used as a second input.

The sum of these two numbers, A and B, appears as $\Sigma1=1$, $\Sigma2=2$, $\Sigma3=4$, and $\Sigma4=8$. If the answer exceeds decimal 15 (binary 1111), a 1 also appears on the C4 line as a Carry Output.

When used only with 4-bit numbers, the C0 input should be grounded. When used as the upper 4 bits on an 8-bit number, the C0 input is connected to the C4 output of the previous (less significant) four stages.

Positive logic with 1 being at high level is used.

Propagation delay is 110 nanoseconds at 10 volts and 300 nanoseconds at 5 volts.

Device is functionally equivalent to the 4008 (CMOS).

74C85

4-BIT MAGNITUDE COMPARATOR

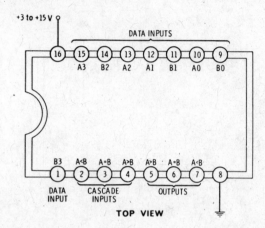

+3 to +15 V

DATA INPUTS

16 15 14 13 12 11 10 9
    A3 B2 A2 A1 B1 A0 B0

B3 A<B A=B A>B A>B A=B A<B
1   2   3   4   5   6   7   8

DATA   CASCADE   OUTPUTS
INPUT   INPUTS

TOP VIEW

This package compares two 4-bit words and provides an output indicating whether they are equal or which is larger.

Usually the input data words to be compared are weighted $A1 = 1$, $A2 = 2$, $A3 = 4$, and $A4 = 8$, while the second word is weighted $B1 = 1$, $B2 = 2$, $B3 = 4$, and $B4 = 8$.

If only 4-bit words are being compared, the $A = B$ Cascade input should be wired *high*. The $A > B$ and $A < B$ Cascade inputs should be *grounded*.

If the two words are equal, the $A = B$ goes high. If $A > B$, the $A > B$ output goes high. If $A < B$, the $A < B$ output goes high. Thus, a high state appears at the proper output; the other two remain low.

To work with 8-bit words, the outputs of the first 4-bit comparison (least significant bits) are connected to the Cascade Inputs of the second stage. The final answer appears as the outputs of the most significant 4-bit comparator, with the proper output going high.

Propagation delay is 100 nanoseconds at 10 volts and 250 nanoseconds at 5 volts. Device is functionally equivalent to the 4063 (CMOS).

QUAD EXCLUSIVE-OR GATE

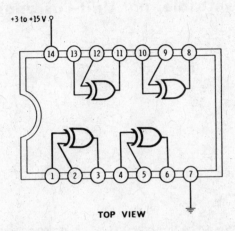

+3 to +15 V

TOP VIEW

The package contains four independent EXCLUSIVE-OR gates. They may be used separately.

On any one gate, when one, *but not both, inputs are high,* the output is *high.* When *both inputs are high* or *both inputs are low,* the output is *low.*

Propagation time is 85 nanoseconds at 10 volts and 180 nanoseconds at 5 volts.

Device is functionally equivalent to the 4030 and 4070 (both CMOS) devices.

74C90

DECADE (÷10) COUNTER (Ripple, not Presettable, not Unit-Cascadable)

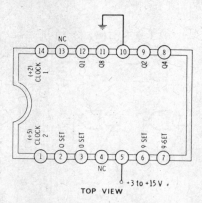

TOP VIEW

This is a divide-by-2 and a divide-by-5 counter in a single package. They may be used together as a divide-by-10, or separately. It ripple counts in the BCD-up direction. Note the unusual supply pinouts.

For a BCD counter, weighted 1-2-4-8, enter via Clock 1, and jumper Q1 to Clock 2. Both 9-Set and both 0-Set inputs must be *grounded* for normal counting. The counter advances on the negative-going clock edge. The clock must be properly conditioned and made bounceless and noise free. If a conventional decade counter is needed, *all set terminals must be held at ground*. The counter may be reset to zero by bringing either or both 0-Set inputs positive. The counter may be preset to 9 by bringing either or both 9-Set inputs positive.

An external jumper must be provided between counter halves. If entry is via Clock 2 and Q8 is jumpered to Clock 1, a counter weighted 1-2-4-5 results, with Q1 as the most significant output and a symmetrical square wave at the output. The circuit is not unit-cascadable. Device is functionally equivalent to the 4510 (CMOS) device.

Typical maximum toggle frequency is 5 megahertz at 10 volts and 2 megahertz at 5 volts.

BINARY (÷ 16) COUNTER
(Ripple, not Presettable)

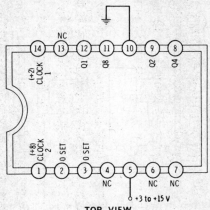

TOP VIEW

This is a divide-by-2 and a divide-by-8 counter in a single package. They may be used together as a divide-by-16, or separately. It ripple-counts in the binary-up direction. Note the unusual supply pinouts.

For a base-16 counter, weighted 1-2-4-8, enter via Clock 1 and jumper Q1 to Clock 2. Both 0-Set inputs must be *grounded* for normal counting. The counter may be reset to zero by bringing either or both 0-Set inputs positive.

The counter advances on the negative-going clock edge. The clock must be properly conditioned and made bounceless and noise free. *Both set terminals must be held at ground for normal counting.*

An external jumper must be provided between counter halves.

Typical maximum toggle frequency is 5 megahertz at 10 volts and 2 megahertz at 5 volts.

Device is functionally equivalent to the 4520 (CMOS).

74C107

DUAL JK LEVEL-TRIGGERED FLIP-FLOP
(With Preclear Only)

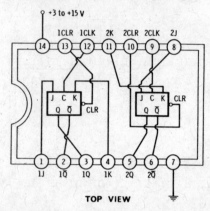

TOP VIEW

Package contains two independent, level-clocked JK flip-flops. This is a clocked logic block. There are two outputs: Q, and its complement $\overline{Q}$. Under certain input conditions, Q and $\overline{Q}$ can change whenever the Clock input goes to a low level. The Q and $\overline{Q}$ outputs do not change for any change in the J and K inputs; the only time they can change is as the input clock goes to a low level.

If J and K are grounded, the clock does *nothing*. If J and K are made positive, the clock changes the output states on Q and $\overline{Q}$, or *divides binarily*. If J is high and K is low, clocking makes Q high and $\overline{Q}$ low. If J is low and K is high, clocking makes Q low and $\overline{Q}$ high.

Information on the J and K inputs can only be changed once immediately after clocking. Further changes can bring about invalid operation. The clock must be conditioned to drop only once and then very rapidly.

The Clear input should be left, or tied, positive for normal operation. If the Clear input is grounded, the flip-flop immediately goes or stays in the state where the Q output is low and the $\overline{Q}$ output is high.

Maximum toggle frequency is 11 megahertz at 10 volts and 4 megahertz at 5 volts.

Device is functionally equivalent to the 4027 and 74C73 (both CMOS) devices.

8-BIT PRIORITY ENCODER

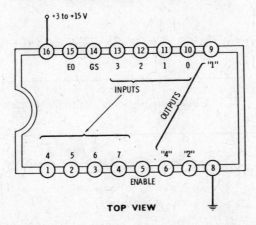

This is a specialized TTL device. It lets you rank eight inputs in order of importance. It produces as an output a binary word indicating the most important input present at any given time.

Selected inputs are placed on the 1-2-3-4-5-6-7 input lines. At the 1, 2, 4 outputs, the binary equivalent of the most significant (largest) input line selected will appear.

The Enable input must be low to get an output.

The EO output will go low if *any* input (one or more) is selected. The GS output will go high if *any* input (one or more) is selected and the Enable input is low.

The package may also be used as a keyboard encoder or as an 8-line-to-3-line encoder. It is expandable. Consult data sheet for more information.

Propagation time is 14 nanoseconds, typical. Device is functionally equivalent to the 4532 (CMOS).

74C150

1-OF-16 DATA SELECTOR

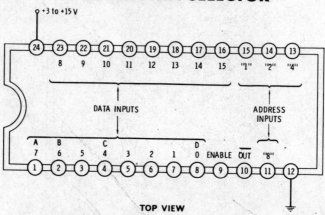

TOP VIEW

This package selects one of sixteen inputs and provides the complement of the selected input as an output. It will also generate any logic function of five or less input variables.

Inputs are selected by applying a code from 0000 through 1111 on the 1, 2, 4, and 8 Address inputs. The complement of the data on the selected input appears as an output.

The Enable input must be *low* for normal operation. Driving it high drives the output high, independently of the condition of the selected input. The Enable input is sometimes called a Strobe input.

For logic function generation, four of the variables are applied to the Address inputs. The selected Data inputs are connected low, high, to the fifth variable, or to the complement of the fifth variable per the desired truth table.

Note that this package inverts the data from input to output.

Select time is 120 nanoseconds at 10 volts and 290 nanoseconds at 5 volts.

For digital signals only, this device is functionally equivalent to the 4067 (CMOS) device.

1-OF-8 DATA SELECTOR

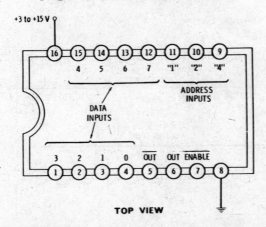

TOP VIEW

This package selects one of eight inputs and provides the data on the selected input or its complement as an output. It will also generate any logic function of four or less input variables.

Inputs are selected by applying a code from 000 through 111 on the 1, 2, and 4 Address inputs. The data on the selected input appear at pin 6; the complement of the selected data appears at pin 5. Pin 5 is faster in responding, as pin 6 is an inverter/follower.

The Enable input (sometimes called a Strobe) must be *low* for normal operation. Driving it high drives the pin-6 output *low* and the pin-5 output *high*, independently of the condition of the selected input.

For logic function generation, three of the variables are applied to the Address inputs. The selected Data inputs are connected low, high, to the fourth variable, or to the complement of the fourth variable per the desired truth table.

Select time is 110 nanoseconds at 10 volts and 240 nanoseconds at 5 volts.

For digital signals only, this device is functionally equivalent to the 4051 and 4097 (both CMOS) devices.

74C154

1-OF-16 DATA DISTRIBUTOR

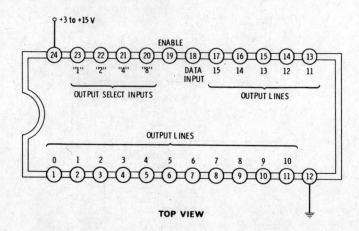

TOP VIEW

This package may be used to provide a 1-low-out-of-16 output or may be used to send input data to one selected output of sixteen, the remaining fifteen staying high.

The output address is selected with the 1, 2, 4, and 8 select lines. For instance, 1 low, 2 high, 4 high, and 8 low selects output No. 6.

If Enable and Data Input are both low, the selected output address goes low. If Enable is low, and a Logic input is provided the Data Input, the selected output address follows the Logic input.

Note that the functions of data selector and data distributor cannot be interchanged. This circuit accepts one input and routes it to *sixteen* outputs.

Select time is 100 nanoseconds at 10 volts and 265 nanoseconds at 5 volts.

Device is functionally equivalent to the 4514 (high output) and the 4515 (low output).

74C155

DUAL 1-OF-4 DATA DISTRIBUTOR

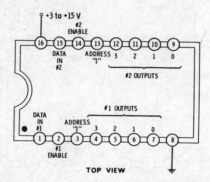

TOP VIEW

This dual package may be used to provide two 1-low-out-of-4 outputs or may be used to send input data to one selected output of four, the remaining three staying high. Both halves of the circuit are identically addressed.

The output address is selected with the 1 and 2 Address inputs. For instance, a 1-low and a 2-low input selects output No. 0 on both sides of the circuit.

If the Data input is made high and the Enable input low on circuit No. 1, the selected output address goes low. If the Data input is made low and the Enable is made low on circuit No. 2, the selected output address goes low.

If the Enable is made low on circuit No. 1, the *complement* of the input data appears at the selected output. If the Enable is made low on circuit No. 2, the input data appear at the selected output.

Note that the two halves of this circuit are not identical. Side No. 1 inverts the data. Side No. 2 does not.

The circuit is converted into a 1-of-8 data distributor by connecting the two Data input lines together and using them as a 4-address line. If both Enables are low, the selected 1-of-8 output goes low. If both Enables are paralleled and fed data, the data are routed to the selected output.

Address select time is 21 nanoseconds. Device is functionally equivalent to the 4555 (CMOS) device.

74C157

QUAD 1-OF-2 DATA SELECTOR

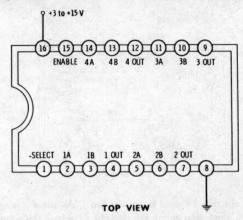

TOP VIEW

This is a four-pole double-throw data selector. All four switches are simultaneously enabled or selected.

If the Enable input is high, all outputs will be low, regardless of the input data. If the Enable input is low, and the Select input is low, the A inputs will provide an output. If the Enable input is low and the Select input is high, the B inputs will provide an output.

Note that this is a data selector and not a distributor. There are two inputs that may be selected to provide one output.

Select time is 80 nanoseconds at 10 volts and 100 nanoseconds at 5 volts.

Device is functionally equivalent to the 4019 (CMOS), and is a similar device to the 74C158, with inverted outputs.

DECADE (÷ 10) COUNTER (Synchronous, Presettable, Unit-Cascadable)

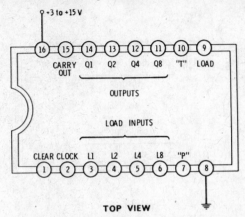

+3 to +15 V

16	15	14	13	12	11	10	9
	CARRY OUT	Q1	Q2	Q4	Q8	"T"	LOAD

OUTPUTS

LOAD INPUTS

CLEAR	CLOCK	L1	L2	L4	L8	"P"	
1	2	3	4	5	6	7	8

TOP VIEW

This is a synchronous, up-only decade (base-10) counter. For normal counting operation, the Clear input is made high, the P and T Enables are made high, and the Load input is left high. The count advances one count synchronously every time the clock goes from the low to the high state. The circuit triggers on positive edges. Outputs at Q1, Q2, Q4, and Q8 are BCD-weighted.

To asynchronously clear to zero, the Clear line is brought momentarily to ground. To load a number in parallel, the desired code is placed on the Load inputs L1, L2, L4, and L8 and the Load terminal is briefly brought to ground. For fully synchronous operation, the Carry Out of the first stage goes to the T Enable of the second. All stages are synchronously driven from the input clock. Refer to data sheet for more design information.

The clock must be properly conditioned to be bounceless and noise free, providing one, and only one, positive edge per desired clocking.

Typical maximum operating frequency is 8.5 megahertz at 10 volts and 3 megahertz at 5 volts.

Device is functionally equivalent to the 4522 (CMOS) device. The 74C160 is the same as the 74C162, except that the Clear is synchronous.

74C161

BINARY (÷16) COUNTER (Synchronous, Presettable, Unit-Cascadable)

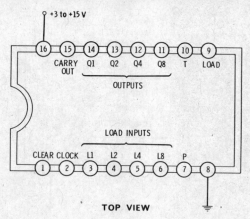

+3 to +15 V

16	15	14	13	12	11	10	9
	CARRY OUT	Q1	Q2	Q4	Q8	T	LOAD

OUTPUTS

LOAD INPUTS

CLEAR	CLOCK	L1	L2	L4	L8	P	
1	2	3	4	5	6	7	8

TOP VIEW

This is a synchronous, up-only binary (base-16) counter. For normal counting operation, the Clear, Load, P, and T pins are made high. The count advances one count synchronously every time the clock goes from the low to the high state. The circuit triggers on positive edges. Outputs at Q1, Q2, Q4, and Q8 are binarily weighted.

To asynchronously clear to zero, the Clear line is brought momentarily to ground. To Load a number in parallel, the desired word is placed on the Load inputs L1, L2, L4, and L8 and the Load terminal is briefly brought to ground. For fully synchronous operation, the Carry Out of the first stage goes to the T Enable of the second. All stages are synchronously driven from the input clock. Refer to data sheet for more design information.

The clock must be properly conditioned to be bounceless and noise free, providing one, and only one, positive edge per desired clocking.

Typical maximum operating frequency is 8.5 megahertz at 10 volts and 3 megahertz at 5 volts.

Device is functionally equivalent to the 4526 (CMOS) device. The 74C161 is the same as the 74C163, except that the Clear is synchronous.

74C164

SHIFT REGISTER, 8 BITS
(Serial Input, Parallel Output)

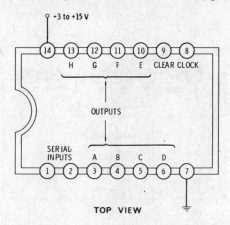

TOP VIEW

This is an eight-stage, shift-right-only shift register. It may be used as a serial-in/serial-out or serial-in/parallel-out register.

For normal operation, one of the Serial inputs is held high and data is sent to the second serial input. Clear is held high. Every negative-to-positive transition (positive edge) of the clock shifts the data one stage to the right. For instance, the data on the input gets shifted into A. A goes to B. B goes to C. C goes to D. D goes to E. E goes to F. F goes to G. G goes to H. H goes to a following stage or is destroyed.

The contents of the register may be cleared to zero by briefly bringing the Clear input low. Both Serial inputs must be *high* to enter a high-input state. Bringing either Serial input *low* forces a low-input condition.

The clock must be bounceless and noise free, producing one, and only one, positive transition per desired shift.

Typical maximum shift frequency is 8 megahertz at 10 volts and 3 megahertz at 5 volts.

Device is functionally equivalent to the CMOS 4015.

74C165

SHIFT REGISTER, 8 BITS
(Parallel Input, Serial Output)

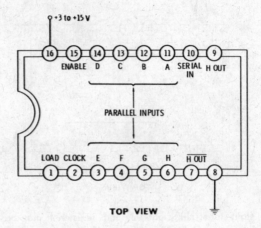

TOP VIEW

This is an eight-stage, shift-right-only shift register. It may be used as a serial-in/serial-out or parallel-in/serial-out register.

For normal operation, the Enable is held *low* and the Load is held high. Every negative-to-positive clock transition (positive edge) shifts the data one stage to the right.

For instance, the data on the Serial input goes into A. A goes to B. B goes to C. C goes to D. D goes to E. E goes to F. F goes to G. G goes to H. H goes to the next stage or is lost. The complement of the eighth stage is also available and provides inverted data.

To parallel-load the register, the Load input is briefly brought to ground while applying the desired input word to the parallel inputs A through H.

Shifting may be inhibited by bringing the Enable input high.

The clock must be bounceless and noise free, producing one, and only one, positive transition per desired shift.

Typical maximum shift frequency is 12 megahertz at 10 volts and 6 megahertz at 5 volts.

HEX "D" MEMORY (Edge-Clocked)

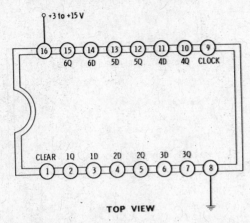

+3 to +15 V

16 15 14 13 12 11 10 9
    6Q 6D 5D 5Q 4D 4Q CLOCK

CLEAR 1Q 1D 2D 2Q 3D 3Q
1 2 3 4 5 6 7 8

TOP VIEW

There are six separate storage latches, each with a noninverted output. All latches are simultaneously updated.

Information at the D inputs is entered into the latches on the ground-to-high (positive edge) of the clock command. The only time that information is entered into this package is on the positive edge of the clock.

The Clear input is normally held high. Briefly bringing it to ground will clear the memory, making all Q outputs low.

This is a fully clocked system and can be used as a shift-register element. Stages can be cascaded. Note that there is no "Follow" mode; storage is updated only on the positive-clock edge.

Propagation delay is 70 nanoseconds at 10 volts and 150 nanoseconds at 5 volts. Maximum update frequency is 12 megahertz at 10 volts and 6.5 megahertz at 5 volts.

Device is functionally equivalent to the CMOS 4174.

74C175

QUAD "D" MEMORY (Edge-Clocked)

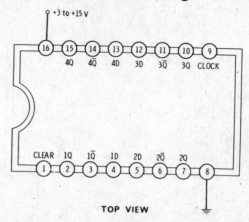

TOP VIEW

There are four separate storage latches, each with normal and complementary output. All latches are simultaneously updated.

Information at the D inputs is entered into the latches on the ground-to-high (positive edge) of the clock command. The only time that information is entered into this package is on the positive edge of the clock.

The Clear input is normally held high. Briefly bringing it to ground will clear the memory, making all Q outputs low and all $\overline{Q}$ outputs high.

This is a fully clocked system and can be used as a shift-register element. Stages can be cascaded. Note that there is no "Follow" mode; storage is updated only on the positive-clock edge.

Propagation delay is 75 nanoseconds at 10 volts and 190 nanoseconds at 5 volts. Maximum update frequency is 10 megahertz at 10 volts and 3.5 megahertz at 5 volts.

Device is functionally equivalent to the CMOS 4175.

DECADE (÷10) UP/DOWN COUNTER
(Carry, Borrow, Presettable, Synchronous)

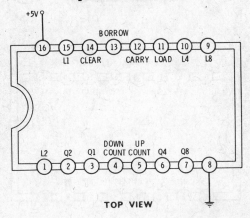

TOP VIEW

This is a synchronous decade (base-10) counter that counts in either direction. Two input clocks are used, and stages are Carry/Borrow cascaded. For a normal up-counting sequence, Load should be high, and Clear should be low.

The counter advances one count on each ground-to-positive transition of the Up-Count input clock. It backs up one count on each ground-to-positive transition of the Down Count input clock. When up-counting, hold the Down-Count input high. When down-counting, hold the Up-Count input high.

To load, the desired word is placed on Load inputs L1, L2, L4, and L8. The Load input is then briefly brought low. To clear the counter, the Clear input is briefly made *positive*. Note that the Clear must be low for normal counting.

Stages are cascaded by connecting Carry to Up Count and Borrow to Down Count.

Maximum operating frequency is 10 megahertz at 10 volts and 4 megahertz at 5 volts.

Device is functionally equivalent to the 40192.

74C193

BINARY (÷16) UP/DOWN COUNTER
(Carry, Borrow, Presettable, Synchronous)

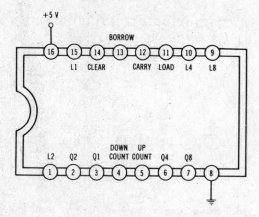

This is a synchronous binary (base-16) counter that counts in either direction. Two input clocks are used, and stages are Carry/Borrow cascaded. For a normal up-counting sequence, Load should be high, and Clear should be low.

The counter advances one count on each ground-to-positive transition of the Up-Count input clock. It backs up one count on each ground-to-positive transition of the Down-Count input clock. When up-counting, hold the Down-Count input high. When down-counting, hold the Up-Count input high.

To load, the desired word is placed on Load inputs L1, L2, L4, and L8. The Load input is then briefly brought low. To clear the counter, the Clear input is briefly made *positive*. Note that the Clear must be low for normal counting.

Stages are cascaded by connecting Carry to Up Count and Borrow to Down Count.

Maximum operating frequency is 10 megahertz at 10 volts and 4 megahertz at 5 volts.

Device is functionally equivalent to the 40193.

7-SEGMENT-TO-BCD CONVERTER, TRI-STATE®

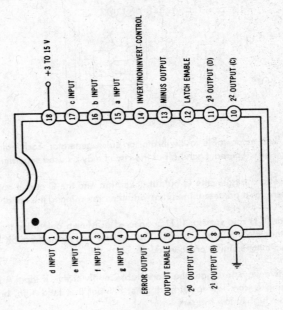

TOP VIEW

This package accepts a 7-segment coded input and converts it to a 1-2-4-8 positive-logic Binary Coded Decimal output code. The device gives the same BCD code for the number 6 with or without a "tail" (segment a lit), or the number 9 (segment d lit), or the number 1 using either segments b and c, or e and f.

A *high* on the Invert Control input (pin 14) selects an active low true decoding at the segment inputs. In addition to four TTL-compatible BCD outputs, an Error and a Minus output are available. The Error output (pin 5) is *low* when a nonstandard 7-segment code appears at the 7-segment inputs. The BCD outputs are forced into a TRI-STATE condition when this occurs or when the Output Enable input (pin 6) is the tied *high*. The Minus output (pin 13) goes *high* whenever a minus (−) code is detected.

Typical propagation delay times are 300 nanoseconds at 10 volts and 500 nanoseconds at 5 volts.

74HC123

MONOSTABLE MULTIVIBRATOR
(Dual, Retriggerable)

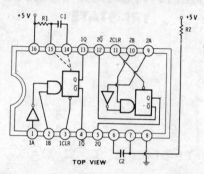

TOP VIEW

This is a dual monostable multivibrator or pulse generator. Each half of the circuit must be triggered. Each half of the circuit may be used separately.

In response to a trigger, the Q output goes high and the $\overline{Q}$ output goes low, staying there for a predetermined time and then returning to the initial state.

A capacitor (C1) and resistor (R1) connected as shown determine the pulse width so that PW = RC (in seconds). The resistor can range from 5K to 25K and the capacitor from 10 pF upward.

There are two ways to trigger the monostable multivibrator. If input A is held *low*, bringing B from low to high triggers. If input B is held *high*, bringing input A from high to low triggers.

The Clear input should remain high. If grounded, it inhibits triggering and returns the circuit to the state with Q low and $\overline{Q}$ high.

The circuit may be retriggered at any time. *Be sure to properly terminate all Trigger and Clear inputs.* Certain forms of clip-on digital testers can upset the operation of this stage, particularly on the Resistor and Capacitor inputs.

Unless very short times or complementary outputs or retriggerability is needed, the 555 is a better choice of monostable multivibrator.

Current per package is 46 milliamperes typical.

QUAD TRI-STATE® DRIVER
(Low Enable)

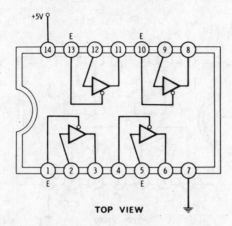

TOP VIEW

This package contains four separate drivers that may be used independently.

Unlike regular CMOS, this package has three possible output states—an output-low, an output-high, and an open-circuit that presents no load either to positive or ground on the output line.

If the TRI-STATE Control input is low, the input gets passed to the output without inversion. If the TRI-STATE Control is high, the output assumes an open-circuit condition.

Propagation delay is 10 nanoseconds. Enabling delay is 6 nanoseconds.

74HC126

QUAD TRI-STATE® DRIVER
(High Enable)

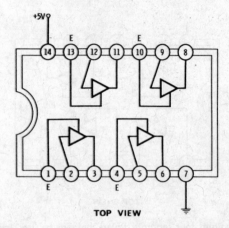

TOP VIEW

This package contains four separate drivers that may be used independently.

Unlike regular CMOS, there are three possible output states—an output-low, an output-high, and an open-circuit that presents no load either to positive or ground on the output line.

If the TRI-STATE Control input is *high*, the input gets passed to the output without inversion. If the TRI-STATE Enable is *low*, the output assumes an open-circuit condition.

Propagation delay is 10 nanoseconds. Enabling delay is 10 nanoseconds.

DUAL 1-OF-4 DATA SELECTOR

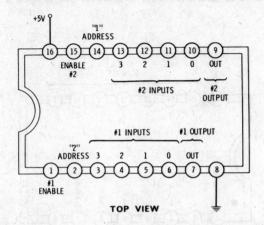

TOP VIEW

This package selects one of four inputs and provides the data on the selected input as its output. There are two separate 1-of-4 selectors with two separate outputs, but their addresses are common.

Inputs to the 1 Address and the 2 Address select the output connection for both sides simultaneously. Input data is not inverted and passed onto the output.

The Enable (or Strobe) must be low to get an output. A high Enable drives the output low, independently of input data.

Note that both halves of this circuit have common address lines, although they have separate inputs, outputs, and enables.

Select time is 26 nanoseconds.

For digital signals only, device is functionally equivalent to the CMOS 4052.

74HC4543

BCD-TO-DECIMAL DECODER/DRIVER

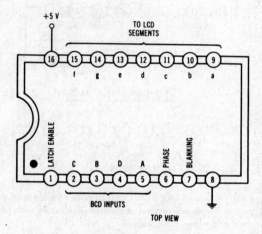

This is a BCD-to-7-segment latch/decoder/driver that can be used as a high-speed decoder, or as an LCD or LED driver. The Phase input (pin 6) controls the polarity of the 7-segment outputs. When low, the outputs are true 7-segment, and when high, they are inverted. For liquid-crystal displays, the Phase input can also be a low-frequency (40 Hz) square-wave backplane signal.

A logic 1 on the Blanking input (pin 7) blanks the display. Data on the BCD inputs are passed onto the 7-segment outputs when the Latch Enable (pin 1) is high, and is latched on the high-to-low transition of the Latch Enable input.

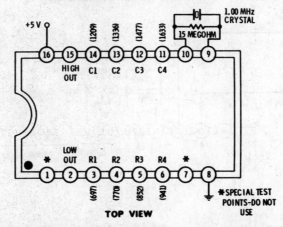

MC14410
(MOTOROLA)

TOUCH-TONE DIALER

This package simultaneously generates two sine waves useful for touch-tone dialing and telephone modem communications.

In operation, a "2-of-8" keyboard *grounds* one of the C inputs and one of the R inputs. For instance, to send the touch-tone "six," R2 and C3 are grounded. This outputs a 770-hertz sine wave on pin 2 and a 1477-hertz sine wave on pin 15.

The outputs are designed to drive a 1K load and may be summed together through series 1K resistors. The peak-to-peak output voltage is about 600 millivolts for the low output and 800 millivolts for the high output.

MC14411
(MOTOROLA)

BAUD-RATE GENERATOR

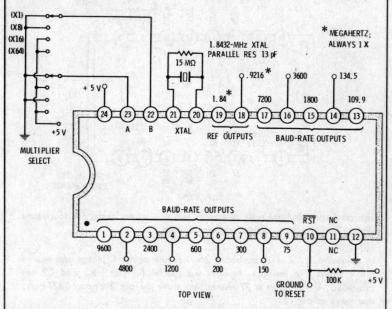

This integrated circuit simultaneously generates 16 reference frequencies useful in tvt, serial interface, and microprocessor applications.

In normal operation, pins 10 and 24 are made +5 V. Pins 22 and 23 are programmed as shown, to output frequencies that are 1X, 8X, 16X, or 64X — those shown on pins 1 through 9 and 13 through 17. Each of these outputs can drive one TTL load or any number of CMOS loads.

Frequency is set by the parallel-resonant 1.8432-MHz crystal and bias resistor on pins 20 and 21. The crystal frequency appears as an output on pin 19, and half that frequency appears at pin 18, independent of the multiplier settings. These outputs are useful for microprocessor clocks, both single and two-phase.

All outputs may be reset to zero by bringing pin 10 low. In normal operation, pin 10 must be held high.

Typical supply current is 2 mA plus one additional milliampere for each output connected to a TTL load. All outputs are symmetrical with a 50-50 duty cycle.

MC14412
(MOTOROLA)

MODEM

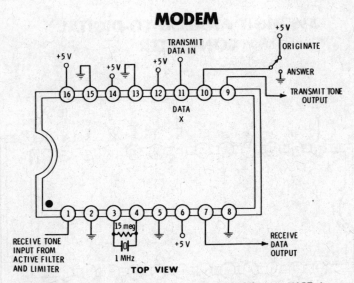

TOP VIEW

This package converts serial data, usually to and from a UART, into tones suitable for "103"-compatible telephone communication.

In the Originate mode, a 0 is transmitted as 1070 hertz and a 1 is transmitted as 1270 hertz. In the Answer mode, a 0 is transmitted as 2025 hertz and a 1 is transmitted as 2225 hertz. Since modems are used in pairs, the receiver responds to the tone group *not* being transmitted.

The transmit input is TTL compatible and responds to 300-baud, or less, data rates. Typical sine-wave output amplitude is 300-millivolts rms into a 100K load.

The receiver input must come from an elaborate and properly designed active filter and symmetrical limiter with controlled group-delay distortion. This input is TTL and CMOS compatible. The Receiver Data output follows this input after demodulation.

More information on modems appears in Motorola Application Note AN731.

Supply current is typically 1.1 milliamperes at 5 volts.

MC14433
(MOTOROLA)

3½-DIGIT ANALOG-TO-DIGITAL CONVERTER

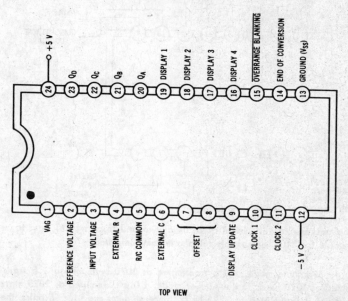

TOP VIEW

This is a single-chip 3½-digit a/d converter with a high input impedance, auto-polarity, and auto-zero. The digital output is a multiplexed 3½-digit BCD code with the most significant digit containing over- and underrange and polarity information, as well as the half digit (blank or "1").

When used as a simple but effective digital voltmeter, the full-scale voltage can be set to either +199.9 or −199.9 millivolts by using a 27K integrator timing resistor between pins 4 and 5 and an external reference voltage of 200 millivolts at pin 2. For a full-scale voltage of +1.999 or −1.999 volts, the external resistor is 470K and the reference voltage must be 2 volts. A pair of 0.1-μF capacitors are used for the integrator (pins 5 and 6) and for the offset correction (pins 7 and 8). Although an external clock may be used at pin 10, the MC14433 contains its own clock so that a single resistor can be used between pins 10 and 11 to set the frequency of the conversion cycle.

A 4511 7-segment decoder/driver can be used to decode the output to a 3½-digit common-cathode display. At pin 15, the overrange blanking signal (normally high) blanks the display whenever the input voltage exceeds the reference voltage at pin 2. See Chapter 10 for more information.

3

Logic

Logic circuits have digital inputs and digital outputs. They accept ones and zeros at their inputs, and they produce new groups of ones and zeros at their outputs. The output one and zero *states* respond to a rule or a set of rules *programmed* into the circuit.

There are two basic types of logic circuits. *Direct*, or *asynchronous*, logic circuits provide immediate outputs as soon as possible after the input information changes. *Clocked*, or *synchronous*, logic circuits do not provide an immediate output. Instead, they wait until a specific time, set by a system clock pulse, and only then do they provide an output change. While more complex, clocked logic allows orderly sequences rather than unchecked races through a circuit. Clocked logic is also more likely to include internal storage or memory that lets the output states be decided both by what is now on the inputs as well as by the past history of the inputs.

This chapter takes a detailed look at direct logic techniques. After deciding what we are going to call a one and a zero, and after giving some simple logic-switch examples, we will look at the common one- and two-input logic functions. We will follow this up with a powerful logic design trick and a quick check into tri-state logic and ultrasimple logic techniques. From here, we will go into advanced logic design methods that use data selectors and programmable read-only memories. These two methods convert virtually any logic problem into a one-package solution that takes only seconds to design and is easy to change. At the same time, these methods totally obsolete any of the traditional "minimization" methods of the fifties and sixties and render them worthless. Finally, we will consider some guidelines on logic design, along with a few examples that include a hexadecimal-to-ASCII converter and a micropower scanning keyboard encoder.

CMOS LOGIC GATES AS SIMPLE SWITCHES

Suppose we have a train of digital pulses we want to turn on or off under digital control. One place this might occur is at the input to a frequency counter, where we would like to measure the number of events that happen in a certain time interval. If we get 745 events in an 0.1-second interval, we get 745 events per unit time, or a frequency of 7.45 kilohertz.

Fig. 3-1 shows several ways to use CMOS as a simple switch. In Fig. 3-1A, a positive control signal passes the input, and a grounded control signal blocks it. The input signal is not inverted. Fig. 3-1B does the same thing, only it inverts the input signal. Fig. 3-1C inverts

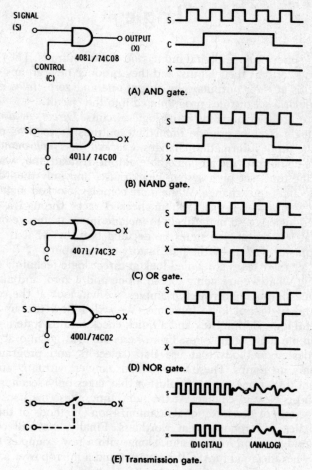

(A) AND gate.

(B) NAND gate.

(C) OR gate.

(D) NOR gate.

(E) Transmission gate.

Fig. 3-1. CMOS logic gates as simple switches.

224

the control input (ground = ON) only, while Fig. 3-1D inverts both the signal and control inputs. Finally, Fig. 3-1E uses a transmission gate to directly pass on the input signal if the control signal is positive and to block the input signal if the control signal is grounded. This final circuit can pass logic in *either* direction as well as handle analog signals.

Note that these circuits might pass partial-width inputs if the control signal is random with the start and finish of each input pulse. More elaborate *synchronization* techniques that use the methods described in Chapter 5 can solve this problem and eliminate count ambiguities.

STATE DEFINITIONS: WHAT IS A ZERO?

All of our CMOS digital logic inputs and most of our outputs accept or provide two output states: a low state near ground and a high state near the positive supply voltage. We call one of these states a *one* and the other a *zero* in order to tell them apart.

Traditionally, in relay and telephone circuits, a zero was always an open contact and a one was always a shorted, or closed, contact. But with CMOS, we have a choice.

If, as in Fig. 3-2, we define the grounded state as a zero and the positive state as a one, we are using *positive* logic. If we define the positive state as a zero and the negative state as a one, we are using *negative* logic. CMOS gates will work equally well with either logic definition, except that *most circuits will provide wildly different results for the two logic definitions.* For instance, as we will see shortly, a positive-logic NAND gate is the same *circuit* as a negative-logic NOR gate.

The names of practically all CMOS logic gates use the *positive-logic* definition. Unlike TTL and some older logic families, this positive-logic definition is also carried over to clocking, presetting,

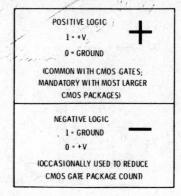

Fig. 3-2. Two logic definitions.

POSITIVE LOGIC
1 = +V
0 = GROUND
(COMMON WITH CMOS GATES;
MANDATORY WITH MOST LARGER
CMOS PACKAGES)

NEGATIVE LOGIC
1 = GROUND
0 = +V
(OCCASIONALLY USED TO REDUCE
CMOS GATE PACKAGE COUNT)

and resetting. A ground-to-positive clock edge does something in clocked logic. A positive state on an input or an enable provides a reset, preset, clear, enable, or whatever. Something unused is normally *grounded*. Also, unlike the earlier logic families, with buffered CMOS, there is essentially no cost or complexity premium between logic blocks that invert and those that do not. Therefore, it is usually a simple matter to stick with the positive-logic definition.

As a general rule, use the positive-logic definition with CMOS. CMOS circuits using medium- and large-scale integration circuits pretty much force you into this. Save the negative-logic definition for use only where it can simplify your circuit, minimize a package count, or help out in some other way.

THE ONE-INPUT LOGIC GATE

The simplest possible direct-logic blocks have one input and one output. The operation of a logic block is usually shown with a chart called a *truth table*. Truth tables simply list the output or outputs obtained for all possible input combinations.

The four possible one-input logic-gate truth tables are shown in Fig. 3-3. In Fig. 3-3A, the output is always a zero regardless of the input, while in Fig. 3-3B, the output is always a one, again regardless of the input. These two possibilities are completely useless.

In Fig. 3-3C, the output is a one when the input is a one, and the output is a zero when the input is a zero. This configuration is called a noninverting *buffer*. Buffers can be used to increase the output drive of a system. They are sometimes used to translate between

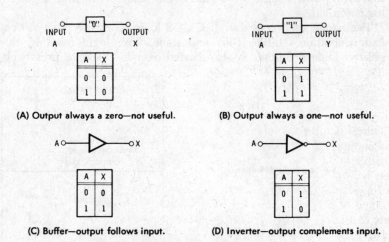

(A) Output always a zero—not useful. (B) Output always a one—not useful.

(C) Buffer—output follows input. (D) Inverter—output complements input.

Fig. 3-3. Four possible one-input, direct logic gates.

226

systems of different supply voltages. For instance, the 4050 is a CMOS buffer that can translate from a higher-voltage CMOS system down to a +5-volt TTL system. Buffers can be used to introduce delays of a few tens of nanoseconds to a few hundreds of nanoseconds. This is sometimes used to let the rest of a system's timing "catch up" to a faster waveform.

You can also resistively couple the output of a buffer back to its own input. This lets you build a simple and economical latch or set-reset flip-flop to use for contact-conditioning and memory circuits. Buffers will also improve the rise time of an input. With external feedback, you can also get a snap-action or *hysteresis* response that minimizes noise effects. Buffers may be combined with simple RC networks for leading- or trailing-edge pulse generators and half monostables. Details on these applications for buffers are given in the next chapter.

The truth table of Fig. 3-3D gives a one when the input is a zero, and vice versa. This is called an *inverter*. Inverters are used to generate the *complement* of a logic signal or to change the definition of a logic signal from positive to negative or back again. Inverting outputs are usually shown on a logic symbol as a small circle at the output of the logic block. A letter identifying a complementary output will usually have a bar above it. For instance, the output of a flip-flop is often called Q. Its complement is $\overline{Q}$, or "Q-bar," or "NOT Q."

Inverters also share the rise-time improving and delay applications of buffers. An inverter, a crystal, and several other parts can be used to build a crystal oscillator. Two inverters cross-coupled form a set-reset memory, a monostable pulse generator, or an astable signal source, depending on whether you use resistors or capacitors for the output-to-input connections. Inverting buffers, such as the 4049 and 4502, are available that can drive ordinary TTL, while the 4069 is a simple hex inverter with normal B-series output compatibility.

THE TWO-INPUT LOGIC GATE

Things get more interesting and more useful when we add a second input to our direct logic block. Our truth table now has four possible states: 00, 01, 10, and 11. Since each of the four output slots has an option of being a zero or a one, there are apparently 2^4, or 16, possible two-input direct logic gates.

Some of these are not too useful. The all-ones and all-zeros outputs of Figs. 3-4A and B are obviously worthless. Fig. 3-4C passes input A but ignores input B. Fig. 3-4D complements input A but ignores input B. Fig. 3-4E passes input B but ignores input A, and

Fig. 3-4F complements input B and ignores input A. We can substitute an inverter or a buffer for any of these last four arrangements.

Almost all of the two-input gates can be converted to buffers or inverters. Usually, this is done by tying two logically similar inputs together, or tying an unused input to +V or ground in a way that doesn't prevent the gate from responding. Fig. 3-5 shows how you can build an inverter or buffer from the most-common two-input logic gates. The inverting gates (those with circles on their symbol output) will form inverters, while the noninverting gates will form buffers or noninverting one-input logic blocks. We will see soon that we can program the EXCLUSIVE-OR and EXCLUSIVE-NOR gates to be either inverters or buffers. However, note that we cannot simply tie two inputs together on these two gates to do this (Fig. 3-5E and F).

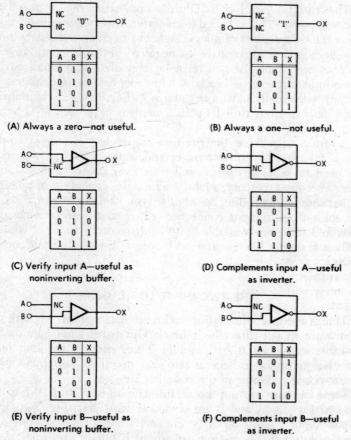

(A) Always a zero—not useful.

A	B	X
0	1	0
0	1	0
1	0	0
1	1	0

(B) Always a one—not useful.

A	B	X
0	0	1
0	1	1
1	0	1
1	1	1

(C) Verify input A—useful as noninverting buffer.

A	B	X
0	0	0
0	1	0
1	0	1
1	1	1

(D) Complements input A—useful as inverter.

A	B	X
0	0	1
0	1	1
1	0	0
1	1	0

(E) Verify input B—useful as noninverting buffer.

A	B	X
0	0	0
0	1	1
1	0	0
1	1	1

(F) Complements input B—useful as inverter.

A	B	X
0	0	1
0	1	0
1	0	1
1	1	0

Fig. 3-4. Six of sixteen possible two-input, direct logic gates.

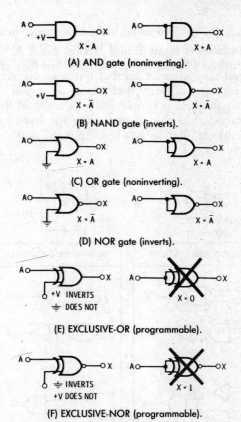

(A) AND gate (noninverting).

(B) NAND gate (inverts).

(C) OR gate (noninverting).

(D) NOR gate (inverts).

(E) EXCLUSIVE-OR (programmable).

(F) EXCLUSIVE-NOR (programmable).

Fig. 3-5. Converting multiple-input gates to inverters or buffers.

The six most-often-used two-input direct-logic functions are the AND, NAND, OR, NOR, EXCLUSIVE-OR, and EXCLUSIVE-NOR gates. Figs. 3-6 through 3-14 give us more details on these logic blocks.

The AND gate is shown in Fig. 3-6. It gives a one output only if both inputs are also ones. The AND gate outputs a zero if either or both inputs are zeros. The AND gate is used for decoding and switching, and whenever we want two or more inputs simultaneously to cause an output.

Fig. 3-6 also shows several ways to build an AND gate. Note that the circuits are different for the positive- and negative-log definitions. Most often, you would use the positive-logic 4081/74C08 AND gate, and you could get by with four AND gates per package. The other arrangements are handy when you have gates left over in other packages and are trying to minimize the total number of packages. The inverters can either be actual inverters or other inverting gates reduced to inverters with the techniques of Fig. 3-5.

The NAND gate of Fig. 3-7 can be built with an AND followed by an inverter. A zero appears at its output if both inputs are ones, and a one appears at its output if either or both inputs are zeros. The circle on the output symbol indicates that this is an inverting gate.

The NAND gate is particularly useful because two of them can be cross-coupled to produce a simple latch or set-reset flip-flop. Two NAND gates connected in this manner form the basis for the more-elaborate logic blocks such as clocked flip-flops, shift registers, and memory elements.

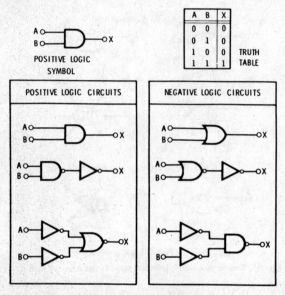

A	B	X	
0	0	0	
0	1	0	
1	0	0	TRUTH
1	1	1	TABLE

POSITIVE LOGIC SYMBOL

POSITIVE LOGIC CIRCUITS

NEGATIVE LOGIC CIRCUITS

Fig. 3-6. The AND gate.

The OR gate of Fig. 3-8 gives us a one output if either or both inputs are ones, and a zero output only if both inputs are zeros. It is used to get an output if *any* input is present. This differs from the AND gate that needs *all* inputs present for an output.

The NOR gate (Fig. 3-9) is the same as an OR gate followed by an inverter. It gives a zero output if any input is a one, and a one output only if both inputs are zeros. The NOR gates can also be cross-coupled to form memory and the more-elaborate clocked logic blocks.

The EXCLUSIVE-OR gate of Fig. 3-10 gives a one output *if either, but not both,* inputs are one. If both inputs are zeros or if both inputs are ones, the output is a zero.

The EXCLUSIVE-OR gate is very different from our earlier gates. Note that the AND, NAND, OR, and NOR gates have a different output for only one-fourth of the truth table, while the EXCLUSIVE-OR gates have ones

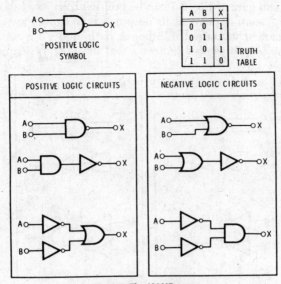

A	B	X
0	0	1
0	1	1
1	0	1
1	1	0

POSITIVE LOGIC SYMBOL

TRUTH TABLE

POSITIVE LOGIC CIRCUITS

NEGATIVE LOGIC CIRCUITS

Fig. 3-7. The NAND gate.

as outputs for half the input states and zeros as outputs for the other half.

Traditionally, EXCLUSIVE-OR gates have been used to perform binary arithmetic. Fig. 3-11A illustrates the rules for the binary addi-

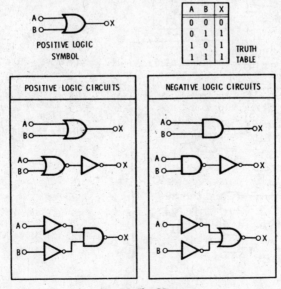

A	B	X
0	0	0
0	1	1
1	0	1
1	1	1

POSITIVE LOGIC SYMBOL

TRUTH TABLE

POSITIVE LOGIC CIRCUITS

NEGATIVE LOGIC CIRCUITS

Fig. 3-8. The OR gate.

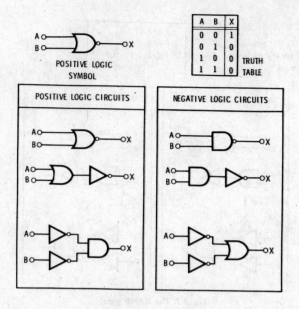

Fig. 3-9. The NOR gate.

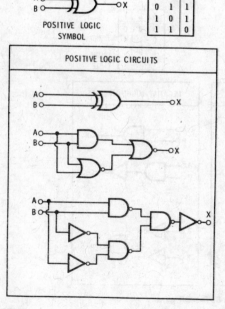

Fig. 3-10. The EXCLUSIVE-OR gate.

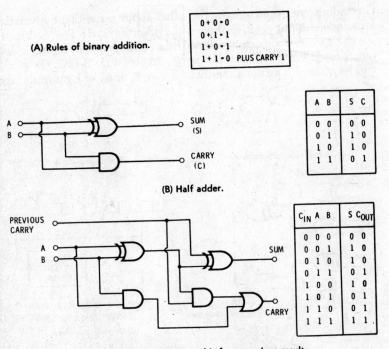

(A) Rules of binary addition.

```
0 + 0 = 0
0 + 1 = 1
1 + 0 = 1
1 + 1 = 0   PLUS CARRY 1
```

A B	S C
0 0	0 0
0 1	1 0
1 0	1 0
1 1	0 1

(B) Half adder.

C_{IN} A B	S C_{OUT}
0 0 0	0 0
0 0 1	1 0
0 1 0	1 0
0 1 1	0 1
1 0 0	1 0
1 0 1	0 1
1 1 0	0 1
1 1 1	1 1

(C) Full adder accepts carry bit from previous result.

Fig. 3-11. EXCLUSIVE-OR gates used for binary addition.

tion of two bits. As Fig. 3-11B shows, we can use an EXCLUSIVE-OR gate to do the addition, helped along by an AND gate to do the carry logic. This is called a *half adder*.

If we are going to add binary bits together, we have to somehow allow for a carry from a previous, or less-significant, two-bit addition. This means we have to double our circuit into the *full adder* of Fig. 3-11C. Bits A and B are added together, and their sum is added to the previous carry to provide the sum and carry as shown in the truth table. Full adders can be cascaded for *parallel* addition of binary bits as is done in a microcomputer, or multiple *serial* trips through the same full adder can be made as is done in a hand calculator. The 4032 and 4038 are typical CMOS serial triple adders.

Binary subtraction is done by adding with complementary numbers, and can use the same circuits. If you need extensive arithmetic capability, you should consider using a stock calculator or microprocessor chip rather than building up a central arithmetic processor out of individual CMOS packages.

Some other uses of our EXCLUSIVE-OR gate are shown in Fig. 3-12. In Fig. 3-12A, we use the gate as a *controllable complementer*.

Grounding one input passes the other input on without inversion. Making one input positive inverts the other input. In Fig. 3-12B, we use the gate to compare two bits. If they are identical, a zero is output. This is extended into the *parity tree* of Fig. 3-12C. Output X will be a zero for an *even* number (0, 2, 4, 6, or 8) of input ones, and

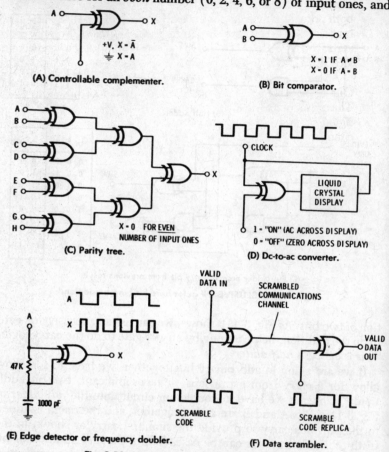

(A) Controllable complementer.

+V, X = Ā
⏚ X = A

(B) Bit comparator.

X = 1 IF A ≠ B
X = 0 IF A = B

(C) Parity tree.

X = 0   FOR EVEN
NUMBER OF INPUT ONES

(D) Dc-to-ac converter.

CLOCK

LIQUID
CRYSTAL
DISPLAY

1 = "ON" (AC ACROSS DISPLAY)
0 = "OFF" (ZERO ACROSS DISPLAY)

(E) Edge detector or frequency doubler.

47K
1000 pF

(F) Data scrambler.

VALID
DATA IN

SCRAMBLED
COMMUNICATIONS
CHANNEL

VALID
DATA
OUT

SCRAMBLE
CODE

SCRAMBLE
CODE REPLICA

Fig. 3-12. Special uses for EXCLUSIVE-OR gates.

will be a one for an *odd* number (1, 3, 5, or 7) of input ones. Parity is a useful way to indicate errors in data transmission systems. When a character is sent, a new bit called a *parity bit* is added to the word. The parity bit is selected to make the total number of ones sent an *even* number. During reception, parity is tested. If an odd number of ones are received, an error has occurred. Simple parity cannot fix errors by itself, but it can request retransmission and show that a mistake was received. The 4531 is a CMOS parity generator/checker.

The dc-to-ac converter of Fig. 3-12D provides a handy way to drive liquid-crystal displays. These displays need an ac drive waveform for long life. A square wave is applied to one side of the liquid-crystal display. This same square wave is routed to a controlled complementer using an EXCLUSIVE-OR gate. If the gate does not invert, both sides of the display bounce up and down together, and the resultant voltage across the display is zero. If the EXCLUSIVE-OR gate inverts, the left side of the display is low while the right side is high, or vice versa, and we end up with a square wave across the display.

An *edge detector*, or *frequency doubler*, is shown in Fig. 3-12E. One of the EXCLUSIVE-OR inputs is fed a delayed replica of the other. The only time these inputs are different is immediately after an input signal *changes*, so we get a one output immediately after each input change. If we select the RC time constant to be relatively short with

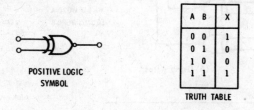

POSITIVE LOGIC
SYMBOL

A	B	X
0	0	1
0	1	0
1	0	0
1	1	1

TRUTH TABLE

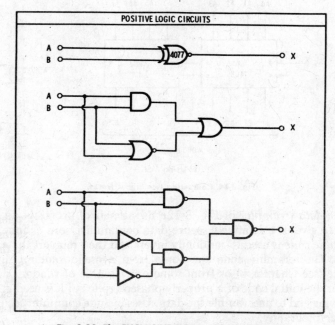

Fig. 3-13. The EXCLUSIVE-NOR or COMPARE gate.

respect to the input frequency, we get a brief pulse on each change of the input. This gives us an *edge* detector. On the other hand, if we set the RC time constant to *one-half* the input frequency period, we end up with a *frequency doubler*. For a given RC setting, the symmetry of the frequency-doubled waveform will change with input frequency, and thus the doubling will be effective over only a limited frequency range. Don't use early 4030s in this circuit because they have a *low* impedance. Use the 4070 or 74C86 instead.

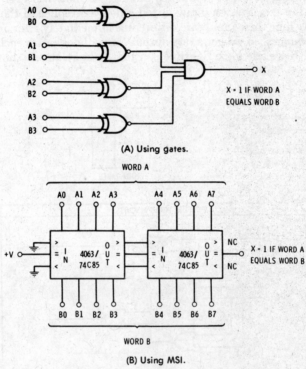

(A) Using gates.

(B) Using MSI.

Fig. 3-14. Comparing two digital words.

The *data scrambler* of Fig. 3-12F uses a pair of EXCLUSIVE-OR circuits to give us a relatively secure data communications channel. A *scramble code* is used to randomly invert and then reinvert the valid data. The scramble code can come from a pseudorandom signal source (see Chapter 6) or from some other source of random data. Either the initial code or a properly phased replica of it is used at the receiving end to unscramble the data. Uses include computer security and cryptography.

The EXCLUSIVE-NOR gate (Fig. 3-13) is also called a *COMPARE* gate. It can be built with an EXCLUSIVE-OR gate and an inverter. It outputs a high state if both inputs are the same, and a low state if both inputs are different.

A *magnitude comparator* (Fig. 3-14) consists of a stack of EXCLUSIVE-NOR circuits that test individual bits of two words that are to be compared. The comparison of each bit pair is ANDed to the results of the other pairs, and a positive output results if all bits of both words are identical. Extra logic can be added to tell which of the two words is larger if they are not identical. The 4063 (or 74C85) is a four-bit comparator. Units may be cascaded together for any word length, as shown in Fig. 3-14B.

OTHER TWO-INPUT GATES

The remaining four possible two-input direct logic gates are shown in Fig. 3-15, along with one possible positive-logic circuit each. These are the A AND NOT B, B AND NOT A, and their complements. While sometimes useful, they simply are not common nor are they available as single gates in most logic families. Later on, Fig. 3-21A shows a B AND NOT A circuit using a 4007.

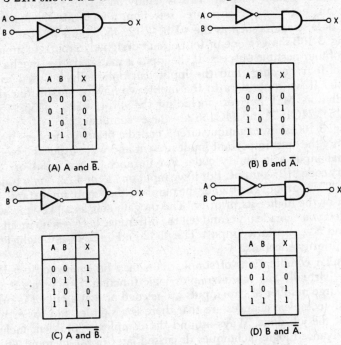

A B	X
0 0	0
0 1	0
1 0	1
1 1	0

(A) A and $\overline{B}$.

A B	X
0 0	0
0 1	1
1 0	0
1 1	0

(B) B and $\overline{A}$.

A B	X
0 0	1
0 1	1
1 0	0
1 1	1

(C) $\overline{A \text{ and } \overline{B}}$.

A B	X
0 0	1
0 1	0
1 0	1
1 1	1

(D) $\overline{B \text{ and } \overline{A}}$.

Fig. 3-15. The remaining two-input direct logic gates.

237

The number of possible gates rapidly gets out of hand as we add inputs. For instance, with three-input direct logic, there are eight slots in the truth table, so there are $2^8 = 256$ possible gates. With four inputs, there are $2^{16} = 65,536$ possible gates. And with single outputs, there are 4,294,967,296 possible five-input direct logic gates.

Very few of these combinations are made available. We have three- and four-input AND, NAND, OR, and NOR gates available. Eight-input NAND and NOR gates do exist, but they are slow and not recommended for use.

More elaborate logic blocks are built up by taking combinations of two-, three-, and, occasionally, four-input gates and combining them to get a more complex MSI function. The most important single combination results when you connect two NAND gates or two NOR gates back to back. This gives an elemental latch or set-reset flip-flop. When these are combined with transmission gates and other latches, we build up the clock logic systems such as JK and D flip-flops, shift registers, counters, and so on.

Fig. 3-16 shows how we can interconnect gates to get some of the more elaborate direct logic functions. In Fig. 3-16A, we have built a one-of-two data selector. This is handy on a shift register to provide an "enter" or "recirculate" selection, and forms the basis for MSI data selectors such as the 4019, 4512, 4539, and 74C157.

Fig. 3-16B shows a one-of-four circuit designed to route an input to one of four possible outputs. If the input is made positive, we have a one-of-four *decoder*, and the input can optionally be used as an enable. If we route data to the input, we have a one-of-four data distributor. MSI devices, including the 4514, 4515, 4555, 4556, 74C154, 74C155, and 74C156 use these techniques.

Fig. 3-16C shows a unique circuit called a *priority* encoder. It will identify the most important input present and will output the weight of that input as a binary code. For instance, if we input on "1," binary code "1" is output. But if we input on "1" and "2," only binary code "2" is output. The higher the number, the "more important" the input or the higher its priority. The OR gate acts as a "key pressed" and lets you cascade units and tell the difference between no inputs at all and a zero-priority input. The 4532 and 74C148 are eight-level MSI priority encoders.

Similar combinations of simple gates may be connected together to get virtually any more-complex logic function. If it seems that a very large number of gate inputs are needed or that a circuit is ridiculously complex, chances are that there is a simpler and easier way to get the job done. Ways around the complexity problem include the advanced logic techniques discussed later in the chapter. Other

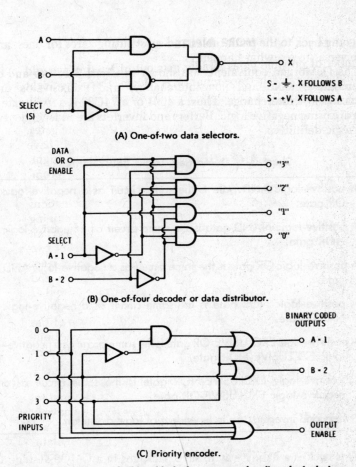

A

B

SELECT
(S)

X

S - ⏚, X FOLLOWS B
S - +V, X FOLLOWS A

(A) One-of-two data selectors.

DATA
OR
ENABLE

SELECT
A - 1
B - 2

"3"

"2"

"1"

"0"

(B) One-of-four decoder or data distributor.

0

1

2

3

PRIORITY
INPUTS

BINARY CODED
OUTPUTS

A - 1

B - 2

OUTPUT
ENABLE

(C) Priority encoder.

Fig. 3-16. Using gates as building blocks for more-complex direct logic devices.

techniques arrange for multiple trips through a serial system, use the overflow or dynamic change of a counter instead of an elaborate state comparison, or the use X-Y or X-Y-Z *matrix* arrays rather than linear or "straight line" arrangements.

A TRICK CALLED DEMORGAN'S THEOREM

Most CMOS logic-gate *circuits* will do two logically different things for us, depending on whether we pick the positive- or negative-logic definition. In general, for every positive way of stating something logically, there is an equally valid negative way of stating the same thing. This is called the *DeMorgan equivalence*. What this means is that a positive-logic NAND gate is exactly the same *circuit* as a negative-logic NOR gate. You can easily prove this yourself

239

by going back to the truth tables and substituting zeros for ones, and vice versa, to see what happens.

The DeMorgan equivalents are shown in Chart 3-1: AND and OR interchange, NAND and NOR interchange, and EXCLUSIVE-OR and EXCLUSIVE-NOR interchange. Thus, a 4001 or a 74C02 is a NAND gate if you are using negative logic. Buffers and inverters are independent of the logic definition.

Chart 3-1. DeMorgan Logic Equivalents

A positive-logic AND gate is the same *circuit* as a negative-logic OR gate.

A positive-logic NAND gate is the same *circuit* as a negative-logic NOR gate.

A positive-logic OR gate is the same *circuit* as a negative-logic AND gate.

A positive-logic NOR gate is the same *circuit* as a negative-logic NAND gate.

A positive-logic EXCLUSIVE-OR gate is the same *circuit* as a negative-logic EXCLUSIVE-NOR gate.

A positive-logic EXCLUSIVE-NOR gate is the same *circuit* as a negative-logic EXCLUSIVE-OR gate.

Buffers and inverters are independent of logic definition.

This trick is a handy way to save packages in a CMOS system. By switching your logic definition, you can use a pair of four-input gates as a NOR gate in one place and a NAND gate somewhere else. Traditionally, when inverting logic was cheaper, you would invert your logic definition at each progressive stage through the system. Remember that the name of the CMOS gate refers to its positive-logic definition, but that the circuit contained inside the package can provide either its "nameplate" function or its negative deMorgan equivalence.

TRANSMISSION-GATE LOGIC (TGL)

We can also do logic with the CMOS analog switches such as the simple 4016 and 4066 and the more-complex MSI 4051, 4052, 4053, 4067, and 4097. We will be looking at applications in detail in Chapter 7. An important advantage of transmission-gate logic (TGL) is that it is bidirectional—just like a relay's contacts—so that we can often freely interchange inputs and outputs of our circuits. This lets us use the same circuit as a data selector or as a data distributor, simply by relabeling the inputs as outputs and vice versa. We can also switch analog signals and other control waveforms with TGL, so long as the signals stay between the limits set by the supply pins on the package.

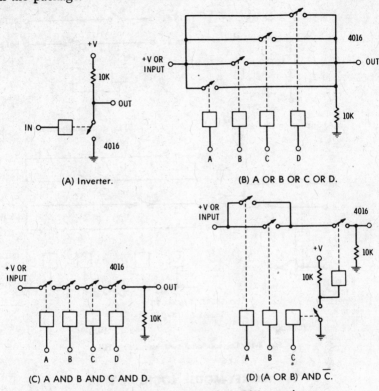

(A) Inverter.

(B) A OR B OR C OR D.

(C) A AND B AND C AND D.

(D) (A OR B) AND $\overline{C}$.

Fig. 3-17. Some transmission-gate logic (TGL) examples.

Fig. 3-17 shows us some "relay-logic" configurations that use positive-logic TGL. In Fig. 3-17A, we use a switch and a resistor to build an inverter. Four parallel switches, arranged as shown in Fig. 3-17B, give us an OR circuit. Four series switches, arranged as shown in Fig. 3-17C, give us an AND circuit, while Fig. 3-17D shows a combination of AND, OR, and inverter logic. Our input can go to +V for conventional logic, or it can be routed to a signal source if the logic is to act on an input signal. When you use TGL, always be sure there is an output load resistor, since any CMOS following a TGL circuit may see an open circuit when all TGL switches are open and will go to an undefined state.

Conventional switch circuits using TGL are shown in Fig. 3-18. For a normally closed contact, you invert your control signal. The inverter can be a conventional CMOS or can be the TGL inverter of Fig. 3-17A.

TGL is particularly useful when you need one unusual logic combination to provide an analog or video output.

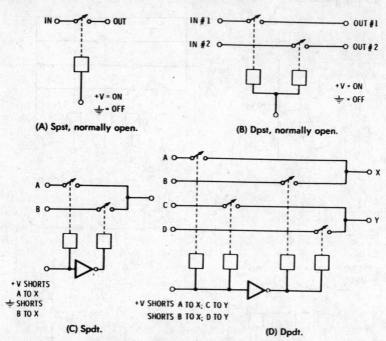

(A) Spst, normally open.

(B) Dpst, normally open.

(C) Spdt.

(D) Dpdt.

Fig. 3-18. Conventional switching using TGL.

MICKEY-MOUSE LOGIC (M²L)

What do you do when you need only one special gate or logic function tucked away somewhere in a corner of your CMOS system? Or suppose you need just one more input or just one more gate. Since all our CMOS gates have open-circuit inputs and since they are very good at improving rise times and "squaring up" sloppy input signals, we can usually put together something very simple that gets the job done. We can call any of these techniques Mickey-Mouse logic, or M²L. Two approaches to M²L are to use ordinary diodes and resistors, or to use the 4007 "do-it-yourself" CMOS gate package.

Fig. 3-19 shows some M²L using diodes and resistors. In Fig. 3-19A, we use a two-diode, one-resistor OR gate to add a second direct-set input to a flip-flop. This can also be used to add inputs to

242

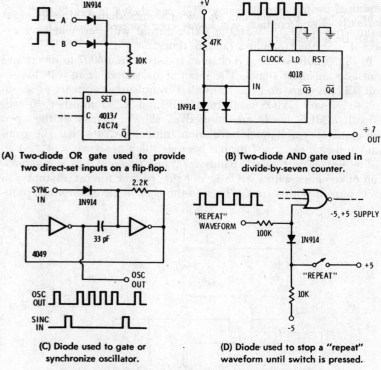

(A) Two-diode OR gate used to provide two direct-set inputs on a flip-flop.

(B) Two-diode AND gate used in divide-by-seven counter.

(C) Diode used to gate or synchronize oscillator.

(D) Diode used to stop a "repeat" waveform until switch is pressed.

Fig. 3-19. Mickey-Mouse logic (M²L) using diodes and resistors.

an OR or a NOR gate. In Fig. 3-19B, we have built an AND gate and used it to force a 4018 into a divide-by-seven counter. We can use the same stunt to add inputs to an AND or a NAND gate. In Fig. 3-19C, we use a diode to gate, or synchronize, an oscillator. When the sync input is made briefly positive, the oscillator is forced into one state and stays there until the sync command goes low. It then acts as a coherent, or gated, oscillator after the sync command goes away. In Fig. 3-19D, a diode is used to prevent a waveform from reaching a CMOS gate until a "repeat" key is pressed. Since the 10K resistor is much smaller than the 100K series input resistor, the input wave-form is too small to cause a response by the NOR gate. Pressing the key reverse-biases the diode and turns it off. This eliminates any loading on the 100K resistor, and the full input swing reaches the gate.

Fig. 3-20 shows some M²L approaches using the 4007. In Fig. 3-20A, we have a B AND NOT A gate used to combine a backspace or clear key with an input timing signal. The output is allowed to go positive only when the key is pressed and the input timing signal is low. Since the output can go low only immediately after an input timing pulse, we are sure the 4007's output will stop when we expect it to, rather than at a random time.

In Fig. 3-20B we use a p-channel transistor in a 4007 to invert and translate an input signal. The input signal comes from a 0- to +5-volt TTL system, and the translated complement appears as a −5-volt to +5-volt CMOS level. Inputs "B" and "C" from other −5-volt, +5-volt CMOS levels are entered as shown, thanks to the open source of the p-channel translation transistor. Note that the companion n-channel transistor has been disabled as shown.

The 4007 is particularly handy for translation and for the combination of switches with CMOS-level logic. In fact, almost anytime you find yourself with a tricky "back-to-the-wall" logic or interface prob-

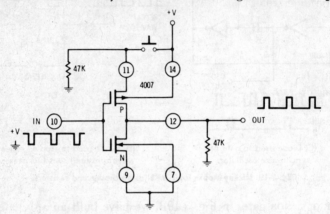

(A) "Backspace" or "clear" gate for a tv typewriter.

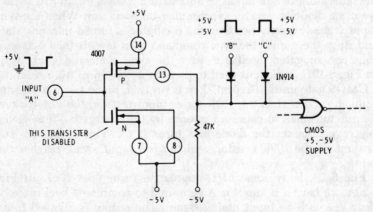

(B) Combined inverter and translator.

Fig. 3-20. M²L using the 4007 "do-it-yourself kit."

lem, the 4007 can help you straighten things out. There are several usage rules for the 4007. All inputs and all outputs must always be between the supply-pin voltage limits. You should never have a completed path of all "ON" switches from +V to ground, except possibly for a brief instant. And, unused transistors should be disabled by shorting them out and tying them to a supply rail. Tie the unused p-channel devices to +V and the unused n-channel devices to ground.

Mickey-Mouse logic should be used with discretion and only in slower portions of systems. It almost always needs extra supply current, and it cuts into potential noise margins. It should never be used above 100 kHz or so. A little bit of M²L goes a long way; after all, we are talking about only a quarter or so for a CMOS gate and maybe twice that for the board space and holes.

TRI-STATE® LOGIC

The outputs of most ordinary CMOS gates are always connected to either +V or ground through the low impedance of an "on" MOS transistor. If we short two logically different outputs together, they can fight each other, one pulling high, and the other pulling low; we end up using lots of supply current to get an undefined output state. Thus:

The outputs of ordinary CMOS gates may not be connected "wired-OR" fashion or otherwise shorted together.

If we want the outputs of several CMOS packages to share a common signal line, or *bus*, we have to use special CMOS devices whose outputs can be floated into a third, or high-impedance, state. Circuits with these special outputs are called *tri-state* logic.

Our third output state is a high-impedance or open circuit, compared with the low-impedance circuit to +V for a positive-logic one and the low-impedance circuit to ground for a positive-logic zero.

Not many CMOS packages have TRI-STATE outputs. The 4502 and 4503/74C126 are a hex inverter and a hex buffer, respectively, which are useful for providing TRI-STATE outputs. Other three-state outputs are present on the 5101 memory, the 4034 bus register, and several other CMOS devices.

Fig. 3-21 shows how we can build a three-state output with a 4007. If the TRI-STATE enable input (pin 6) is high, the lower n-channel transistor turns ON. The TRI-STATE enable input low also turns ON the uppermost p-channel transistor. The inverter in the middle now has supply and ground connections so that it is free to output a high state for a low input and vice versa. If the TRI-STATE enable is grounded, both the lowermost n-channel transistor and the uppermost p-channel transistor turn OFF, and the inverter cannot source or sink output current. With enable high, the circuit behaves as an ordinary inverter. With enable low, the output floats in a high-impedance state. CMOS devices that provide three-state outputs have similar internal circuitry.

Bus-oriented systems are particularly handy to minimize interconnections, especially in computers and microprocessors. In Fig. 3-22, a single bidirectional data bus can send data either from right to left or vice versa, depending on which output is enabled.

Note that A and B have to take turns gaining bus access, or they will fight each other, which leads to this rule:

On any TRI-STATE system, one and only one data source can be enabled at any particular time.

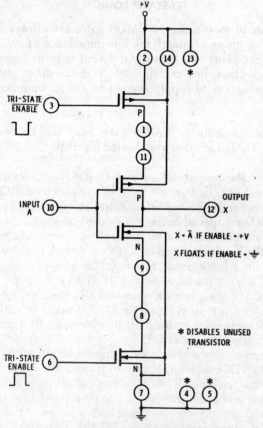

Fig. 3-21. TRI-STATE inverter built with a 4007.

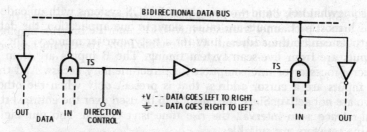

Fig. 3-22. Three-state logic used in bidirectional data bus system.

Most often, in computer systems we use multiple three-state outputs with 8 bus lines for an 8-bit word, or 16 bus lines for a 16-bit word, and so on. Note that only a single enable line is needed in these multiple-bus systems.

Fig. 3-23 shows how we can use three-state outputs to eliminate the need for data selectors when several data sources are to drive a common output. The source select switch is normally a 1-of-n CMOS decoder rather than the switch as shown. Again, in 8- or 16-bit systems, we will have 8 or 16 output lines, but we will need only one enable line for each input source selected.

Fig. 3-24 shows one way we might save one set of drivers in a CMOS-only system. This is done by floating the output lines on resistors that are driven by one of the inputs. In Fig. 3-24, the C inputs set the data outputs unless the A inputs or the B inputs are enabled. If the A or B inputs are enabled, they swamp the C resistors and take over control of the output lines. This arrangement

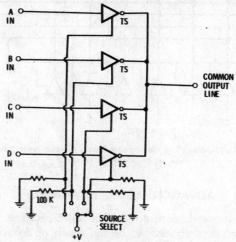

Fig. 3-23. Three-state logic used to eliminate data selectors or multiple-input switching.

is somewhat noisy and works only on CMOS systems with unloaded bus lines. The C inputs are rather slow. In one application, the data outputs can be the address lines for a tv-typewriter memory. The A inputs are from live-scan system timing. The B inputs are from a microprocessor or minicomputer for direct-memory access. And the C inputs are a cursor address that is present only when the other two are not active. Since the cursor can be used over the entire vertical-retrace scan interval, the rise time isn't critical, and our high-value resistors are suitable.

In general, three-state CMOS outputs cannot drive very much load capacitance without slowing down a lot, so TRI-STATE and bus-oriented systems using CMOS should be limited to lower speeds and to reasonable amounts of load capacitance. If a lot of devices are to be driven, or if the bus is to go off-board, or if terminated bus lines are used, heavier TTL-style bus drivers, such as the 26S10, the 3443, the 75138, or the 8838 are called for. Any of these drivers will need considerably more continuous supply power than an all-CMOS driver, but they have much greater drive ability.

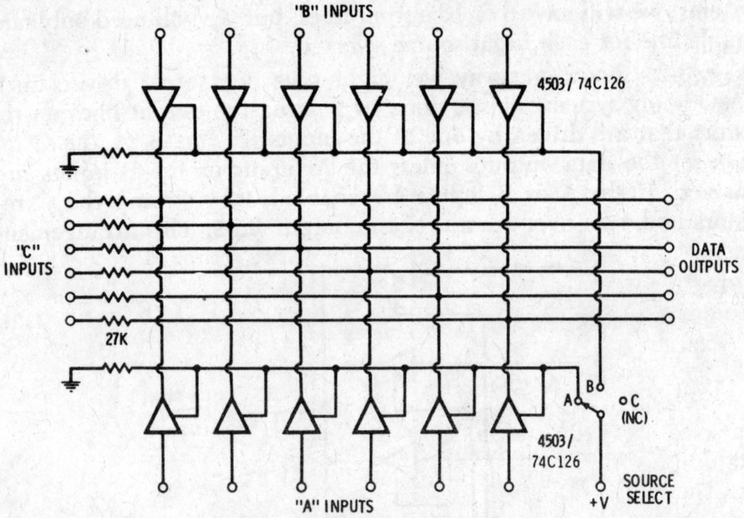

Fig. 3-24. "Floating" or swampable inputs (C) can sometimes save one set of TRI-STATE drivers.

ADVANCED LOGIC TECHNIQUES

Whenever we need more outputs or more-complex functions, or whenever there doesn't seem to be too much of an arithmetical or logical relationship between inputs and outputs, more powerful logic tools are needed. While simply stacking great quantities of gates together can do these jobs, there are much quicker, cheaper, and saner ways to do the same thing. One obvious way is to check the available MSI for a stock device that does the needed job or that can be readily adapted.

In the past, there have been elaborate procedures and sets of rules for determining ways of connecting gates so that the minimum number of gates are used to do a given job. These methods include Karnaugh mapping, Venn diagrams, Boolean algebra, state-sequence diagrams, and related horrors. Ironically, these "minimization" methods do not minimize design time, the number of IC *packages* in use, the system cost, or the ability to handle changes. Today, these techniques are worthless for most CMOS circuit designs.

Instead, we use a concept called *redundant circuit design,* or simply, *redundancy.* We pick one or more ICs that can be made to do an incredible variety of things, and then we restrict these universal or redundant ICs to doing only the job we want. Our restrictions can be done by external connections, by software commands, or by internal fusing operations.

There are lots of advantages to this route. The first advantage is ultraquick and ultralow cost design. You work directly with a truth table, and your logic design time is usually measured in seconds rather than weeks. A second advantage is the ease of changes. If something turns out wrong, a simple change of an IC, a software word, or an input connection is all that is needed for repair. The third advantage is low circuit cost, made possible both by mass production of identical ICs and by simplified PC board layouts.

Redundant ICs useful for advanced logic problems include the *data selector,* the *read-only memory* or ROM, the *programmable logic array* or PLA, and the *microprocessor.* Let's take a closer look at these powerful logic-design tools.

DATA-SELECTOR LOGIC

Data selectors offer a simple, one-package solution to an unusual truth table involving three, four, five, or even six variables and one output. They are particularly handy when there isn't any logical relationship between the input and the output, such as in musical notes, character generation, computer op code, and so on. A data selector is similar to a large selector switch that can pick one of a number of inputs and present that input as an output. With CMOS, we have a choice of conventional one-way selectors, such as the 4019/74C157, 4512, and 4539, or the two-way analog switches that include the 4051, 4052, 4053, 4067, and 4097.

Fig. 3-25A shows a typical four-variable, one-output truth table. Note that there is no obvious way to solve this truth table with a few gates and there is no apparent one-package, "gates-only" solution. In Fig. 3-25B, we have simply used a one-of-sixteen data selector to sample each line of the truth table and to input ones and zeros as needed to do the job. The design takes only seconds. Changes are trivial, and only a single package is needed.

Can we do even better? Fig. 3-25C shows how we can use only an *eight*-input data selector by employing an input-*folding* technique. Three of our variables (A, B, and C) go to pick the input positions. The fourth variable, its complement, a one, or a zero goes to various inputs as needed. Usually you pick a signal whose complement is available or easy to generate for variable D. The rest can often go in any order as needed.

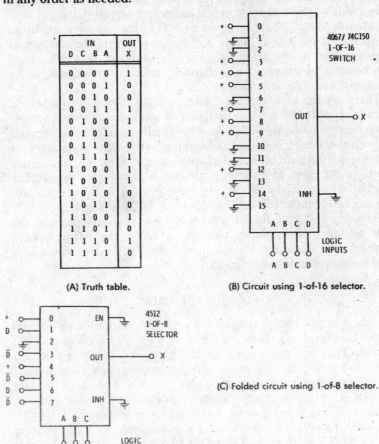

(A) Truth table.

(B) Circuit using 1-of-16 selector.

(C) Folded circuit using 1-of-8 selector.

Fig. 3-25. Data selector logic. Circuits shown provide one-package solution to 65,536 possible four-input, direct logic problems.

Now, we look at corresponding *pairs* of lines in our truth table having identical A, B, and C values. For our truth table, the two 000 cases are both ones, so we hard-wire a one to this input. In the 001 case, the output is a zero when D is a zero, and a one when D is a one, so we put D on this input. In the 010 pair, the output is a zero regardless of D, so we hard-wire a zero to this input. For the 011 pair, the output is a zero when D is a one and vice versa, so we put the complement of D, or $\overline{D}$, here.

We continue down the truth table, checking pairs of values that have an identical A, B, and C code and inputting ones, zeros, D's, or $\overline{D}$'s as needed. With our folding technique, we can build any of the 65,536 direct-logic, four-input gates by using only an eight-input selector. With one 16-input selector, we can build any of the over five billion five-input, direct-logic gates.

Chart 3-2 shows the available CMOS data selectors and the sizes of logic problems they can handle.

While you can go on to use bigger and bigger selectors for larger problems, there is an elegant way to minimize even further the size

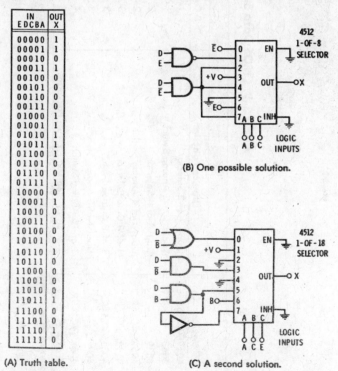

(A) Truth table.

(B) One possible solution.

(C) A second solution.

Fig. 3-26. Data-selector logic using "residue" technique. One-of-eight selector and simple input gating solves any of 4,294,967,296 five-input, direct logic problems.

Chart 3-2. CMOS Data Selectors

Device	Device Function	Input Variables			Truth Tables per Selector Package
		1-0 Only	Folded	Residue	
4019/74C157	Quad 1-of-2 Selector	1	2	3	4*
4053	Triple 1-of-2 Analog Switch	1	2	3	3
4539	Dual 1-of-4 Selector	2	3	4	2
4052/74C153	Dual 1-of-4 Analog Switch	2	3	4	2*
4512	1-of-8 Data Selector	3	4	5	1
4051/74C152	1-of-8 Analog Switch	3	4	5	1
4097	Dual 1-of-8 Analog Switch	3	4	5	2*
4067/74C150	1-of-16 Analog Switch	4	5	6	1
4067 × 2	1-of-16 Analog Switch, Two Packages	5	6	7	1/2

*Common Select Inputs

of the selector in use. Unfortunately, it is a method that only a logic freak could love. It involves using the data selector for most of the truth table and ending up with some "residue" two-input logic problems that can be handled with gates. As the example in Fig. 3-26 shows, we can use an eight-input selector as before and have both D and E variables left over. Our inputs are then ones, zeros, D, $\overline{D}$, E, $\overline{E}$, or any of the remaining two-input logic functions. Since we have lots of options on which two variables go to the input, it pays to pick one that has the simplest possible gating. (There ar eight versions of each of ten options for a total of eighty combinations. Ten of these will need different logic input combinations, while the rest will just need to have their input pins interchanged.)

While these residue techniques are lots of fun to play with, read-only memories are often simpler and cheaper.

A CMOS Cameo

Square-Wave Oscillator

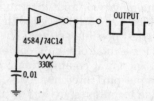

Frequency is approximately 1 kHz for values shown.

Chapter 4

Multivibrators

The term *multivibrator* is rather dated, but it is still probably the best one to use for the two-state CMOS circuits in this chapter. These circuits become two-stated by cross-coupled feedback between a pair of inverting gates or a pair of inverters. Depending on how we use this feedback, the circuit can have two stable states, two stable states dependent on an input signal, one stable state, or zero stable states.

Circuits with two stable states give us latches, memories, flip-flops, and contact conditioners. Two stable states that depend on an input signal give us *Schmitt triggers* and other snap-action circuits for input conditioning and rise-time improvement. One stable and one unstable state lets us build a monostable, pulse generator, or edge detector. And two unstable states let us build an oscillator or signal source.

Similarly, we can feed back around one inverter or inverting gate. This gets us into *linear* CMOS operation, useful for amplifiers and very simple crystal-oscillator circuits.

Multivibrators and linear feedback are the key to memory circuits, clocked logic, pulse shapers, input conditioners, and signal sources. CMOS is particularly good for these uses, since it has open-circuit, nonloading inputs and also since it is extremely easy to bias into a linear operating mode. The open-circuit inputs let us work our pulse circuits from either supply rail with equal ease, and let us use high-value timing resistors along with small capacitors in low-frequency applications.

255

BISTABLE CIRCUITS

Bistable, or two-stable-state, circuits are built by cross-coupling inverters or inverting gates with direct-wire connections. Bistable circuits are used for memories, latches, storage, and contact conditioning, but their most important use is their combination into master-slave pairs for the *clocked-logic* blocks of the next two chapters.

Cross-Coupled Inverters

Suppose that we connect two inverters so that the output of the first is connected to the input of the second and vice versa as shown in Fig. 4-1. Assume that output A happens to be low. A low input to inverter B produces a high B output that reaches around and holds the input to inverter A high. The high input on inverter A gives us a low A output, and the circuit will apparently stay in this particular condition, or *state*, as long as supply power remains applied and we don't otherwise change the circuit.

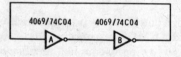

Fig. 4-1. Two cross-coupled inverters.

What if the A output happens to be high? The same sequence occurs, except in reverse order. Making output A high produces a low output B, which in turn holds output A high. Once again, the circuit is stable. There are two stable states to this circuit. Traditionally, we call it a *bistable multivibrator.* Today, it is more likely to be called a latch or a one-bit memory.

Mechanical Triggering

Our latch would be much more useful if we could force it into a desired state. In Fig. 4-2A, we have added two push buttons to momentarily short an output to ground. This momentary short flips or holds the latch into the state the push button is pulling low. Short circuits on our CMOS outputs are tolerable, but they are not good practice as they will introduce TTL-like current glitches or spikes onto the power supply. This is particularly true of heavier inverters such as the 4049 and the 4502.

We can eliminate these current glitches by adding resistors as in Fig. 4-2B. We have also arranged our two switches into a spdt (single-pole, double-throw) push button. The resistors isolate the temporary shorts from the output and thus greatly minimize the size of the transition glitch.

This particular arrangement is called a *bounceless push button*. The latch flips the instant that the switch "makes," and holds until the instant the switch closes the normally closed contacts. Debouncing circuits like this one are absolutely essential anytime a mechanical contact is to be used with clocked logic, because any contact noise or bounce can otherwise be read as multiple input signals.

The latch output that is high when the button is pressed is called a Q output; its complement is called the $\overline{Q}$ output, sometimes called "not-Q" or "Q-bar."

Fig. 4-2C shows a hex contact conditioner that uses the six buffers in a 4050 noninverting buffer for six individual spdt push buttons or switches. Note that two cascaded inverters are the same as a non-inverting-buffer logic block, but that we lose the optional $\overline{Q}$ output. We can interchange the switch contacts to get a $\overline{Q}$ output instead of a Q, but we can't get both at once.

Electronic Triggering

The simplest way to electronically trigger the latch is to use inverting two-input gates instead of inverters. Fig. 4-3 shows a pair of NOR gates connected as a set-reset flip-flop and also shows the symbol and truth table for this flip-flop. This circuit obeys the following rules:

- If both inputs are grounded, the flip-flop stays the way it was.

- If the Set input is made positive, the flip-flop goes or stays in the state with Q positive and $\overline{Q}$ grounded.

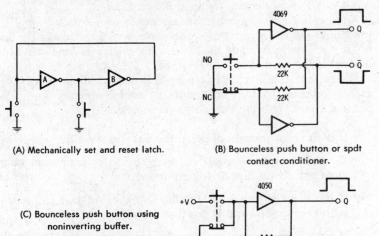

(A) Mechanically set and reset latch.

(B) Bounceless push button or spdt contact conditioner.

(C) Bounceless push button using noninverting buffer.

Fig. 4-2. Mechanical triggering and bounceless push button.

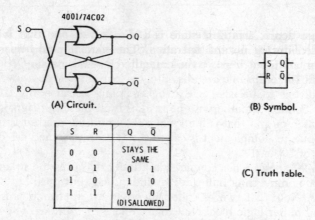

(A) Circuit.

(B) Symbol.

S	R	Q	Q̄
0	0	STAYS THE SAME	
0	1	0	1
1	0	1	0
1	1	0	0
		(DISALLOWED)	

(C) Truth table.

Fig. 4-3. Set-reset flip-flop built with NOR gates.

- If the Reset input is made positive, the flip-flop goes or stays in the state with Q grounded and Q̄ positive.

- If both Set and Reset are simultaneously made positive, the flip-flop temporarily goes into a state with *both* Q and Q̄ grounded. The last input to go to ground will decide the final state of the flip-flop.

This last condition is called a *disallowed state*. It is normally avoided. For proper operation of a set-reset flip-flop, *the set command and the reset command must never be present at the same time*. If this ever happens, a disallowed state will result. The *last* input to go away decides the final flip-flop state.

An easy way to remember the operation of a NOR flip-flop is to imagine that the Set input passes its "one" across the top of the symbol to the Q output, while the Reset input passes its "one" across the bottom of the symbol to the Q̄ output.

We can also build a Set-Reset flip-flop with NAND gates, as shown in Fig. 4-4, but this time our inputs are complementary ones called Set and Reset. This flip-flop obeys these rules:

- If both inputs remain positive, the flip-flop stays the way it was.

- If the S̄et input is grounded, the flip-flop goes or stays in the state with Q positive and Q̄ grounded.

- If the R̄eset is grounded, the flip-flop goes or stays in the state with Q grounded and Q̄ positive.

- If both S̄et and R̄eset are simultaneously grounded, the flip-flop temporarily goes into a state with both Q and Q̄ positive. The last input to go positive will decide the final state of the flip-flop.

258

Once again, this final state is a disallowed one that is usually avoided during normal operation. The NAND flip-flop operates by taking an input zero, complementing it, and passing it directly across the symbol to the respective Q or $\overline{Q}$ output.

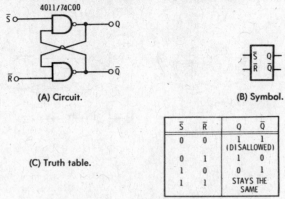

(A) Circuit.

(B) Symbol.

(C) Truth table.

$\overline{S}$	$\overline{R}$	Q	$\overline{Q}$
0	0	1 (DISALLOWED)	1
0	1	1	0
1	0	0	1
1	1	STAYS THE SAME	

Fig. 4-4. Set-reset flip-flop built with NAND gates.

The 4043 is a quad NOR flip-flop, while the 4044 is a quad NAND flip-flop. Clocked-logic flip-flops, such as the 4013/74C74 and 4027/74C73, can also be used as set-reset devices by using only their direct inputs. More details on this in the next chapter.

Edge Triggering

Most often, we want to set or reset our flip-flops on the start or finish of some other waveform. This is called *edge triggering*. It lets us use input commands of any length and greatly minimizes the disallowed-state problems.

Edge triggering is done by differentiating the leading or trailing edge of an input with a resistor-capacitor network. Fig. 4-5 shows how we edge-trigger the NAND and NOR flip-flops with positive or negative signal edges. In the circuits of Figs. 4-5B and C, input inverters are used to invert the input command and make it compatible with the triggering. Although we have shown identical edge triggering on each input pair, we can combine them any way we like.

The time constants shown generate approximately a 20-microsecond pulse when fed with a fast-rise or fast-fall waveform. These time constants can be shortened for faster applications, or they can be lengthened whenever signals with poorer rise time are expected. The time constant is equal to the product of the resistor and the capacitor. Resistor values in K and capacitor values in nanofarads (.001 μF or 1000 pF = 1 nanofarad) will give the time constant in microseconds. The time constant must be long enough to accept an input (at least ten times the input rise time if possible), but short enough that the triggering command goes away before the flip-flop is to be returned to the other state. Otherwise, a disallowed-state condition temporarily crops up.

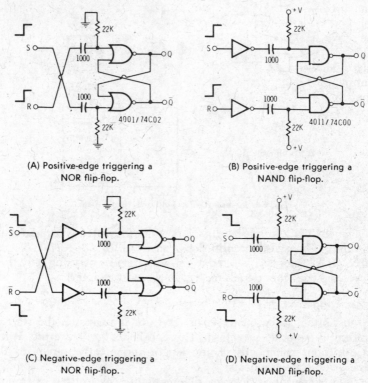

(A) Positive-edge triggering a NOR flip-flop.

(B) Positive-edge triggering a NAND flip-flop.

(C) Negative-edge triggering a NOR flip-flop.

(D) Negative-edge triggering a NAND flip-flop.

Fig. 4-5. Edge triggering.

Counting and Shifting

There is no ultrasimple way to get a single set-reset flip-flop to alternate its states, binarily count, or shift information to the next flip-flop down the line. You will *always* get into steering problems, preferential states, and unchecked races if you try this.

For these more-complex counting, shifting, and synchronizing circuits, you have to use the *clocked logic* of the next two chapters. The key to clocked logic is to use alternating *pairs* of set-reset flip-flops in a master-slave arrangement. Only half the pair can change at a time, giving you orderly steering and one-state-at-a-time operation. *Never* attempt these more-complex functions with a circuit having only one flip-flop per stage.

260

Hold-Follow Latches

Fig. 4-6 shows two versions of a different type of bistable circuit called a *hold-follow* flip-flop, or a *storage latch*. There are two inputs—a *data* input and a *control* input. The control input decides whether the flip-flop is to keep an old piece of data or follow the changing input. In Fig. 4-6A, we use NOR logic. Grounding the control input causes the flip-flop to follow the data input, since a data one sets Q to a one, and a data zero, after inversion, sets $\overline{Q}$ to a one.

Fig. 4-6B shows how to do the same thing with NAND gates. The only difference is that the control sense is reversed, with a ground control input holding previous information and a positive control input following the data input.

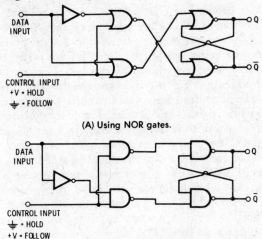

DATA INPUT

CONTROL INPUT
+V - HOLD
⏚ - FOLLOW

(A) Using NOR gates.

DATA INPUT

CONTROL INPUT
⏚ - HOLD
+V - FOLLOW

(B) Using NAND gates.

Fig. 4-6. Hold-follow storage latches.

Latches of this type are handy for data storage, particularly for holding numbers to be digitally displayed on readouts. Like the earlier set-reset flip-flops, these circuits **cannot** be cascaded. The 4042 and 4508 are single and dual four-bit hold-follow latches, respectively, while the 4514 and 4515 combine latches with a 1-of-16 decoder. The 4511 is a similar latch and seven-segment readout decoder-driver.

"Nonvolatile" Memory Techniques

An ideal memory should be instantaneous, super small, and nearly free. In addition, it should be able to read or write data at system speeds but still be able to hold information for very long periods of

TO MAKE A CMOS SYSTEM "NONVOLATILE":

1. BRING AND HOLD ALL SYSTEM INPUTS EXACTLY TO GROUND.
2. UNLOAD ALL OUTPUTS THAT ARE SOURCING CURRENT.

-- THEN --

3. SMOOTHLY SWITCH FROM MAIN POWER TO A BACKUP BATTERY.

TO POWER UP A "NONVOLATILE" CMOS SYSTEM:

1. SMOOTHLY SWITCH FROM THE BACKUP BATTERY TO MAIN POWER.

-- THEN --

2. RECONNECT OUTPUT LOADS AS NEEDED.
3. REACTIVATE INPUTS AS NEEDED.

Fig. 4-7. CMOS systems can simulate nonvolatile storage if these rules are followed.

time in the absence of supply power. At this writing, no such device exists.

The PROMs of the last chapter are nonvolatile, but either they are not changeable at all or they take a long time to make changes. Newer erasable and rewritable nonvolatile RAMs (random-access memory) are just being introduced, but they are expensive and still take a relatively long time (milliseconds per block-written bit) to write data. And magnetic core, while nonvolatile with proper power-down and write-after-destructive-read circuits, takes too much in the way of support to be practical in small, low-cost, low-power systems.

CMOS offers a way to fake nonvolatility in a simple and low-cost manner. In fact, the faking is so good that it may even further set back the development of a true nonvolatile RAM. The key to the process is that CMOS needs almost zero supply power when inputs are not changing and outputs are not loaded. All you do is carefully switch over to a holding battery during power-down times. Since the battery need only provide a few microamperes of current, it will last as long as its normal shelf life. Nicads (nickel-cadmium) or other rechargeable batteries are usually good for this application as they can be automatically charged during power-up times and hold the CMOS data for power-down times.

These "nonvolatile" techniques are good for applications such as computer data storage, electronic checkbooks, and remote data acquisition in caves, volcanoes, glaciers, satellites, and other hard-to-get-to places. And, while a backup battery may seem like a cop-out, there are lots of natural and nonelectronic systems that use the same idea for nonvolatile storage. Two obvious examples are the genetic code and the human brain.

Fig. 4-7 shows what we have to do to make a properly designed CMOS system "nonvolatile" and then switch it back to system power. To go to battery power, you have to prevent inputs from changing and you have to unload outputs. Then you are free to switch over to battery power, but this has to be done smoothly without any loss of power. Fig. 4-8 shows an automatic changeover circuit. Closing

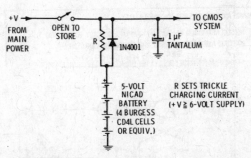

Fig. 4-8. Automatic changeover circuit for "nonvolatile" CMOS operation.

the switch applies system power. It also trickle charges the Nicad battery stack at a rate controlled by resistor R. Opening the switch causes the battery to take over.

There are a few more usage rules to consider if our CMOS system is going to absolutely minimize battery drain in the hold mode. All input changes must be prevented. The simplest way to do this is to design the circuit so that all inputs are low during the hold mode and so that all drive power is disconnected from the inputs. While our outputs are still free to sink current to ground if we want them to, we must prevent them from sourcing any current from the positive power supply, since the battery has to provide this current. Tristate outputs are a very good choice, but any method that keeps all outputs floating or low during hold times will work.

We also have to be sure that every CMOS input in the system goes either solidly to ground or solidly to +V, for being within just a fraction of a volt from ground or +V causes the standby current to sharply increase. This also means that any resistors in the circuit draw zero current during power-down times, and that no CMOS stage is purposely biased into an active region for use as an oscillator or timer.

Once all these restrictions are met, it is a simple matter to build a "nonvolatile" system that can hold data for months or even years, without attention. Let's look at two examples.

A Model-Railroad-Track Turnout

The turnouts, or track switches, in a model railroad layout are often pulse operated. Thus, there is no way of immediately knowing which state they happen to be in, particularly when everything is turned off. You can apply continuous power or you can add contacts to feed back information concerning the switch state to the main panel, but this takes a lot of wire.

CMOS offers another way to indicate the state of the switch, as shown in Fig. 4-9. We build a "nonvolatile" NOR flip-flop memory,

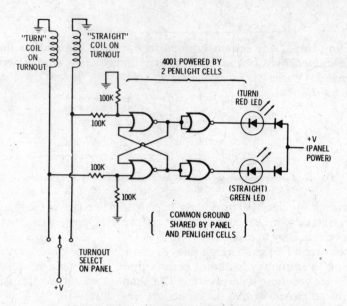

Fig. 4-9. "Nonvolatile" CMOS memory used as an indicator for a model-railroad-track turnout.

permanently powered by two penlight cells. Switching sets or resets the memory and lights the red or green indicator LEDs as shown. The diodes, the LEDs, and the coils on the turnout are powered from the main panel, while the memory itself has its own permanently on battery supply. The battery need provide only the CMOS leakage current, along with an ultrabrief switching transient of a few mils for a few nanoseconds every time the switch is changed. Only one backup battery is needed for all the switch memories in the system. The diodes in series with the LEDs reduce the reverse-bias leakage current to a very low value during power-down times.

A 256 × 8 Microcomputer Memory

Fig. 4-10 shows how we can build a "nonvolatile" memory for a microprocessor or hobbyist computer. It is organized as 256 words of eight bits each and uses two 5101 RAMs. All of the power-down circuitry needed is built into these CMOS memories.

Each RAM covers four of the eight bits per word. In normal operation under system power, the read ($R/\overline{W}$) input is left high and an address is sent to the eight address pins, along with input data to the eight input-data pins—four per IC. Briefly bringing the $R/\overline{W}$ input low will load that information into the addressed location. To read the data, simply address the location and leave the $R/\overline{W}$ input high. The addressed data will appear on the output lines. As

264

with any RAM, it is extremely important to keep the address lines stable immediately before, during, and immediately after a write command. Otherwise, other locations might get "flashed" and can change their contents.

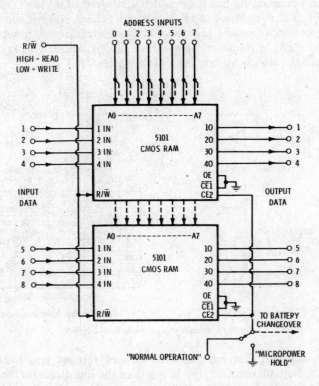

Fig. 4-10. A 256-word, 8-bit-per-word "nonvolatile" CMOS microcomputer memory.

The second chip enable, or CE2, controls the automatic change-over from system operation to low-power "nonvolatile" holding. To hold data, connect CE2 to ground and then switch over to your battery supply. Grounding CE2 automatically disconnects all inputs and all outputs. Supply current needs drop to 15 microamperes per chip, or 30 microamperes total, which is easily handled by a stack of four Nicad button cells, such as Burgess CD4L cells or their equivalent.

SCHMITT TRIGGERS

Schmitt triggers are bistable circuits that are driven by an input. They provide a snap-action response with a *hysteresis,* or dead band.

They are useful for squaring up slowly rising or noisy inputs, for contact debouncing, and for general input interface and conditioning.

Fig. 4-11 shows how you can build your own Schmitt trigger by using a noninverting buffer or gate. There are two resistors, an *input* resistor and a *feedback* resistor. If our feedback resistor were very large, our input would always switch, without any snap action, at the threshold or transition point of the gate or buffer. This is usually one-half the supply voltage, or a voltage of $+V/2$.

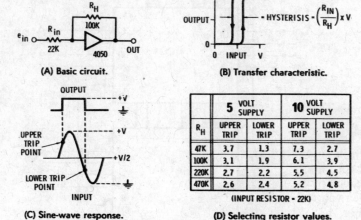

(A) Basic circuit.

(B) Transfer characteristic.

(C) Sine-wave response.

(D) Selecting resistor values.

R_H	5 VOLT SUPPLY		10 VOLT SUPPLY	
	UPPER TRIP	LOWER TRIP	UPPER TRIP	LOWER TRIP
47K	3.7	1.3	7.3	2.7
100K	3.1	1.9	6.1	3.9
220K	2.7	2.2	5.5	4.5
470K	2.6	2.4	5.2	4.8

(INPUT RESISTOR = 22K)

Fig. 4-11. Schmitt trigger made from a buffer.

But, suppose that our positive-feedback resistor was somewhat smaller, but still considerably larger than the input resistor. Suppose further that both our output and our input happened to be low. If our input voltage starts rising from a low value, it will have to go well *above* $+V/2$ before any triggering can occur, because of the positive-feedback resistor acting as a step-down voltage divider and reducing the gate or buffer input voltage below the actual input voltage.

When we finally reach an *upper trip point* that is above $+V/2$ but corresponds to the trigger voltage at the gate input, the output will go positive. This output will reach around through our feedback resistor and drive the input positive as well, reversing the sense of the voltage divider. The positive feedback rapidly causes the circuit to change state. Even if there is some noise on the input, once this change starts, it rapidly finishes itself. Similarly, when the input voltage is high and falling, the reverse action occurs, and we have to go down to a *lower trip point* that is *below* $+V/2$. This time, when

266

the threshold is reached, the snap action goes in the opposite direction, again because of the feedback resistor.

The amount of dead band, or hysteresis, you get between the trip points is simply the resistor ratio, R_{in}/R_H, times the supply voltage. Thus, with a 10-volt supply and a feedback resistor that is ten times the value of the input resistor, we get a 1-volt hysteresis. This means our upper trip point will be +5.5 volts for rising inputs and our lower trip point will be +4.5 volts for falling inputs. Input voltages between these two values are in the dead band and will be ignored unless they drop below the lower trip point or above the upper trip point.

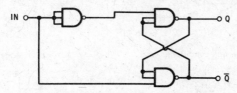

Fig. 4-12. Schmitt trigger made from NAND gates has higher input impedance.

The rest of Fig. 4-11 shows the transfer characteristic of this circuit, its sine-wave response, and the amount of dead band you will get for various resistor and supply values. This is a handy circuit, particularly when you only need to condition one or two inputs, and when you want complete control over the amount of hysteresis.

Fig. 4-12 shows us a CMOS-only Schmitt trigger that is faster, needs no resistors, and has an open-circuit input impedance. It is based on the difference in input threshold voltages needed for different combinations of inputs on a three-input gate. With a three-input NAND gate, it takes a higher input voltage to produce a low output when all three inputs are driven together than when only one input is driven and the other two inputs are separately controlled. So, it takes a higher input voltage to set our flip-flop on a rising input than it does to reset it on a falling one. The dead band varies with the device and supply voltage. Typically, hysteresis is one-third the supply voltage. This can be dropped to one-sixth the supply voltage by driving only two inputs on the first stage and tying the third input high.

The same basic idea is used in two ready-to-go CMOS Schmitt triggers, the 4093 quad two-input NAND version, and the 4584 hex inverter Schmitt trigger. These devices have open-circuit inputs and pinouts identical to common inverters and gates, except that they provide an automatic snap-action input response.

Fig. 4-13 shows some applications for Schmitt triggers. In Fig. 4-13A, we use the circuit as a rise-time improver. Extremely slow

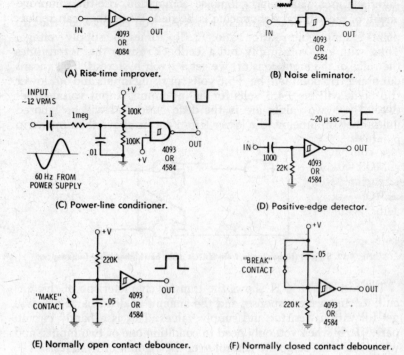

(A) Rise-time improver.

(B) Noise eliminator.

(C) Power-line conditioner.

(D) Positive-edge detector.

(E) Normally open contact debouncer.

(F) Normally closed contact debouncer.

Fig. 4-13. Using CMOS Schmitt triggers.

signals and ultralow-frequency sine waves can be converted to fast-rise outputs this way. In Fig. 4-13B, we do the same thing to eliminate noise on a system input. The noise must be less than the dead band for complete rejection, and it can still jitter the time location of an input change. In Fig. 4-13C, we have a power-line conditioner that gives a sharp 60-hertz output pulse once each power-line cycle for use as a frequency reference. The input resistors and capacitors are both a noise filter and a way of centering the trigger bias to the middle of the dead band. Because of the hysteresis, the duty cycle of the output normally will not be exactly 50/50, but the frequency will exactly follow the input reference.

In Fig. 4-13D, we have a positive-edge detector. This is easily converted to a negative-edge detector by returning the input resistor to +V instead of ground. Two contact debouncers or conditioners are shown in Figs. 4-13E and F, for normally open and normally closed single contacts. These circuits operate by discharging or recharging an RC network whose time constant exceeds *three times* the worst-case bounce time.

Schmitt trigger circuits are recommended for most CMOS input interface uses, particularly if noise elimination, rise-time improvement, or mechanical debouncing is needed. While they can replace conventional gates, Schmitt triggers tend to be slower.

ASTABLE CIRCUITS

Neither of the two states is stable in an *astable multivibrator*. So, these circuits will continually flip back and forth, *free-running*, to form an oscillator or signal source.

There are several good ways to build CMOS astables. We can use conventional resistor-capacitor flip-flop charging circuits. With CMOS we can get by very nicely with a single resistor and a single capacitor in an ultrasimple, sure-starting circuit. Or, we can bias our CMOS as a linear amplifier and build a high-stability crystal oscillator, again in an ultrasimple circuit. We also have options of building CMOS voltage-controlled oscillators and other types of signal sources using transmission gates, Schmitt triggers, and non-CMOS integrated circuits. Let's find out more about these concepts.

A Simple Astable

The basic CMOS astable circuit is shown in Fig. 4-14. This circuit uses two inverters, a resistor, and a capacitor. If the output of inverter B happens to be high, its input must be low, and the resistor (R_T) will charge the capacitor in the positive direction.

When the input voltage gets to the threshold point of inverter B, output B starts to go low and also drives input A low, which in turn drives output A high. The charge on our capacitor can't change instantaneously, so the right side of the capacitor jumps positive as well. This sudden jump, which is in the direction that input B is going, provides positive feedback, snapping the circuit into the other state.

Output B is now low, its input is high, and this time the resistor charges the capacitor in the negative direction. Charging continues until the threshold is once again reached. At the time the threshold is reached, the circuit snaps back into its original state. Our circuit

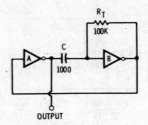

Fig. 4-14. CMOS astable oscillator.

continues to alternate states, with the resistor first charging the capacitor in the positive direction and then in the negative direction.

Our output is shown taken off at the A inverter. A complementary output is also available at the output of inverter B. However, if you use A-series CMOS, this output may have limited amplitude and rounded edges. For this reason, *B-series devices are preferred for all astable and monostable uses.* B-series Schmitt-trigger devices are particularly handy for very low frequency applications.

An Improvement

Our basic astable is a very simple and very effective signal source. Unlike traditional transistor astables, the circuit self-starts, since we have only one RC network. A minimum supply voltage of 4.5 volts is recommended for most CMOS astable circuits.

Our simple circuit would perform even better if it were not for the input protection diodes. If both diodes are present, they make the frequency dependent on the supply voltage and cause the corners of output A to be rounded. Worse yet, if a zener protection network is used instead of the two-diode protection (see Fig. 1-15), we get a clamping on one state only—and an asymmetrical output waveform.

The way around these protection problems is shown in Fig. 4-15. We add a new resistor as shown. This unloads the RC network from the clamping effect of the protection diodes. It will give us a nearly square, or 50/50, duty cycle, and will minimize the rounding of output A and make the frequency almost totally independent of power-supply variations.

Normally, you make the input resistor *ten times* the timing resistor to minimize the effect of the protection diodes. Since our CMOS inputs are normally open circuits, this resistor has no effect on operation. The upper limit is set by the resistance getting so large that it can't charge the 5-picofarad input capacitance of inverter B in a time that is reasonably small compared with one cycle of the frequency of operation.

The waveforms of our improved astable circuit appear in Fig. 4-16. Inverter B switches at one-half the supply voltage, or $+V/2$, so

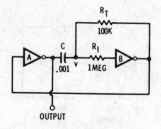

Fig. 4-15. Improved version of the CMOS astable oscillator.

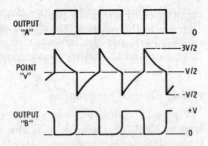

OUTPUT "A"

POINT "V" — 3V/2, V/2, -V/2

OUTPUT "B" — +V, 0

Fig. 4-16. Waveforms of improved CMOS astable oscillator.

the input voltage to inverter B swings the supply voltage from a maximum of +3V/2 to a minimum of −V/2. The output rounding shown is typical of A-series devices but is minimized or eliminated entirely with B-series devices.

The initial charging voltage across R_T in either direction is 3/2 the supply voltage, so the threshold is set at the 2/3, or 67%, point on the charging curve. The 63% point on a charging curve is called the *time constant*, and is equal to the RC product (resistance in K times capacitance in nanofarads equals time in microseconds).

So, one charging cycle apparently takes 67/63 of a time constant, or about 1.1 RC. Since there are two chargings involved, the total period has to be double this, or 2.2RC, and the frequency will be 1/2.2RC, or:

$$\text{Period} = 2.2 \text{ RC}$$
$$\text{Frequency} = 1/2.2 \text{ RC}$$

Curves of frequency versus R for different values of C appear in Fig. 4-17. Resistors can range from 3K to several megohms, and capacitors can vary from 50 picofarads, or so, upwards. Below 1000 picofarads, the frequency will be somewhat lower than our formula

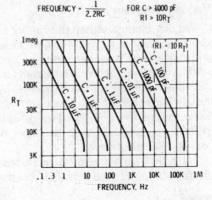

$$\text{FREQUENCY} = \frac{1}{2.2RC} \quad \text{FOR C > 1000 pF} \quad RI > 10R_T$$

Fig. 4-17. Component values for improved CMOS astable oscillator.

R_T

1meg, 300K, 100K, 30K, 10K, 3K

.1 .3 1 10 100 1K 10K 100K 1M

FREQUENCY, Hz

predicts. This is shown by the closer spacing between the 100- and 1000-pF curves. These curves are approximate and assume an input resistor ten times the feedback resistor. Schmitt-trigger circuits will have somewhat higher frequency values than those shown in Fig. 4-17.

Higher Output Frequencies

Fig. 4-18 shows a 6-megahertz astable using a pair of inverters from a 4049. This astable will also work with 4502 inverters. These higher frequencies require faster, higher-current devices and tend to be dependent on the device, supply voltage, and load capacitance. The circuit shown uses the RCA 4049 with a 10-volt supply. Selected ICs may be needed.

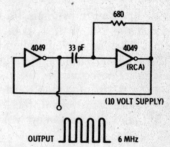

Fig. 4-18. Higher output frequencies require high-current inverters.

Another approach to high-frequency operation is to build a *ring oscillator* of three or five cascaded inverters. The propagation delay through the devices sets the operating frequency, which can be as high as 20 megahertz. The frequency may be lowered by placing a capacitor across one of the inverters. This circuit configuration is also dependent on the supply voltage, the load, and the type of device, and should be used only with a regulated power supply and hand-picked devices.

Varying Frequency

We can change either the timing resistor or the capacitor to vary the frequency of any of our CMOS astable circuits. Fig. 4-19 shows a circuit whose frequency may be smoothly varied over a 10:1 range with a potentiometer. Frequency may also be switch-stepped in decades by changing the capacitor by tens. The duty cycle remains nearly 50/50 for all frequency settings. The recommended resistance values range from 3K to 10 megohms. Capacitors can range from 50 picofarads upwards, with the most uniform and predictable results above 1000 picofarads.

Note that *higher* resistance values are associated with the *lower* frequencies, and vice versa. If you use a calibrated dial and an ordi-

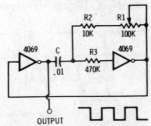

C SETS <u>RANGE</u> OF OPERATION
R2 SETS <u>MAXIMUM FREQUENCY</u>
R1 + R2 SETS <u>MINIMUM FREQUENCY</u>
R3 ISOLATES PROTECTION DIODES

Fig. 4-19. Varying output frequency of astable circuit.

nary linear potentiometer, the range will be very cramped at one end. You can get a more uniform calibration either by using a reverse-taper log pot or by using a normal audio-taper log pot and putting the dial on the pot shaft and the pointer on the panel.

Symmetry

The output duty cycle of the circuits we have looked at so far is pretty much 50/50, meaning that the output is high for one-half the time and low for the other half. There will be slight variations from device to device, particularly at higher frequencies and if the protection diodes are not properly isolated.

Sometimes we want to purposely unbalance the duty cycle for special uses. Fig. 4-20 shows some possibilities. In Fig. 4-20A, we have added a diode and a resistor to the charging network. The connection shown puts R1 in the circuit when output B is high and leaves it out when B is low. This means that our capacitor charges in the positive direction through the parallel combination of R1 and R2, but that it charges in the negative direction only through R2. The result is an output that is high for a greater time than it is low. The ratio of low time to high time is set by the ratio of the parallel R1–R2 combination to R2 alone. Duty cycles beyond 100:1 are possible. If you want the opposite duty cycle, simply reverse the diode.

In Fig. 4-20B, we have an astable circuit whose duty cycle can be varied over an extremely wide range, and pretty much independent of frequency. Our pot setting determines two resistances for us, R1 and R2. Resistors R1 and R2 add together to equal the end-to-end resistance of the pot. Resistor R2 charges the capacitor in the positive direction, and R1 charges it in the negative direction. We can get any duty cycle we want, ranging from narrow positive pulses, through 50/50, to narrow negative pulses.

You can improve the circuit further as shown in Fig. 4-20C. Here, R4 sets a duty-cycle limit in either direction, independent of pot settings. This guarantees starting even if the pot is at one of its

273

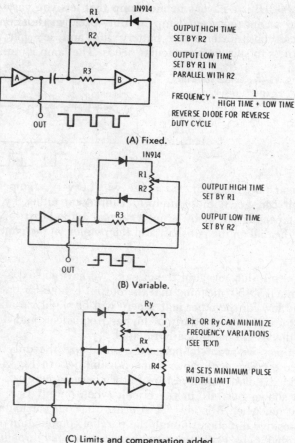

OUTPUT HIGH TIME
SET BY R2

OUTPUT LOW TIME
SET BY R1 IN
PARALLEL WITH R2

$$\text{FREQUENCY} = \frac{1}{\text{HIGH TIME} + \text{LOW TIME}}$$

REVERSE DIODE FOR REVERSE
DUTY CYCLE

(A) Fixed.

OUTPUT HIGH TIME
SET BY R1

OUTPUT LOW TIME
SET BY R2

(B) Variable.

Rx OR Ry CAN MINIMIZE
FREQUENCY VARIATIONS
(SEE TEXT)

R4 SETS MINIMUM PULSE
WIDTH LIMIT

(C) Limits and compensation added.

Fig. 4-20. Changing symmetry or duty cycle.

limits. Usually, variation of duty cycle will be almost totally independent of frequency. If a slight difference remains, a resistor, *either* R_x or R_y, may be added to make the frequency identical at both ends of the pot rotation. Normally, the value of the selected resistor will be five times the pot value, or higher.

Some applications of asymmetric circuits appear in Fig. 4-21. Fig. 4-21A is the master clock for a tv typewriter. It runs at 1.06 megahertz but has a very narrow, positive duty cycle. The positive portion of the waveform is used to load a video shift register with information from a character generator. The circuit is also gated with the diode as shown. By varying the off time, you set the horizontal-scan and horizontal-blanking rates. This lets you lock the tvt scan to the power line with suitable extra circuitry.

Fig. 4-21B is a caver's helmet lamp that lets you vary the brightness of an incandescent lamp by duty-cycle modulation. This circuit can both extend the battery life and let you adjust the brightness to suit your particular needs. The lamp is rapidly turned on and off, with the on time being set by the duty cycle of the astable. The A-series devices are used with three transistors paralleled at the output to get enough drive current to power the output transistor or Darlington pair. Note that this is essentially a "lossless" dimmer, compared with the power you would waste if you simply put a high-wattage potentiometer in series with the lamp.

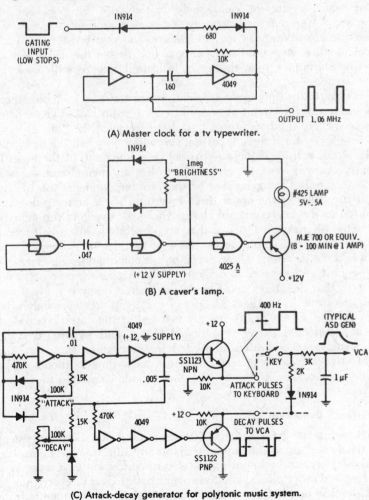

(A) Master clock for a tv typewriter.

(B) A caver's lamp.

(C) Attack-decay generator for polytonic music system.

Fig. 4-21. Asymmetric circuit applications.

Similar duty-cycle modulation techniques can let you control the brightness of a light-emitting-diode (LED) display by blanking the drive waveforms as needed. Switching-mode power supplies also use this principle for voltage regulation.

An attack-sustain-decay generator for a *polytonic* (more than one or two notes at a time) electronic music system is shown in Fig. 4-21C. Each note to be controlled is routed through a *voltage-controlled amplifier* (vca), whose gain is set by the charge on a capacitor. Changing the charging rate sets the attack, while changing the discharge rate sets the decay of each individual note. If we had only one vca, we could easily do this with two pots, one for the attack and one for the decay. To work with multiple vca's, we need an electronic way to simulate two variable resistors for each vca. Further, we would like to somehow disable the decay setting during the attack time to ensure full amplitude regardless of the two settings.

We do this by duty-cycle-modulating two resistors, switching them in and out of the circuit as needed. A 3K resistor looks like a 3K resistor if there is current through it 100 percent of the time. It looks like a 6K resistor if there is current through it only half the time, and like a 30K resistor if there is current only one-tenth of the time. If there is zero current, the resistor looks like an open circuit.

Attack pulses are generated by the variable-symmetry astable circuit consisting of the upper three inverters. These attack pulses are routed to the keyboard and charge the vca capacitors through the selected 3K resistor. Decay pulses are generated with a half-monostable technique (more on this later in the chapter), by using the lower three inverters. These decay pulses continuously discharge all capacitors at a selected decay rate. Note that attack dominates decay, so that the attack cycle begins as soon as a key is pressed and full amplitude is reached. The decay cycle can take over only after key release. Decay pulses are always shorter than attack pulses. A true exponential decay is provided. Extra half monostables and thresholds can be added as needed for percussion, snubbing, and other two-step decay effects. For faster decay times, a diode can be added across the 3K resistor with the cathode connected to the key side.

"Perfect" Symmetry

How can you get a "perfect" 50/50 symmetry, totally independent of the device and its duty cycle, without any adjustments or compensation? This can be important in electronic music, waveshape synthesis, and other places where we want both halves of everything to be identical or where we want to eliminate all even harmonics from the output waveform.

276

The key to "perfect" symmetry is to put a binary divider or divide-by-two counter on the output of an astable circuit. The output will change only on the leading edge of each astable cycle, independent of the duty cycle. We end up with 50/50 symmetry and an output frequency *one-half* that of the astable circuit. Fig. 4-22A shows a type-D flip-flop binary divider added to an astable circuit to get a 50/50 output. We will be finding out more about these devices in the next chapter.

A general-purpose laboratory square-wave generator is shown in Fig. 4-22B. Here, we cascade a stack of divide-by-ten circuits on the output of a 10:1 variable-frequency astable oscillator. These divide-by-ten circuits are detailed in Chapter 6.

Note that we use the "2" output each time rather than the "8" output, since the "2" output has a 50/50 duty cycle, while the "8" output has an 80/20 duty cycle. We get a switch-selected output range of 10 hertz to 100 kilohertz. To extend the range farther, one additional binary divider, such as a 4013 or 74C74, could be added to the

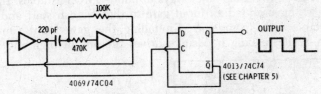

(A) Binary flip-flop added to output halves frequency and gives perfect 50/50 duty cycle.

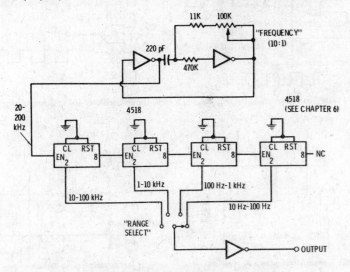

(B) General-purpose lab square-wave generator.

Fig. 4-22. "Perfect" symmetry is gained by dividing down the output frequency.

final "8" output for a 1-hertz to 10-hertz range. Or, if we like, any number of additional divide-by-ten circuits can be cascaded on the output.

We also have the option of simply switching timing capacitors instead of all those divide-down circuits. But the dividers are actually cheaper than capacitors (particularly the low-frequency ones) and much easier to switch, and the range-to-range tracking and calibration are automatically preserved without needing precision components. With CMOS, not much extra current is for the divider chain, since practically all the supply current will go into the astable and the loaded output driver.

By replacing the astable with a crystal oscillator (circuit is shown in Fig. 4-25), we can build a time base for a digital counter or other instrument that will provide time gates of 0.01, 0.1, and 1 second. A one-chip example of this is provided in the 7207.

Other Astables

Fig. 4-23 shows some other ways to build CMOS astables. In Fig. 4-23A, we have added an input gate so that we can start and stop our astable upon command. Grounding the gate input runs the astable, while making it positive drives the output high and stops

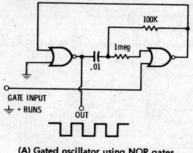

(A) Gated oscillator using NOR gates.

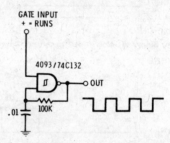

(B) Gated oscillator using Schmitt trigger.

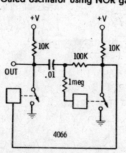

(C) Transmission-gate astable—low sinking impedance to ground.

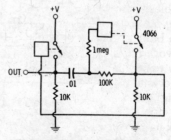

(D) Transmission-gate astable—low sourcing impedance to +V.

Fig. 4-23. Other astable circuits.

oscillation. Note that the gating input has been applied to the side of the astable with resistor feedback. This automatically moves this gate out of its linear region during times when there is no oscillation. Besides reducing supply current, this "turn-it-off" move guarantees the direction in which the output will go when the input gating is released.

Incidentally, note that the other NOR gate is reduced to an inverter by grounding one input, rather than tying the two inputs together. When you use multiple-input gates reduced to inverters, you will get the threshold closest to "half way up" if you change only one input. Multiple parallel inputs can shift the threshold and change the duty cycles somewhat.

In Fig. 4-23B, we have used a Schmitt trigger as an astable. This is a very effective and simple circuit. The capacitor voltage swings from the upper trip point to the lower trip point and back again, so it never gets near the protection diodes. The frequency for a given capacitor will be around six times higher than that shown in Fig. 4-17. We can gate the oscillator with the other input. Placing this input at $+V$ runs the circuit; grounding it forces the output high. The first output pulse will be delayed for about three periods after release of the gate. Internal delays in the circuit limit operation to a megahertz or less.

In Figs. 4-23C and D, we have used transmission gates as astables. These take two more resistors and continuously draw supply current. They have the advantage in Fig. 4-23C of being able to sink lots of current to ground and in Fig. 4-23D of being able to source lots of output current from $+V$.

CRYSTAL OSCILLATORS

When you need extreme stability and freedom from calibration or adjustments, you can go to a simple and effective CMOS crystal oscillator circuit. These circuits work somewhat differently than the astables we have just looked at. We bias an inverter, or an inverting gate, into its linear region so that it behaves as an amplifier. We then build a feedback network that includes a crystal. The feedback network routes a percentage of the output back to the input where it is reamplified, forming an oscillator whose frequency is set by the crystal.

Fig. 4-24 shows the basic biasing schemes for linear operation. In Fig. 4-24A, we simply provide a 10-megohm resistor from output back to input, *making sure that there is no other dc path to the input*. With this negative-feedback resistor, the inverter, or inverting gate, will automatically bias itself to an output voltage of *one-half* the supply voltage, or $+V/2$. It will behave as a high-gain amplifier, swinging almost to the supply limits in either direction with very good linearity.

The gain you get will depend on the device you use. Most A-series devices end up with a gain of about 30 or so. But triple-buffered, inverting B-series CMOS will give you gains of 10,000 or even more. One exception is the 4069B which isn't buffered.

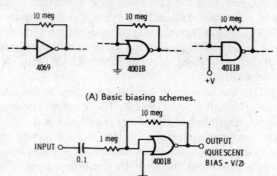

(A) Basic biasing schemes.

(B) Gain-of-ten ac amplifier.

Fig. 4-24. Linear biasing of inverting gates.

A typical gain-of-ten ac amplifier is shown in Fig. 4-24B. It is a handy circuit if you need a small amount of audio or other linear gain in your system. With B-series devices, the gain is set by the ratio of the feedback resistor to the input resistor (10 megohms/ 1 megohm = 10). The lower frequency limit is set by the input capacitor (around 2 hertz for values shown). Your upper cutoff frequency will be several megahertz, and this can be reduced as needed by adding a capacitor across the feedback resistor. In Chapter 7, we will be looking at some much better CMOS linear amplifier circuits using the 3130.

Fig. 4-25 shows how we can convert our amplifier into a crystal oscillator. A network called a *pi network*, which includes the crystal, is connected from output to input. Component values are selected to provide a 180-degree phase inversion from the gate's output back to its input, and a loss somewhat less than the gain of the amplifier.

The crystal used is specified to have its *parallel* resonant frequency delivered to a 20-picofarad load, if the circuit frequency and the frequency marked on the crystal are to be the same. Varying this 20-picofarad load capacitor will shift the frequency slightly for exact calibration. A variation of one kilohertz per megahertz in either direction is typical.

The math involved in the pi network is extremely involved. Full details appear in RCA application note ICAN-6539.

Our input RC network serves two purposes. It helps set up the proper loss and phase shift in the pi network. At the same time, it acts as a low-pass filter that discourages the crystal from running at a third or fifth overtone, or other higher frequency.

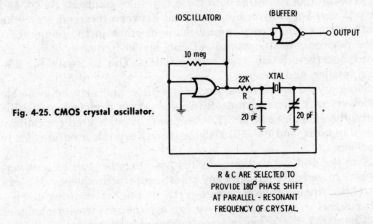

Fig. 4-25. CMOS crystal oscillator.

R & C ARE SELECTED TO
PROVIDE 180° PHASE SHIFT
AT PARALLEL - RESONANT
FREQUENCY OF CRYSTAL.

The output buffer (either a 4001 or a 74C04) is used to give a full-amplitude, square-wave output. If we use a 4001A as shown in Fig. 1-28, we can drive an ordinary TTL gate as an output load. This circuit works best at supply voltages above 4.5 volts.

Finding a low-cost crystal from a source with good delivery can be quite a problem. There are literally hundreds of crystal manufacturers. Chart 4-1 lists a few I've had reasonable luck with.

Chart 4-1. Crystal Sources

I. High Frequency (1 to 6 MHz):

JAN Crystals
2400 Crystal Drive
Ft. Myers, FL 33901

C.T.S. Knights
400 E. Reimann Avenue
Sandwich, IL 60548

International Crystal
10 North Lee Street
Oklahoma City, OK 73102

II. Low Frequency (10 kHz to 240 kHz):

Statek Corp.
1233 Alvarez Avenue
Orange, CA 92668

There are two basic low-cost types of crystals that you should consider using. Parallel-resonant, AT-cut crystals ranging from 1 to 6 megahertz are one choice. The other choice is the tuning-fork style, low-frequency crystals available in a dozen stock frequencies from 10 kHz through 240 kHz. Both these types are compact, cheap, and easy to get. Should you need a crystal between these ranges, or an oddball lower frequency not available as stock in the tuning-fork-style devices, use a standard AT-cut, higher-frequency crystal instead, and then divide down with counters. This is usually far simpler, smaller, and cheaper.

Some crystal frequencies are very popular and may be available at lower cost. These include the digital-watch frequency of 32,768 hertz, the 100-kilohertz and 1.0-megahertz crystals used for calibration standards, and the 3.579545-megahertz crystals used in color-tv receivers.

A crystal and a countdown divider used together can be a powerful way to generate virtually any adjustment-free frequency. We can get either square-wave outputs directly or sine waves using on-chip synthesizers in special ICs, or we can use the sine-wave techniques in Chapter 7. Let's look at some examples of crystal oscillator/divider techniques.

Baud-Rate Generators

Data communications between microprocessors, computers, tv typewriters, teletypewriters, and other *peripheral* devices are often done in a *serial*, or one-bit-at-a-time, manner, using standard UARTs (Universal Asynchronous Receiver-Transmitter) or their dedicated microprocessor serial I/O equivalents.

To be understandable at the receiving end, the information must be sent out at a predetermined rate called the *baud rate*. For error-free operation, the baud rate usually has to be set and held to well within one percent. Ideally, this should be done without any adjustments.

Standard, slower baud rates include 110, 150, 300, 600, and 1200 bits per second. Depending on how your serial interface is set up, we will need these frequencies or their second ($2 \times$), sixteenth ($16 \times$), or sixty-fourth ($64 \times$) harmonics as references. These higher-frequency references let us do things like test for error, sample in the middle of a bit, and generate sine waves for cassette storage.

Fig. 4-26 shows how we can use a CMOS 4060 crystal oscillator/divider to generate *all* of these frequencies, using one crystal for the 110-baud outputs and a second for everything else. The 4060 can divide the crystal frequency by as high as 2^{14}, or 16,384. The frequency you get depends on the crystal frequency and binary divider output you pick. For instance, suppose we want the 300-baud, $64 \times$

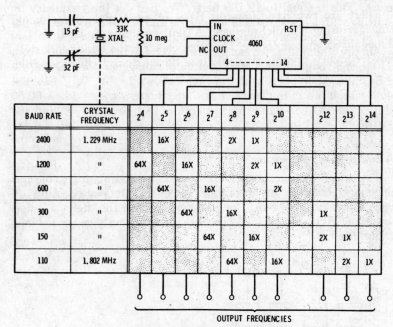

BAUD RATE	CRYSTAL FREQUENCY	2^4	2^5	2^6	2^7	2^8	2^9	2^{10}		2^{12}	2^{13}	2^{14}
2400	1.229 MHz		16X			2X	1X					
1200	"	64X		16X			2X	1X				
600	"		64X		16X			2X				
300	"			64X		16X				1X		
150	"				64X		16X			2X	1X	
110	1.802 MHz					64X		16X			2X	1X

OUTPUT FREQUENCIES

Fig. 4-26. A baud-rate generator.

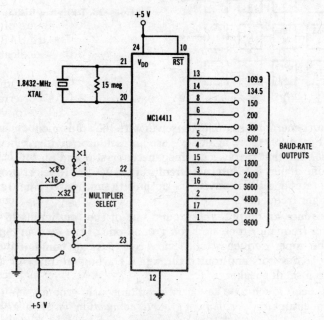

Fig. 4-27. An MC14411 baud-rate generator.

clock. This is equal to 19,200 hertz. We can get this frequency by starting with a clock of 1.229 megahertz and dividing it by 2^6, or 64. All of the outputs are square waves with a 50/50 duty cycle.

On the other hand, we can generate 13 reference frequencies from 110 to 9600 baud using a MC14411 baud-rate generator and a single 1.8432-megahertz crystal in the circuit shown in Fig. 4-27. Like the 4060 circuit, all output frequencies are square waves with a 50/50 duty cycle. Other baud-rate generators include the 4702 and the 5016.

A Touch-Tone® Generator

The touch-tone frequencies are used as dialing tones by the telephone company. A pair of tones is sent simultaneously, using the code shown in Fig. 4-28. For instance, to send the digit "5," sine waves of 770 hertz and 1336 hertz are simultaneously generated and transmitted.

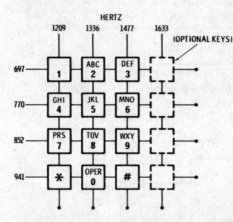

Fig. 4-28. Touch-tone frequencies.

We can generate these tones ourselves with the adjustment-free circuit in Fig. 4-29. Once again, we use an oscillator/divider scheme; this time, one that involves a 1-megahertz crystal and an MC14410 touch-tone dialer. The internal circuits of the MC14410 select proper division rates and then convert the outputs to synthesized sine waves of the right frequency.

To produce an output, you ground one of the *row* inputs to get the lower-frequency tone, and you ground one of the *column* inputs to get the upper-frequency tone. These two tones are combined with the two 1K resistors and routed through a Darlington transistor amplifier to a small speaker.

The special touch-tone keyboard simultaneously grounds one row and one output *to a common ground connection* when a key is

284

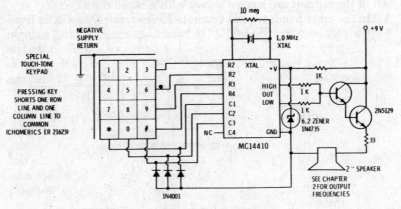

Fig. 4-29. A touch-tone generator.

pressed. Note that the keyboard ground or common is the only battery-return connection. Power for the chip is obtained through the three diodes, one of which becomes forward-biased if any column line is grounded. This way, no power is used until a key is pressed.

You can use this circuit as an automatic acoustically coupled touch-tone dialer or for remote control and signalling uses.

Equally Tempered Music

Our oscillator-divider scheme is also ideal for many electronic music applications. Fig. 4-30A shows how we can combine a CMOS 2-megahertz oscillator with a non-CMOS 5024 divider that automatically generates all twelve notes of the top octave. Lower-octave notes are easily obtained by binarily dividing each output as needed. The 4024 binary ripple counter is ideal for this use.

Whenever we are using a short keyboard synthesizer or otherwise want to get by with fewer vca's (voltage-controlled amplifiers) and other downstream electronics, we can translate down our top octave to any octave we want, as shown in Fig. 4-30B. Here, we use a 4024 to divide down our crystal so that we get references of normal, down one octave, down two octaves, and down three octaves. One of these references is selected and routed to the 5024, which now generates the selected octave rather than a top one.

This same technique is handy for pitch references, which are useful for tuning other instruments. But, remember that a properly designed tuning aid has to have a sine-wave output so that your ear doesn't trick you into mistuning the instrument. Also, most tuning aids applied to a piano will result in disaster, as a piano keyboard is "stretched" away from exact frequencies so that nonharmonic overtones will sound correctly.

285

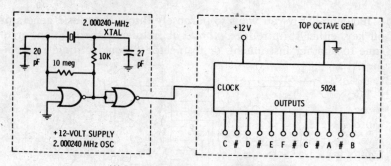

(A) Basic circuit generates top octave only.

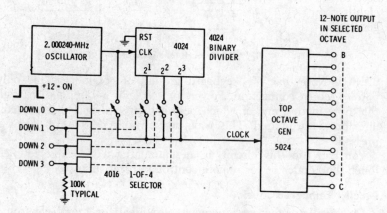

(B) Octave translators added.

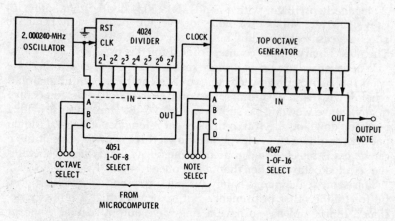

(C) Microprocessor interface produces any of 96 notes under digital command.

Fig. 4-30. Equally tempered note generators for electronic music.

Incidentally, if you want to purposely throw the music generator off key, either to introduce *vibrato* or frequency modulation, or to tune to another instrument, or to include a nonelectronic randomness, you break the input connection and add an *offset oscillator* that shifts the crystal frequency as much as needed.

For complete microprocessor or computer control, we would like to be able to output a seven- or eight-bit computer word that our top octave circuit could convert to a desired note in a desired octave. This is done in Fig. 4-30C. An input one-of-eight selector made from a 4051 picks the octave to be output, while a 4067 set up as a one-of-twelve selector picks our output note. Thus, three bits are used to select the octave and four more to select the output note. An eighth computer bit can be used to key an attack-sustain-decay generator.

VOLTAGE-CONTROLLED OSCILLATORS

Yet another type of astable is the *voltage-controlled oscillator* (vco), sometimes also called a voltage-to-frequency converter or V/F. You can build a very simple and effective, ultrawide-range vco with one-half of the 4046 phase-locked loop, as shown in Fig. 4-31. A capacitor and a resistor, C and R1, set the operating frequency. This set frequency will vary from zero for a grounded control-voltage input to *twice* the value shown in Fig. 4-31B for a control-voltage input of +V. The control input is a very high impedance. We can easily sweep 100:1 or 1000:1 in frequency with this circuit.

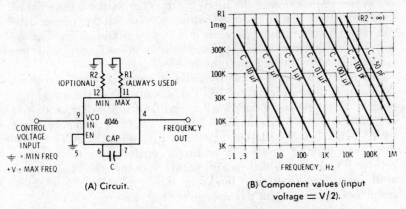

(A) Circuit.

(B) Component values (input voltage = V/2).

Fig. 4-31. Voltage-controlled oscillator using a 4046.

If we want a more limited operating range, we can add a second resistor, R2, as shown in Fig. 4-32. This resistor provides us with a frequency *offset*. Resistor R2 raises both the minimum and maximum frequency in direct proportion to its size with respect to R1.

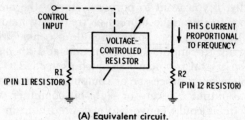

(A) Equivalent circuit.

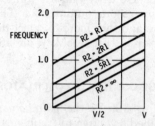

(B) Effect of adding R2.

Fig. 4-32. Adding a frequency offset to a 4046 vco.

For instance, if R2 is twice R1 and our original circuit went from 0 to 10 kilohertz, adding R2 with a value of 20K will change the range from 5 to 15 kilohertz.

Some simple applications for vco circuits appear in Fig. 4-33. Fig. 4-33A is a siren. Pressing the input button charges the 2-microfarad capacitor. The vco starts at zero frequency, to which the speaker cannot respond, and rapidly increases in a typical siren fashion. Releasing the push button causes the note to decay in frequency, with the time being set by the 2.2-megohm resistor.

We can turn the 4046 off by making pin 5 positive. A pulsed oscillator is shown in Fig. 4-33B. This circuit gives us a tone-burst output. The tone is produced when enable pin 5 is low, and it is not produced when enable pin 5 is high. The tone frequency is set by R1 and C1, while the burst rate is set by R3 and C2. The burst duty cycle may be changed by adding a resistor and a diode across resistor R3, as we did back in Fig. 4-20A. Note that the 4046 is operated in a fixed-frequency mode in this application. For fixed-frequency operation, tie the voltage-control input to +V.

You can build a two-tone alarm by changing the vco input from ground to +V and using an offset resistor, R2, to select the ratio of the two notes you are going to produce. Another trick, shown in Fig. 4-33C, is to use the EXCLUSIVE-OR gate available elsewhere in the 4046 as an outside-world inverter. In this case, the gate is used as half of the astable that sets the rate at which the two tones change.

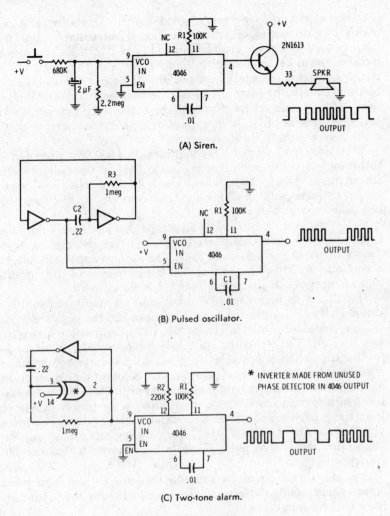

(A) Siren.

(B) Pulsed oscillator.

(C) Two-tone alarm.

Fig. 4-33. Some simple 4046 vco circuits.

More Accuracy

The 4046 is ideal for utility vco applications and is reasonably stable. This is particularly true in the closed-loop applications we will be looking at in detail in Chapter 7. This is a simple, cheap, and easy-to-use circuit. But the 4046 is not stable enough for ultracritical vco applications such as you might need for a digital voltmeter or an electronic music-tone source.

Two somewhat more stable circuits are the CMOS 555 timer and the Intersil 8038 sine-square-triangle generator.

If we need very wide range, good linearity, stability, and the ability to go down to zero frequency for a zero-voltage input, we have to turn to fancier vco techniques. Typically, these will use a modified form of *dual-slope integration*, or else they will use a charge-balancing scheme. Chips available to do just this include the Siliconix LD110, the Motorola 14433, the Raytheon 4151, and the Intersil 8052, among others. Some of these chips are expensive and may need special codes, large capacitors, stable references, or other restrictions that limit their ease of use.

A "semidiscrete" vco, or V/F, that uses very low cost parts and is surprisingly good in performance over its 3½- or 4-decade range is shown in Fig. 4-34A. An input voltage of 0 to −2 volts produces an output frequency of 0 to 20 kilohertz. The circuit uses a V/F technique called *charge balancing*.

We let the input continuously fill a capacitor while we empty the capacitor in discrete charge packets, which are determined by a constant current for a constant time. The average capacitor voltage is continuously forced to zero, so that the frequency of the current-time charge packets linearly follows the input current.

The 4558 is an improved 1558 dual linear op amp made by Raytheon. Half of it is operated as an integrator, and the negative input voltage continually tries to charge the .047-μF capacitor in a positive direction. The second half of the 4558 is operated as a comparator that trips a 555 monostable that generates a 10-microsecond pulse every time the capacitor goes positive. This constant-time pulse turns on a constant-current generator comprised of the field-effect transistor, which in turn reaches around and removes a packet of charge from the capacitor and drives the output back negative. The more negative the input voltage, the higher the input current; so, the more often the charge pulses have to be removed to balance the capacitor charge.

You calibrate the circuit by adjusting the input current for a given input voltage, while the zero-adjust control balances the input currents for zero input.

One application of the circuit is the digital thermometer shown in Fig. 4-34B. It uses a Corning TSR5 temperature sensor and has a resolution better than a tenth of a degree. This particular sensor costs less than a dollar in quantity and is linear over the 0°C to 100°C range of the instrument.

The input circuit is a simple bridge driven by two identical and matched constant-current generators. Any bridge unbalance is caused by temperature, and is converted to an output of 10 kHz at 100°C and 0 kHz at 0°C. Negative temperatures are handled by interchanging the input leads.

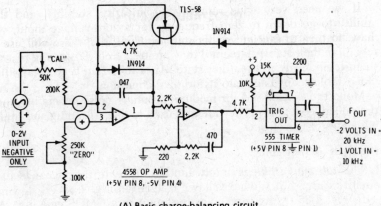

(A) Basic charge-balancing circuit.

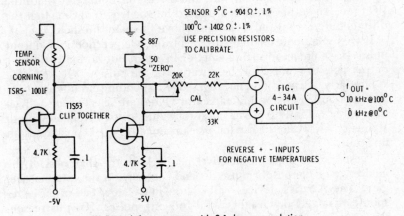

(B) Digital thermometer with 0.1-degree resolution.

Fig. 4-34. A 3½-decade voltage-frequency converter.

You calibrate the circuit with two precision resistors equal to the 5° and 100° sensor values. Measuring the output frequency for a tenth of a second gives you a direct digital reading in tenths of a degree.

The effects of hum and noise on the inputs of any V/F can be minimized by measuring the output frequency for an exact number of whole power-line cycles, so that the hum tends to average itself to zero. The use of a 0.1-second time gate, derived from the power line with dividers, is one good way to do this.

When advanced features such as automatic zero and automatic input polarity are needed, use the fancier V/F chips previously mentioned. Incidentally, one precision, low-cost analog voltage divider useful for digital voltmeter inputs is the Analog Devices AD1802.

291

MONOSTABLE CIRCUITS

Monostables have one stable state and one unstable state. They remain in their stable state until they are triggered. Triggering places them temporarily in the other state. After a certain time delay, the monostable snaps back into its original condition.

Monostables are used for medium-accuracy pulse generators and time-delay generators. They are also used to detect leading and trailing edges of waveforms.

The Half Monostable

The *half monostable* is simply another name for the edge detector circuit of Fig. 4-5. We have shown it again in Fig. 4-35.

In Fig. 4-35A, a resistor holds the input to a CMOS gate low. A positive edge is RC-coupled to it, and drives the input positive for the RC time. An output pulse is produced for the length of time that the differentiated input pulse remains above the gate threshold.

If we use a noninverting gate, or buffer, we get a positive output pulse. If we use an inverting gate, or inverter, our normally positive output goes to ground for the duration of the pulse.

Similarly, we can build negative-edge responding circuits as shown in Fig. 4-35B. Resistor and capacitor values are selected for a given time as shown in Fig. 4-35C. Resistor values can go from 10K up to 10 megohms, while the capacitors can range from 100 picofarads on up.

These circuits have no positive feedback, so the sharpness of the final edge of the output pulse depends entirely on the internal gate gain. The A-series devices will give a fairly slow end-of-pulse response that gets downright awful for long pulses. For this reason, the *B-series devices (except the 4069) or Schmitt triggers should be used for all half-monostable circuits.* This is especially true if the half monostables are cascaded. Schmitt triggers should be used for very long pulse times.

There are several usage limits for half monostables:

- The input waveform must stay in its post-trigger state for a period of time longer than the on time.

- The input waveform must rise and fall faster than ten times the on time. It must also be noise free and bounceless.

- The input waveform must go into the pretrigger state long enough to let the RC network recover fully.

- The gate must have enough gain (B-series except 4069) or snap action (Schmitt triggers) to get a sharp end-of-pulse transition.

- The circuit may not be retriggered until after full recovery.

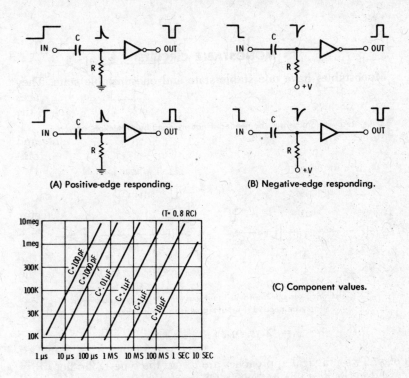

(A) Positive-edge responding.

(B) Negative-edge responding.

(T- 0.8 RC)

(C) Component values.

Fig. 4-35. Half-monostable circuits.

Some refinements in half monostables are shown in Fig. 4-36. If an input protection diode doesn't happen to be present in the direction we need it, external diodes can be added as shown in Fig. 4-36A to shorten the recovery time.

The high current recharging through an internal or external diode puts a load on the nontriggering input edge. This gives a droopy or sloppy *input* response. You can eliminate this extra loading with a series resistor as shown in Fig. 4-36B. This works with either internal or external diodes. It slows the recovery time but improves the input waveform. The extra resistor should be used whenever the input signal source is fed to other places besides this one particular circuit. For fast recovery with this resistor in place, always use the largest practical timing resistor in combination with the smallest reasonable capacitor.

True Monostables

True monostables use internal feedback to get their end-of-pulse snap action. You can accomplish this by immediately setting a flip-flop and later resetting it after a resistor-capacitor determined time

293

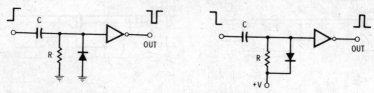

(A) Recovery diodes may be needed if not provided by protection network.

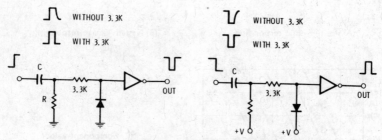

(B) Isolating resistor improves input waveshaping during recovery; may be needed if input signal is used elsewhere.

Fig. 4-36. More half monostables.

delay. Two practical approaches are to use the type-D flip-flops discussed in the next chapter or CMOS monostable circuits such as the 4528. Several possibilities are shown in Fig. 4-37.

In Fig. 4-37A, a positive input edge immediately sets the 4013 type-D flip-flop. This makes the Q output high and the $\overline{Q}$ output low. The high Q output starts charging capacitor C through timing resistor R until the voltage on C gets high enough to reset the flip-flop to its original state. As soon as the circuit flips, the capacitor is rapidly discharged through the diode. The circuit then rests to await a new trigger pulse.

The circuit recovers quite fast, particularly for lower values of C. If a sharp waveform is needed on the Q output, a 3.3K, or higher, resistor should be added in series with the diode. This will give a longer recovery time but a cleaner Q fall time. This circuit should not be retriggered until recovery is complete. Triggering too soon will give a shorter-than-expected output pulse.

Sometimes we would like to have a *retrigger* capability in which the time delay starts over again for each trigger pulse. If the trigger pulses are close enough together, the output continues until the fixed delay after the *last* triggering event.

Actually, there is no such thing as a truly retriggerable monostable because *some time* must *always be provided to discharge the timing capacitor at some point in the cycle*. We can make this time

294

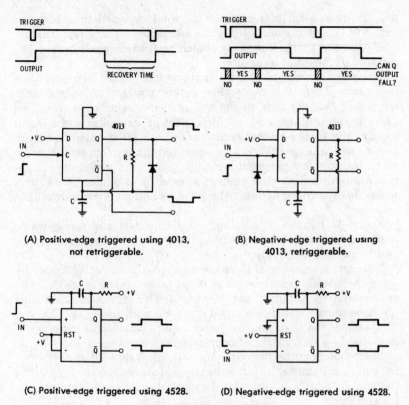

(A) Positive-edge triggered using 4013, not retriggerable.

(B) Negative-edge triggered using 4013, retriggerable.

(C) Positive-edge triggered using 4528.

(D) Negative-edge triggered using 4528.

Fig. 4-37. Full monostables.

very brief, which puts limits on capacitor size, or we can make this recovery time depend on how long we stay in the pretrigger state, or we can delay ever getting *any* output on initial triggering till after the capacitor is discharged to some known value. Regardless of what you do, there is always some time in which the output pulse either cannot start after triggering or end before triggering. These times *must* be allowed for in all retriggerable circuits.

Fig. 4-37B shows a retriggerable monostable using a 4013. When the input clock is low, the capacitor is discharged rapidly to ground. When the input clock or trigger goes positive, the capacitor is released, Q goes high, and timing starts in the usual way, with R charging C until reset. If the input remains high during this time, the monostable cycles out and awaits a new input-low state to reset the capacitor. If an input-low state arrives before timing out is complete, the capacitor is reset and the cycle continues. Note once again that *the output cannot end during the entire time the input trigger is low.* Normally, you use a low-duty-cycle clock that makes the in-

295

put low time a small fraction of the total expected time between triggers. The narrowness of this clock depends on how fast you can discharge the timing capacitor, which in turn depends on the impedance of the input source.

You retrigger the circuit by bringing the input high and then low again. *There is a delay time before you get an output after triggering that depends on the size of your timing capacitor.* This delay is almost a microsecond for a 1000-pF capacitor and a 5-volt supply, and can be several *milli*seconds with large, microfarad size capacitors. Note that this is an unexpected thing to happen—usually when you trigger a monostable circuit, you expect something to happen promptly rather than stalling around for a while. The 4528 is unusable if you need both long time delays and an immediate output response.

Fig. 4-33D shows us a falling edge triggered and retriggerable monostable. You get this by driving the − input, making reset high, and grounding the + input.

You can make either of these last two circuits nonretriggerable by using the output to lock out new input pulses. For Fig. 4-33C, you cross-couple the $\overline{Q}$ output to the − input. This gives you a fixed output time during which new triggers are ignored. In Fig. 4-33D, you cross-couple the Q output to the + input, which also gives you a monostable with a fixed but nonretriggerable output time.

The problem of the 4528 delaying an output till well after triggering can be minimized by using the largest feasible resistor for a given time delay. This will minimize the capacitor size and minimize the turn on delay.

The 4098 is a much improved dual monostable that replaces the 4528. While the pinouts are identical, the delay-on-triggering limitation is missing in the 4098, and pins 1, 8, and 15 are internally connected together for common grounding. One limitation remains with the 4098 in that the RESET pulse, if used, must be very long for large values of timing capacitor.

The reset input on the 4528 should normally remain high. If it is grounded, it will immediately terminate any output pulse. Reset can also be used to inhibit any output pulses during power-up times. This is done by connecting a large-value capacitor from the reset input to ground and a high-value timing resistor to +V. This causes a time delay during power up that prevents any output pulses. Values for any of the monostable circuits can be roughly approximated from Fig. 4-35B.

A half monostable was shown in Fig. 4-21C as the decay portion of our electronic music generator. We purposely used a half monostable here, since we never want the decay pulses to be longer than the attack pulses. A half monostable automatically terminates if the

296

input reverts to the other state. Since A-series devices were used in this circuit, several inverters are cascaded to get acceptable output rise and fall times.

Electronic Dice or Roulette Run Down

How would you build a run-down circuit for a pair of electronic dice (like those in Chapter 6) or for a roulette wheel?

To get the best effect, we would like a signal source that starts at a high frequency, slows down progressively, and then stops cold. There are two details that are extremely important. Whenever the output stops, it must stop completely and permanently so that a late pulse doesn't jump things later on. Second, the final pulse must be a clean one and not a glitch of some sort. This prevents the dice or roulette counters from jumping into a preferential state because of a sloppy or too brief clock command.

One possible circuit is shown in Fig. 4-38. We use a 4046 as a vco, set to run at a frequency twice the fastest rate we want our dice or wheel counters to go. Pressing the spin button charges the 1-μF capacitor, and the 4046 immediately jumps to a very high output frequency. It then progressively slows down as the 10-megohm resistor discharges the timing capacitor. It will keep slowing down until the capacitor completely discharges or until the spin button is once again pressed.

To get the output to stop, we build a retriggerable monostable using half a 4013 driven by the complement of the vco output. As long as the frequency is above that set by the RC values on the monostable, the $\overline{Q}$ output stays low and a second D flip-flop is allowed to continue to divide the vco output by two.

When the frequency first drops below the magic value that lets the monostable time out, the monostable resets the flip-flop counter and holds it that way for all successive vco clock pulses. Thus, the output is held, even though the 4046 continues to produce lower and lower output frequencies. This is a good example of how the vco, monostable, and clocked-logic circuits of the next chapters can be combined to solve a system problem.

There are two important details of this circuit. Be sure you use the $\overline{Q}$ output on the last 4013, since it is positively held clamped by the reset input, and be sure the inverter and its delay go on the monostable side, and not on the divide-by-two side of the 4046 output.

THE CMOS 555 TIMER

A device that has a wide range of astable (signal source) and monostable (pulse source) possibilities is the low-power CMOS ver-

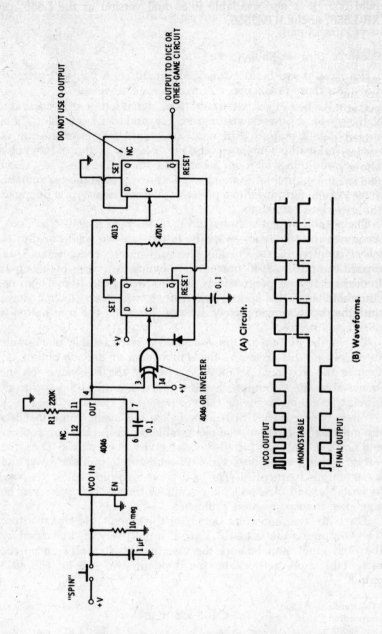

Fig. 4-38. Rundown circuit for electronic dice or roulette.

298

sion of the 555* timer, manufactured as either an L555, XR-L555, or ICM7555. It is also available in a dual version as the L556, the XR-L556, or the ICM7556.

The 555 timer is unique in that it simply, cheaply, and accurately serves as a free-running *astable multivibrator*, square-wave generator, or signal source, as well as being useful as a pulse generator and serving as a solution to many special problems. It may be used with any supply voltage from 2 to 18 volts and can source or sink up to 100 mA at the output.

In 555 timer circuits, the timing capacitor always has one end connected to ground, and a positive-only current is applied. This lets you use large electrolytic capacitors for long timing periods. The input is also a very high impedance, which lets you use high-value resistors and small capacitors for a given time constant. You can easily get a 1000:1 frequency range out of a single capacitor by simply changing the series resistor.

The astable or signal-source connection appears in Fig. 4-39. Assume that the output is high (1), and the charge on the capacitor is low. The capacitor begins charging through R1 and R2 in series toward the positive supply voltage (+V). When the voltage across the capacitor gets to two thirds of the supply voltage, an internal "threshold" comparator senses this and flips the internal circuitry to the other state. The output now goes low (0) and the capacitor

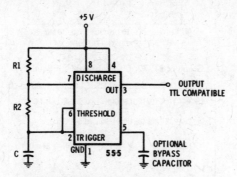

C CHARGES THROUGH R1 AND R2 IN SERIES.
C DISCHARGES THROUGH R2 ONLY

OUTPUT IS POSITIVE WHILE C IS CHARGING.
OUTPUT IS GROUNDED WHILE C IS DISCHARGING.

Fig. 4-39. A 555 timer connected as an astable multivibrator or signal source.

---

* The standard 555 and 556 timers, which are pin-for-pin compatible with the corresponding CMOS versions, can be also used directly with CMOS circuits without any additional interfacing circuitry. However, the CMOS version consumes less than 5% of the power required by the standard 555 timer.

discharges through R2. Discharge continues until the capacitor voltage reaches one third of the supply voltage. At this instant, an internal "trigger" comparator senses the low capacitor voltage and flips the circuit back to its initial state. The cycle continuously repeats, and the output is a rectangular waveform. The output is high while the capacitor is charging and low while the capacitor is discharging.

Fig. 4-40 shows how to choose the timing resistors. These can range from 1K (minimum value—R1 or R2) through 100 megohms (maximum value—R1 and R2 in series). This gives a potential adjustment of 100,000:1. Best results can be obtained with capacitors of 100 pF or larger. The maximum operating frequency is around 500 kHz. The minimum operating frequency is limited only by the size and leakage of the capacitor you use.

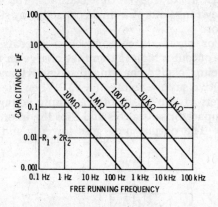

(A) Graph of R1, R2, C, and operating frequency.

CHARGE TIME (OUTPUT HIGH):	$0.693 (R_1 + R_2) C$
DISCHARGE TIME (OUTPUT LOW):	$0.693 (R_2) C$
PERIOD:	$0.693 (R_1 + 2R_2) C$
FREQUENCY:	$\dfrac{1.44}{(R_1 + 2R_2) C}$

LIMITS: MAX $R_1 + R_2$ -- 3.3 meg
  MIN R_1 OR R_2 -- 1 K
  MIN RECOMMENDED CAPACITANCE: 500 pF
  MAX CAPACITANCE -- LIMITED BY C LEAKAGE

DUTY CYCLE: $= \dfrac{\text{TIME HIGH}}{\text{TIME LOW}} = \dfrac{R_1 + R_2}{R_2}$

(B) Design equations.

Fig. 4-40. Choosing component values for a 555 astable multivibrator.

By making R2 large with respect to R1, we can get an essentially symmetrical square-wave output. For instance, if R1 is 1 kilohm and R2 is 1 megohm, the difference in the charging and discharging resistance is only 0.1%, and good symmetry results. Any symmetry from 50% through 99% can be obtained by a selection of the ratio of R1 and R2. Since the high output time always charges through two series resistors (R1 +R2), it is always longer than the time that the output is low.

Similar to the technique shown earlier in Fig. 4-22, Fig. 4-41 shows how we can achieve a perfectly symmetrical output by adding a clocked flip-flop binary divider to the output. The output of the clocked flip-flop will be one half of the frequency of the 555 timer and the symmetry will be 50/50, independent of the ratio of R1 to R2. This technique works with any signal source and may be used where a constant or exact symmetry is needed.

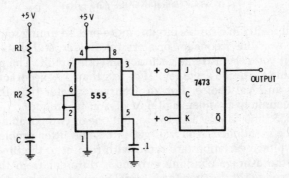

Fig. 4-41. A perfectly symmetrical output of half frequency, independent of the ratio R1 to R2, is obtained with a clocked flip-flop binary divider added to the output.

We can extend the duty cycle below 50% by using the circuit of Fig. 4-42. Two diodes are added to make the charge and discharge paths independently controllable. This will give us any selected-output duty cycle, at a slight loss in temperature stability. The basic circuit of Fig. 4-39 is independent of supply-voltage effects because of the trip points of the internal comparators, which are proportional to the supply voltage. Thus, *long-term* supply variations are ignored, and temperature variations are only 50 parts per million per degree Celsius, meaning that an 18 °F temperature change is needed to shift the frequency 0.05%.

Even though the circuit is stable, it is not always suitable for precision frequency work. For instance, setting the frequency to extreme accuracy takes a multiple-turn potentiometer and high stable components for the capacitor and resistors R1 and R2. While the 555 timer does make stability and a reasonably accurate frequency a

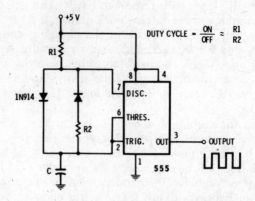

Fig. 4-42. Two diodes added to the basic 555 astable multivibrator give wider control over the duty cycle.

practical reality in a simple circuit, more precise timing sources, such as the power-line frequency or a crystal divider, still should be considered if you need extreme accuracy in your circuit. The problem is in defining "extreme accuracy." For less than 1% accuracy, the 555 timer is an ideal choice. For an accuracy greater than 0.1%, you should definitely consider digital or crystal techniques.

The 555 timer is very easy to use, but one or two possible problems should be mentioned. You will get into some jitter problems if the supply-voltage variations are rapid with respect to the timing cycle, so that the average charging current is different from the instantaneous voltage at the time the internal comparator changes the state of the output.

Bypassing the Control input (pin 5) to ground with a capacitor (typically 0.1 μF) helps a lot, but supply filtering or regulation should be used to hold the supply voltage constant over the timing interval. One place this problem can show up is in battery-powered CMOS instruments, where the different multiplexed (LED) display numbers shift the supply voltage and introduce potential jitter. This specific problem is eliminated by using the fastest possible display multiplexing and by adding a regulator or having bypassing between the 555 timer and the rest of the circuit.

Ac hum can also cause instability if large resistors and a timing capacitor with long leads is used. The cure is to locate the timing resistors and capacitor as close to the 555 timer as possible.

Incidentally, you can intentionally frequency modulate the output frequency by capacitively coupling a signal to the Control input. Operation, however, is nonlinear and the range limited, but this technique is handy for such things as switching-mode power-supply designs and for adding vibrato to electronic music.

302

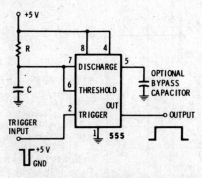

C CHARGES TO 2/3 SUPPLY VOLTAGE AFTER TRIGGERING VIA R
C DISCHARGES RAPIDLY TO GROUND AT END OF TIME CYCLE.

OUTPUT IS POSITIVE FOR TIME DURATION

TRIGGER MUST REMAIN POSITIVE UNTIL INITIATED BY
DROPPING TO GROUND

TRIGGER PULSES MUST BE NARROWER THAN OUTPUT TIME PULSE.

Fig. 4-43. A 555 timer connected as a monostable multivibrator or pulse generator.

When you use the 555 timer as an astable multivibrator, the Reset input (pin 4) should be connected to the positive supply voltage to prevent any noise problems. When the Reset pin is low, the timer is turned off.

The 555 timer can be connected as a monostable vibrator as shown in Fig. 4-43. An external trigger command applied to the Trigger input (pin 2) starts the action. This is normally done by leaving the Trigger input positive and briefly pulling it to ground to initiate the timing sequence.

When the input is triggered, the output goes positive. The timing capacitor then charges positive until it reaches two thirds of the supply voltage. At this instant, the internal comparator flips the circuit over. The output goes to ground, and the capacitor is rapidly discharged to ground. Unlike the astable multivibrator, the cycle does not repeat itself. An external command must be sent again to the Trigger input to initiate a new time delay.

The design chart and key equations are shown in Fig. 4-44. As with the astable circuit, the time width is independent of temperature and supply-voltage variations that are long with respect to the timing cycle. Supply variations, such as hum, glitches, or digit bobble, that are short compared to the timing cycle can introduce jitter or instability and must be suitably filtered out. Bypassing pin 5 (typically 0.1 μF) is also recommended.

The Reset input (pin 4) may be used to hold the output low or to stop a timing cycle after it begins. This is done by bringing the input

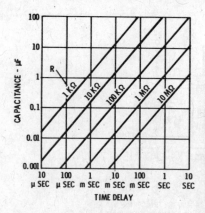

ON TIME: T - 1.1 RC
MAX R: 3.3 meg
MIN R: 1 K
MINIMUM RECOMMENDED C: 500 pF
MAXIMUM C: LIMITED BY LEAKAGE
MAXIMUM RECOMMENDED DUTY CYCLE:
80% WITH LARGE R; 50% WITH SMALL R.
MAXIMUM RECOMMENDED TRIGGER PULSE WIDTH:
1/4 - ON TIME.

(A) Graph of ON TIME versus R and C.   (B) Design equations.

Fig. 4-44. Component values for a 555 monostable multivibrator.

to ground for the length of time you wish to inhibit the operation. When not used, the Reset input should be tied to the positive supply. Many circuits that specifically use the 555 timer are described in *The 555 Timer Applications Sourcebook with Experiments* by Howard M. Berlin (published by Howard W. Sams and Company).

THE 4047 MULTIVIBRATOR

Besides the 555 timer, there is another integrated-circuit CMOS multivibrator—the 4047, which can be wired for both monostable and astable operation.

Fig. 4-45 shows the connections necessary for monostable operation. Unlike the 555, we can have either positive or negative triggering. For positive triggering, an external trigger command applied to the positive trigger input (pin 8) starts the action. This is normally done by leaving the trigger input at ground and briefly pulling it to the positive supply voltage to initiate the timing sequence. For negative triggering, the external trigger command is applied to the negative trigger input (pin 6) to start the action. This is done by leaving the trigger input positive and then briefly pulling it to ground to initiate the timing sequence. In either case, the minimum trigger pulse width is typically 500 nanoseconds for a 5-volt supply, 200 nanoseconds for 10 volts, and 140 nanoseconds for 15 volts.

When triggered, the complementary outputs Q and $\overline{Q}$ change for a time width that is dependent upon R and C, as shown by the graph and design equations in Fig. 4-46. For reliable operation, the resistance of the timing resistor should be between 10K and 1 megohm, while the timing capacitor can be any practical value

304

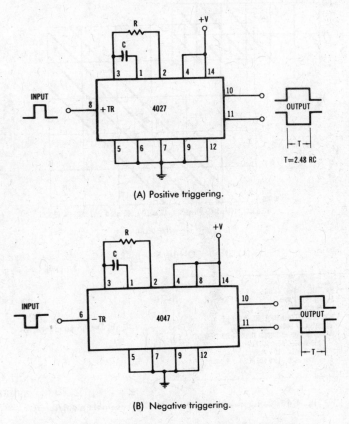

(A) Positive triggering.

(B) Negative triggering.

Fig. 4-45. Two 4047 monostable multivibrator circuits.

greater than 0.001 μF. The output time width can then be as small as 25 microseconds.

The astable connections for the 4047 appear in Fig. 4-47. There are two output frequencies. The oscillator output at pin 13 is twice that of the complementary outputs at pins 10 and 11. Unlike the complementary outputs, however, the oscillator output will not necessarily have 50/50 symmetry.

Fig. 4-48 shows how to choose the timing components. These can range from 10K through 1 megohm for R. This gives a potential adjustment of 100:1. Best results can be obtained with capacitors that are 100 pF or larger. The maximum reliable operating frequency is around 500 kHz.

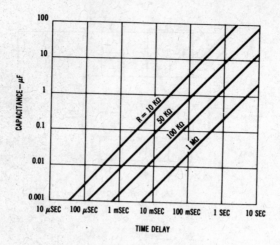

(A) Graph of ON TIME versus R and C.

ON TIME: T=2.48 RC
MAX R: 1 meg
MIN R: 10 KΩ
MINIMUM RECOMMENDED C: 0.001 µF
MAXIMUM C: ' IMITED BY LEAKAGE
MAXIMUM RECOMMENDED TRIGGER PULSE WIDTH:
    500 nSEC - 5-V SUPPLY
    200 nSEC - 10-V SUPPLY
    140 nSEC - 15-V SUPPLY

(B) Design equations.

Fig. 4-46. Component values for a 4047 monostable multivibrator.

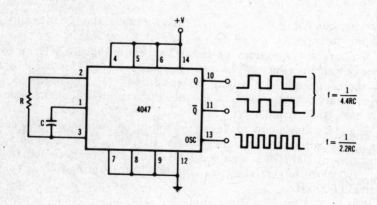

Fig. 4-47. A 4047 astable multivibrator with three outputs.

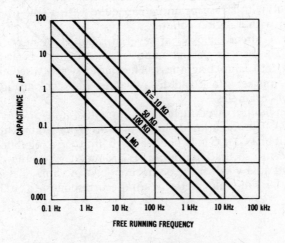

(A) Graph of R, C, and operating frequency at pin 13.

OUTPUT FREQUENCY (PIN 13): $\dfrac{1}{2.2RC}$

    DUTY CYCLE: 50-50 SYMMETRY, NOT GUARANTEED

OUTPUT FREQUENCY (PINS 10 OR 11): $\dfrac{1}{4.4RC}$

    DUTY CYCLE: 50%

LIMITS: MAX R ....1 meg
       MIN R ....10 K
       MIN RECOMMENDED CAPACITANCE ....100 pF
       MAX CAPACITANCE ....LIMITED BY C LEAKAGE

(B) Design equations.

Fig. 4-48. Component values for a 4047 astable multivibrator.

DUTY-CYCLE INTEGRATORS

There are several places where monostables and other multivibrators often crop up that can be collectively called *duty-cycle integrators*. They all use a duty-cycle change to produce an output of some sort, and all are based on the rule:

If the amplitude of a waveform stays constant, its average value will be proportional to its duty cycle or symmetry.

We can do our averaging with the mechanical inertia of a meter movement, with a resistor and a capacitor, with a true op-amp integrator, or with a digital counter.

We have already looked at several examples of duty-cycle integrators. The caver's helmet lamp integrated duty cycle and converted it into lamp brightness. The music system used duty-cycle integration to set attack and decay values. Let's look at Fig. 4-49 for some more examples.

We can build a dwell meter for automotive tune-ups, as shown in Fig. 4-49A. We condition an input pick-off signal and use it to swing the output of a CMOS gate between two limits on a regulated supply. This is set up so that the output is grounded when the points are closed and at +V when the points are open. A value of R is selected so that the full-scale meter reading corresponds to 100% point

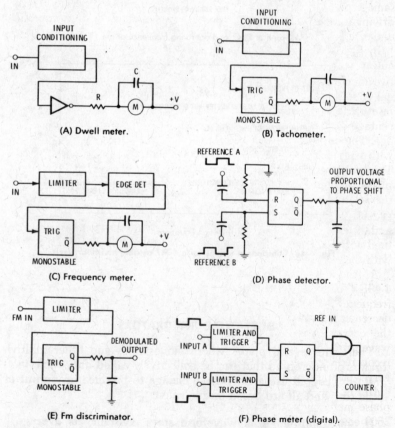

(A) Dwell meter.

(B) Tachometer.

(C) Frequency meter.

(D) Phase detector.

(E) Fm discriminator.

(F) Phase meter (digital).

Fig. 4-49. Duty-cycle integration circuits. Regulated power supply must be used.

closure, and a zero reading matches 0% point closure. Capacitor C is selected to minimize meter bobble at low-idle speeds.

An analog tachometer is built by triggering a monostable and then measuring the duty cycle that results between triggerings, as in Fig. 4-49B. The average meter current will be linearly proportional to the input pulse rate, which, in turn, is easily related to engine rpm by considering the number of cycles and cylinders. Normally a 30% or 40% maximum monostable on time is selected to provide for full recovery and linear operation at full scale. We will be looking at a more complex digital tachometer in Chapter 7.

A frequency meter is really the same thing as a tachometer; only the input conditioning and calibration differ. Input signals (Fig. 4-49C) are limited and edge-detected to trip a monostable. The average duty-cycle current is then proportional to frequency. This same type of circuit is easily converted to a capacitance meter by driving it from an astable whose frequency depends on the capacitor under test.

In Fig. 4-49D, we have built a phase detector that provides an output voltage that is proportional to the phase difference between two identical input frequencies. The leading edge of reference signal A sets the flip-flop, and the leading edge of reference signal B resets it. The time difference between the two is integrated as a percentage of the total period, and determines the average output voltage. The output voltage varies from zero when the two are in phase with each other, to nearly +V when signal B lags by nearly 360°. This particular circuit is normally set up to work around 180° of phase shift, with a +V/2 output that can swing higher or lower as the relative phase changes. We can do something similar with an EXCLUSIVE-OR gate,, only this time the optimum phase angle is 90° and the circuit repeats every 180°. Symmetrical input signals must be used with the EXCLUSIVE-OR phase detector, but any duty cycle works with the set/reset system.

A frequency-modulation (fm) discriminator is shown in Fig. 4-49E. The monostable is usually set to a period of one half the center frequency of the incoming signal. As the frequency increases or decreases, the duty cycle changes in proportion. After filtering away the carrier, we end up with an audio or other demodulated waveform. The linearity of such a system is outstanding, but the recovered audio output is very low for small-deviation fm systems. A very well filtered and regulated power supply is usually essential.

In Fig. 4-49F, we have converted our phase detector into a digital phase meter by adding a gating waveform whose frequency equals 3600 counts during the input period. The resulting count per cycle will equal the phase shift expressed in tenths of a degree. Usually you lock this reference to the 3600th harmonic of the input frequency. If the input frequency is too high, input scalers can be used.

For all duty-cycle techniques, holding and regulating the power supply to a constant value is absolutely essential because the total energy in a pulse is set by the product of its duty cycle and its amplitude. To get an output that is only proportional to duty cycle, the amplitude must remain constant.

SOME GUIDELINES

The circuits of this chapter can solve many timing and interface problems, particularly in smaller and simpler applications. With larger CMOS systems, there are some definite limits on how many of these circuits we should use to solve a given set of circuit problems.

In general, it is best to design any digital system to have a single main clock or reference frequency and then derive all possible pulse widths and other frequencies from that main source. Our main clock can be a crystal standard or locked to some outside-world signal like the power line or other reference.

This "lock-everything-to-everything-else" design eliminates system adjustments, minimizes the discrete-parts count, and makes us much more independent of CMOS device-to-device variations. Most often, a totally locked-together system will rely heavily on the clocked-logic circuits discussed in the next chapters.

As a general rule, use the astable and monostable circuits of this chapter only when they solve or simplify a problem. Try to eliminate adjustments whenever possible, and try to eliminate any circuits that may depend on device-to-device variations. Always try to minimize the number of discrete components (resistors, capacitors, and, particularly, pots) except in those cases when using a few "outside-world" components clearly saves you money, simplifies the circuit, or gains you simplicity on a pc layout.

When you do use the circuits of this chapter, remember to use the doubly or triply buffered B-series or Schmitt-trigger devices wherever you can. Limit your supply voltage to between 4.5 and 15 volts because lower values will cause marginal operation and higher values can cause excess heating. Performance will generally be better in the 9- to 12-volt range, compared with 5-volt operation.

The system accuracy you need determines other limits. Astables and monostables should be reserved for low-accuracy applications having tolerances of five percent or higher. Ordinary astables are fine for electronic games, toys, and other wide-range, easy-to-calibrate systems. Whenever you need more accuracy or stability, or whenever misadjustment can cause operating problems, use more accurate frequency and time references digitally derived from a master clock or other stable source. Among the places where astables should NOT be used are baud-rate generators, modems, clocks, and counters.

We can sum up the foregoing guidelines with this general rule:

For any particular digital logic system, always use the most stable, most accurate, most adjustment-free circuit you can possibly afford.

Now, on to those clocked-logic circuits.

AUTOMATIC KEYBOARD REPEAT

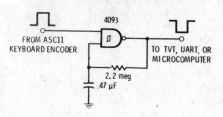

Briefly pressing a key produces a single output pulse. Holding key down produces a single pulse, a wait of one second, and then repeating pulses at a three-per-second rate.

5

Clocked Logic: JK and D Flip-Flops

In *clocked, synchronous,* or *"step-by-step"* logic, the outputs of logic blocks don't change immediately after their inputs change. Instead, the logic block waits until a specific time which is set by a waveform on a *clock* input. Only then are output changes allowed. There are two essential steps to the clocked-logic process. In the *setup* step, inputs decide what the logic block is going to do. In the *clocking* step, the logic block actually does what it was told to do and provides an output.

There are lots of advantages to clocked logic. First and foremost is the orderliness of the process. Logic signals move one, and only one, stage at a time. This lets us move data from one logic block to another without unchecked races and "domino" effects, in which a logic one or zero goes several stages beyond where it was first headed. What is equally important is that the *output* of a logic block can now determine, or at least influence, its own *next* output condition, without any preferential states or wild oscillations taking place.

Clocked logic also internalizes the variable processing delays from logic block to logic block. As long as the *slowest* block completes its internal operations before the next clock arrives, all outputs of all stages will be valid and predictable at the instant of clocking. Therefore, we automatically know when to look for valid data. As a side benefit, most modern clocked logic is *edge sensitive*. This edge sensitivity eliminates any need for resistors and capacitors to determine a leading or a trailing edge of a logic signal.

Clocked logic is used in virtually all advanced electronic systems. This is particularly true if counting, shifting of data, or storage of characters is needed. In this chapter, we will first look at a "do-it-yourself" clocked-logic block and will then check into the 4013 type-D flip-flop and the 4027 JK flip-flop. These devices are extremely useful by themselves as the detailed applications catalog found later in this chapter will show you. The same operating principles will be important as the basic building blocks of the "heavier" counters and registers covered in the next chapter.

CMOS CLOCKED LOGIC

Most CMOS logic blocks are clocked on the *positive edge* of the clock. This is the ground-to-positive transition of the clock input. Clocking is defined in a positive-logic sense for most CMOS devices.

There are a few exceptions to this positive clocking rule. Binary ripple counters such as the 4020, 4024, 4040, and 4060 are clocked on the negative edge or positive-to-ground transition of the clock. This lets you cascade binary stages for longer count lengths. A few CMOS counters give you a choice of clock polarity, which can be set by a logic signal on a separate pin. The 4518 and 4520 dual-decade and dual-hexadecimal counters are the most important of these devices. They give you a choice of positive-edge clocking for synchronous counting systems or negative-edge clocking for cascaded ripple counting. The 4192 and 4193 have a pair of normally high clock inputs. One clock input is brought low for an up count; the other for a down count. Clocking occurs on the trailing positive edge of the pulse.

Except for a few easy-to-live-with setup and hold-time limitations (details in Chapter 6), *only the input conditions that exist at the instant of the clocking transition or edge affect the output.* Inputs can change regardless of whether the clock is high or low, eliminating the "one-swallowing" problems that plagued early TTL-level clocked flip-flops.

There is one important clock restriction that remains with CMOS and applies to just about any logic family:

In any clocked-logic system, the clock must cycle only once, noiselessly and bounce-free, per intended output change.

This means that all our clocking signals must be clean. In particular, clocking commands that come from the outside world or from mechanical push buttons MUST be properly conditioned to give you one, and only one, clean transition per desired output change.

With CMOS, it also pays to keep the clock rise time as fast as possible. Five microseconds is a normal worst-case maximum clock transition time. If possible, make your clock signals have much faster rise times than this. Slower rise times may let one stage output a new state before the next stage has a chance to complete clocking. This mixes old and new data and generates garbage. In large CMOS systems, it pays to avoid *clock-slew* problems by deriving all clock signals from the same source or from parallel sources with identical delay.

A CLOCKED-LOGIC BLOCK

One inherent feature of clocked logic is that it takes *two* regular flip-flops or storage devices to build one clocked-logic block. One of these flip-flops takes care of accepting and setting up the input information, while the second flip-flop actually carries out the intended operation and holds the result as an output.

In early clocked logic, one of these two storage elements consisted of a diode-capacitor memory or "steering network." More recently, the stored charge in a base-emitter junction of a transistor has served the same purpose, with the presence or absence of a stored charge representing a one or a zero. Today's CMOS, along with many other logic families, uses two distinct flip-flops—an input or *setup* storage device called the *master,* and an *output* flip-flop called the *slave*. On the clocking edge, the content of the previously setup master flip-flop is transferred to the slave. The slave flip-flop then provides the final output and between-clocking storage of output data.

Once again, the all-important purpose of the two-step process is to give an orderly one-stage-at-a-time shift of data between clocked-logic blocks, and to count or binarily divide without getting into preferential state and hang-up problems. Let's see what kind of trouble we can get into by trying to make an ordinary set-reset flip-flop binarily divide. In other words, we will make it try to alternate states on every input command.

Fig. 5-1A repeats the NOR logic set-reset flip-flop of the last chapter. With both set and reset low, the flip-flop holds the last state it was put into, with Q and $\overline{Q}$ providing complementary outputs. A high on set drives output Q high and output $\overline{Q}$ low, while a high on reset does the opposite. Driving both set and reset high at the same time gives a disallowed state, and the last input to go to ground decides the final result.

In Fig. 5-1B, we have converted the NOR flip-flop into a sort of clocked set-reset flip-flop. We did this by adding AND gates to the inputs, controlled by a new *clock* line. When clock is low, inputs are ignored. When clock is high, inputs are accepted. We can now set up what the flip-flop is going to do while the clock is low, and actually carry out the operation by briefly bringing the clock high.

So far, so good. This is a useful clocked-logic block. We can obviously make it binarily divide by cross-coupling output $\overline{Q}$ to set and output Q to reset (Fig. 5-1C). Now every time the clock goes high, the flip-flop will change state, since it was told to go to the opposite state. Right? Well, not quite.

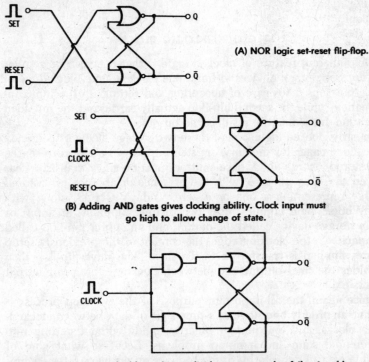

(A) NOR logic set-reset flip-flop.

(B) Adding AND gates gives clocking ability. Clock input must go high to allow change of state.

(C) An attempt at building a binary divider or counter that fails miserably.

Fig. 5-1. Steps toward clocked-logic flip-flops.

It is true that the instant the clock goes high, the outputs change state. But what if the clock *stays* high? These new output states reach around and change the input, which changes the output, which changes the input, and so on. What you really end up with is a complicated and unpredictable gated oscillator that runs while the clock is high and stops in one state or the other while the clock is low. This is hardly what we had in mind.

We might try to solve the problem by picking a clock pulse just wide enough to let one, and only one, change take place. But this will make our clocked logic dependent on loading, device, temperature, and supply voltage. The point is that *there is no reliable way to let a single clocked flip-flop count or shift information.* That is why we have to use two separate storage elements or master-slave *pairs* of flip-flops for workable clocked logic.

AN ALTERNATE-ACTION PUSH BUTTON

Fig. 5-2 shows an *alternate-action* push button that does work reliably. It changes its output state every time the button is pushed. At the same time, it provides debouncing and contact conditioning.

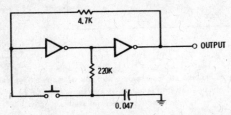

Fig. 5-2. Alternate action (push on-push off) bounceless push button. Resistor and capacitor form temporary storage for "steering."

While this circuit looks almost as simple as that of Fig. 5-1C, there is a crucial difference. Here we have two storage devices—a master capacitor and a slave flip-flop. The capacitor remembers what the new state is going to be. When the button is pressed, the capacitor voltage is transferred to the slave flip-flop. No race or oscillation is possible since the capacitor can't recharge as long as the button is pressed, and after the button is released, no problem remains. This is a low-frequency circuit ideally suited to manual button pressing.

A MASTER-SLAVE CLOCKED LOGIC BLOCK

Fig. 5-3 replaces the capacitor master with a conventional flip-flop. What we have done here is use two of the clocked NOR flip-flops again. The first, or master, flip-flop accepts data only when the clock is low; the second, or slave, flip-flop accepts data only when the clock is high.

Now, when the clock is low, the master flip-flop can accept data and will remember the last input to go high. When the clock goes high, the input flip-flop is *disconnected* from the set and reset inputs and is no longer allowed to change state. But, with the clock high, the slave flip-flop is enabled, and the content of the master is transferred to the slave and appears as an output immediately after the clock goes high.

Even if we cross-coupled the outputs back to the inputs of cascaded stages, a wild race couldn't result because the next flip-flop down the line would not be enabled at any particular instant.

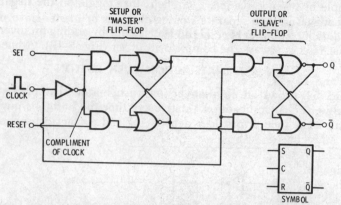

Fig. 5-3. The key to reliable clocked logic is the use of master-slave pairs of clocked flip-flops. Only one flip-flop is active at any time, eliminating unchecked races and preferential states.

Our circuit is said to clock on the positive edge because that is the time an output apparently appears. In reality, clocking is continuous, with the low clock state accepting data into the master and the high clock state transferring the master contents to the slave.

If we wanted a negative-edge clocked flip-flop instead, we would move the inverter so that the first stage is active with clock high and so that the second stage is active with clock low. Note that with either system, inputs can change virtually at any time without "1-swallowing" or similar problems.

We call this particular circuit a clocked RS flip-flop, and unlike the circuits in Fig. 5-1, it is a genuinely useful building block without unchecked-race or preferential-state problems. Just sitting there by itself, this circuit can't binarily divide, and it still has disallowed input conditions when both set and reset are high, but we can fix these limitations.

It is an easy matter to convert the clocked RS flip-flop into the more-useful and more-common clocked-logic blocks, as shown in Fig. 5-4. These more common flip-flops are the type-T flip-flop, the type-D flip-flop, and the JK flip-flop.

The T in the type-T flip-flop of Fig. 5-4A stands for *toggle*. By adding two external *feedback* leads from output Q to reset and from output Q̄ to set, we tell the flip-flop to *change* state each time. This modification alternates states each clocking. Since the output changes state each positive clock transition, you get only half as many positive transitions in the output. This gives you a square wave of one-half the input clocking frequency. The T-type flip-flop is not available separately as a CMOS package since it is easy to convert D-type and JK flip-flops into binary dividers. The 4024 is an

318

example of seven cascaded T flip-flops that toggle on the negative clock edge.

A *data*, or *delay*, or *type-D* flip-flop is built by adding an inverter so that reset is always the complement of set (Fig. 5-4B). A one on the D (data) input is stored in the flip-flop on the positive clock edge and appears at the Q output. A zero similarly applied is clocked in and appears at the $\overline{Q}$ output. The D-type flip-flop is useful in storing or delaying one bit of information. It is the key to the shift registers of the next chapter. We will see that shift registers store data and move information on an orderly one-stage-at-a-time basis. We can convert a D-type flip-flop to a T-type flip-flop by externally feeding back the $\overline{Q}$ output to the D input.

The most versatile and universal clocked flip-flop is the JK flip-flop of Fig. 5-4C. The extra gates on the input make the JK flip-flop into a type-D flip-flop if the inputs are different. They make the JK flip-flop into a type-T flip-flop if the inputs are both high. Finally, if both J and K are low, the *same* state is clocked back into the flip-flop, making it *appear* to do nothing.

The JK flip-flop is a universal one that can store data, binarily divide, or do nothing, all depending on the conditions of the J and

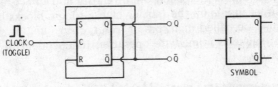

(A) Type-T flip-flop can only binary divide or alternate states.

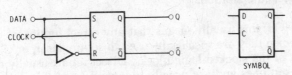

(B) Type-D flip-flop shifts or stores information.

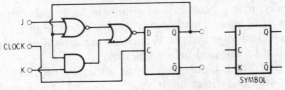

BOTH J & K INPUTS LOW INHIBITS,
BOTH INPUTS HIGH ALTERNATES,
DIFFERENT INPUTS SHIFTS OR STORES.

(C) JK flip-flop shifts, stores, binary divides, or does nothing.

Fig. 5-4. Converting a clocked RS flip-flop into other clocked flip-flops

319

K inputs. There are no disallowed states or disallowed combinations of J and K logic. When all this versatility is needed, the JK flip-flop is the obvious choice to use, particularly for subtle timing sequences.

However, the type-D flip-flop is often in a shorter package, is slightly cheaper, uses somewhat less power, and often has a simpler pc board layout. So the type-D flip-flop is most often the best choice to use, and it is a good policy to save the fancier JK versions for those applications where you definitely need the do-nothing or inhibit option of having both inputs low.

DIRECT INPUTS

After we have gone to all the trouble of making our clocked-logic block operate only when clocked and only when we want it to without any unchecked races or disallowed-state conditions, we usually go back and add some new *direct* inputs that let us immediately set or reset the flip-flop into some state *independently* of the clocked inputs. We can use these direct inputs to initialize a flip-flop into a certain state, to reset a group of counting flip-flops to zero, or to preset or "jam" a certain count or word into a register or latch.

These new inputs are called the *direct-set* and *direct-reset* inputs. Similar direct inputs on the fancier clocked-logic blocks of the next chapter may be called *load, preset, reset, clear, jam,* or some other name that suggests immediate operation independent of the clock.

Note that:

All direct inputs to a clocked-logic block must be disabled during clocked operation.

In CMOS, this usually means that any direct inputs are held *low* except when they are specifically used to set up, clear, or change the contents of the clocked-logic blocks. Direct inputs usually dominate the clocked ones and are usually independent of the clock level or the conditions on the clocked inputs.

Generally, it is a good rule to edge-couple or pulse direct inputs when they are used. This keeps a steady direct high from hanging up your clocked-logic system. When you use direct-logic inputs, they always must be released before clocking.

Since the direct inputs behave as ordinary set-reset unclocked flip-flops, only one direct input should be used at a time. If you try using both direct inputs at once, you will get a disallowed state condition, just as you did with the simpler flip-flops of the last chapter. There is, of course, no reasonable way to let direct inputs shift or binarily divide without problems—this is why we went to a clocked-logic block in the first place.

THE 4013 DUAL TYPE-D FLIP-FLOP

With CMOS, we can use transmission-gate techniques to greatly simplify the internal design of clocked-logic blocks. Let's take a detailed look at the 4013 dual type-D flip-flop and the 4027 dual JK flip-flops to see how they work and how transmission gates simplify the logic for us.

The logic diagram for half of a 4013 appears in Fig. 5-5A, with the actual pinouts given in Chapter 2. While we could use AND gates for clocked logic with CMOS (just as we did in the earlier logic blocks), the CMOS transmission gate set up as a spdt switch greatly simplifies things for us.

Assume that the direct-set and -reset inputs are low. This reduces our master flip-flop to a pair of inverters that can be cross-coupled, and does the same for the slave. Assume further that the clock is low. The slave flip-flop is cross-coupled through its transmission-gate switch so that it remembers a previous answer and outputs it via the buffered Q and $\overline{Q}$ outputs. These inverting buffers prevent outside loading from affecting the state or speed of operation. With the clock low, the master flip-flop is *not* cross-coupled. Instead it *follows*

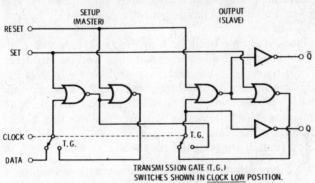

(A) Logic diagram for half a 4013.

CLOCKED INPUTS:			
D	CLOCK	Q	$\overline{Q}$
0	⌐	0	1
1	⌐	1	0
$\overline{Q}$	⌐	CHANGES	

DIRECT INPUTS:			
R	S	Q	$\overline{Q}$
0	0	CLOCK OPERATION	
0	1	1	0
1	0	0	1
1	1	1	1
		(DISALLOWED)	

(B) Truth tables.

Fig. 5-5. A 4013 dual type-D flip-flop.

321

the data input. It will keep following the data input and remembering its instantaneous value as long as the clock is low.

As the clock suddenly goes high, the two spdt transmission gates switch to the other side. This now cross-couples the master flip-flop, disconnects the master from the D input, and forces the master to remember the last value on the D input at the instant the clock went high. Since the D input goes nowhere when the clock is high, anything new on the D input after the positive clock edge occurs is ignored.

When the clock goes high, it also breaks the cross-coupling on the slave flip-flop, turning the slave into a pair of inverters that reflect the state of the master. Thus, with the clock high, the master is holding data and ignoring any new D inputs. The slave is simply passing on (without remembering) the contents of the master directly to the outputs.

What happens when the clock returns to low? From the outside world, it appears that nothing happens. The switches flip over to the other side. This cross-couples the slave output so that it now remembers the data for us independently of what the master is doing. The master is now released and allowed to follow new input data. So, while a rather dramatic internal change takes place on the falling clock edge, no outputs change, and things externally appear to stay as they were.

The clock rise time must be fast. Five microseconds is the usual limit. The clock input must be conditioned and bounce-free. A slow rise or fall time can cause switching problems where old and new data can get mixed. *Note that the fall time is as important as the rise time for proper operation.* Both must be fast and clean. Note also that this circuit is fully static. It can remain in the clock-high or clock-low states indefinitely.

We can summarize the rules for the 4013* as follows:

- Both direct inputs must be low for normal clocked operation.
- If the D input is high, the flip-flop goes or stays in the state with Q high and $\overline{Q}$ low on the positive edge of the clock.
- If the D input is low, the flip-flop goes or stays in the state with Q low and $\overline{Q}$ high on the positive edge of the clock.
- If the D input is cross-coupled to the $\overline{Q}$ output, the flip-flop changes to the other state on the positive edge of the clock, behaving as a binary divider.
- If the direct-set input is made high by itself, the flip-flop will immediately go or stay in the state with Q high and $\overline{Q}$ low.

---

* Except for the internal logic diagram, a 74C74 is functionally equivalent to the 4013.

- If the direct-reset input is made high by itself, the flip-flop will immediately go or stay in the state with Q low and $\overline{Q}$ high.

- If the direct-set and direct-reset inputs are simultaneously made high, a disallowed state results with both Q and $\overline{Q}$ high, independently of the clock and D inputs. This state is normally avoided. The last direct input to go low decides the final result.

- Both direct inputs must be returned to ground before effective clocking can resume.

- The clock must be bounceless and noise-free with rise and fall times faster than five microseconds.

The truth tables in Fig. 5-5B summarize these rules.

THE 4027 DUAL JK FLIP-FLOP

A JK flip-flop has two advantages over a D-type flip-flop: it can be made to binarily divide under external control, and it can appear to do nothing (not change) despite repeated clockings. These extra performance features are obtained at the cost of a somewhat larger and more expensive IC that takes slightly more supply power and usually requires a more complex pc layout. The JK flip-flop is important where full performance is needed, such as in sequencers, odd-length walking-ring counters, divide-by-three circuits, fully synchronous counters, and some other special uses.

The logic diagram of one half a 4027 is shown in Fig. 5-6A. It is simply the type-D flip-flop circuit repeated with extra gates added to the input. These gates respond to a J input, a K input, and an internal feedback line that monitors the *present* Q output. Since each flip-flop has one new input, we end up with a total of 16 pins, compared with the 14 of the 4013 and 74C74 dual type-D flip-flops.

Suppose both J and K inputs are low when we bring the clock from the low to the high state. What happens? The low K input disables the AND gate, holding its output low. The low J input is ignored by the NOR gate, and the present Q output is inverted twice and presented to point "D." On clocking, the old state of the flip-flop is *reentered.* To the outside world, it looks like nothing at all happened. If J and K are both low, clock commands appear to be ignored.

What happens if the J and K inputs are both high? This will disable the NOR gate and enable the AND gate. The Q output is inverted once and sent to "D." Clocking will change the flip-flop to the other state. We alternate states or binarily divide when the J and K inputs are both high.

If J is high and K is low, the AND gate is disabled, and a one unconditionally appears at "D" and is loaded. Similarly, if J is low and K is high, a zero unconditionally appears at point "D." This zero

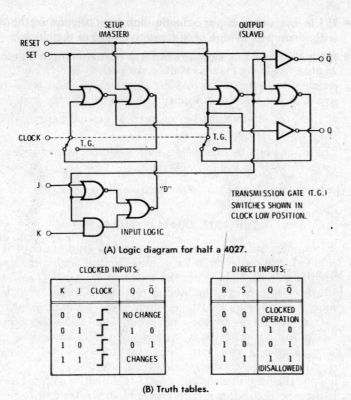

(A) Logic diagram for half a 4027.

TRANSMISSION GATE (T.G.)
SWITCHES SHOWN IN
CLOCK LOW POSITION.

CLOCKED INPUTS:

K	J	CLOCK	Q	Q̄
0	0	⌐	NO CHANGE	
0	1	⌐	1	0
1	0	⌐	0	1
1	1	⌐	CHANGES	

DIRECT INPUTS:

R	S	Q	Q̄
0	0	CLOCKED OPERATION	
0	1	1	0
1	0	0	1
1	1	1	1
		(DISALLOWED)	

(B) Truth tables.

Fig. 5-6. A 4027 dual JK flip-flop.

results as a "don't-care" condition. If Q is high, it goes through the AND gate, is inverted once, and ends up a zero. If Q is low, it goes through the NOR gate, is inverted twice, but *still* ends up a zero. Either way, having J low and K high loads a zero.

Our JK flip-flop acts like a type-D flip-flop if the inputs are different. If both J and K are low, the circuit appears to ignore clock pulses. When both J and K are high, the flip-flop binarily divides.

We can summarize the rules for the 4027* as follows:

- Both direct inputs must be low for normal clocked operation.

- If J is low and K is low, no apparent output change takes place on the positive edge of the clock.

- If J is high and K is low, the flip-flop goes or stays in the state with Q high and Q̄ low on the positive edge of the clock.

---

* Except for the internal logic diagram, a 74C73 is functional equivalent to the 4027.

324

- If J is low and K is high, the flip-flop goes or stays in the state with Q low and $\overline{Q}$ high on the positive edge of the clock.
- If J is high and K is high, the flip-flop changes output states, binarily dividing on the positive edge of the clock.
- If the direct-set input is made high by itself, the flip-flop will immediately go or stay in the state with Q high and $\overline{Q}$ low.
- If the direct-reset input is made high by itself, the flip-flop will immediately go or stay in the state with Q low and $\overline{Q}$ high.
- If the direct-set and direct-reset inputs are simultaneously made high, a disallowed state results with both Q and $\overline{Q}$ high, independently of the clock and D inputs. This state is normally avoided. The last direct input to go low decides the final result.
- Both direct inputs must be returned to ground before effective clocking can resume.
- The clock must be bounceless and noise-free, with rise and fall times faster than five microseconds.

The truth tables in Fig. 5-6B summarize these rules.

An easy way to remember the operation of the direct inputs is that if you do nothing to them (keep them low), they do nothing. On the type-D flip-flop, the D input is passed across the flip-flop to the Q output on clocking. The same thing happens to the JK flip-flop with different J and K inputs. Do nothing to J and K (keep them low), and the flip-flop does nothing. Do everything to J and K (both high), and you get a binary divider.

AN APPLICATIONS CATALOG

We have put together a catalog of many of the different things that can be done with the basic clocked-logic type-D and JK flip-flops. These techniques are useful with the 4013/74C74 and 4027/74C73 dual flip-flops in simpler circuits. They also form the basis for the fancier clocked-logic blocks in the next chapter.

Most often, you will probably want to use only a few 4013s or 4027s in your circuit, since the fancier MSI blocks cram much more performance in a single package. If you find yourself using lots and lots of JK or type-D flip-flops, try to find an MSI substitute or a different approach that will simplify the job for you. On the other hand, it is a very rare CMOS circuit that doesn't have two or three 4013s and maybe a 4027 tucked away in a corner somewhere to pick up some loose ends that the MSI can't handle directly. So, it pays to be aware of all the different things you can do with these basic clocked-logic blocks.

Binary Counters

Binary counters are probably the oldest use for clocked flip-flops. We can get a single stage to divide by two by cross-coupling $\overline{Q}$ to D on a 4013/74C74 or by making both J and K high on a 4027/74C73. The output alternates states, giving a square wave with a 50/50 duty cycle and a frequency of one half the input clock frequency.

We can *cascade* binary counters as shown in Fig. 5-7. This lets us count to numbers higher than two or lets us divide an input clock by a higher ratio. If the output of one divide-by-two (Fig. 5-7A) is connected to a second counter so that its output clocks the second stage, we end up with the divide-by-four counter of Fig. 5-7B. Add another stage, and we have the divide-by-eight counter of Fig. 5-7C. Additional stages mean more possible count states and a higher division ratio. Four stages is a particularly interesting combination. By itself, it can represent 16 different things, count to 16, or scale an input frequency by 16. But, if we properly tamper with the count sequence, we can shorten the divide-by-sixteen into a divide-by-ten, or decimal, counter and do "by tens" counting and arithmetic.

These binary counters are called *ripple* counters. One stage has to change completely before the next stage can start changing. Note

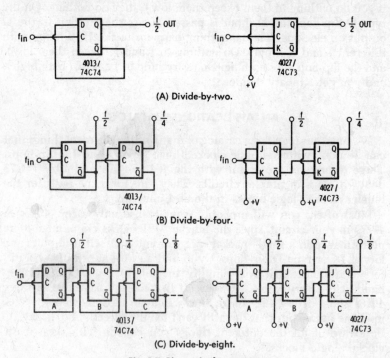

(A) Divide-by-two.

(B) Divide-by-four.

(C) Divide-by-eight.

Fig. 5-7. Binary ripple counters.

326

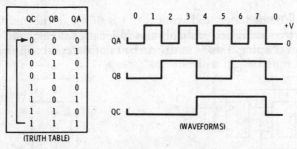

(A) Normal, **add**, or up counter. Stages change on high-to-low transition of previous stage.

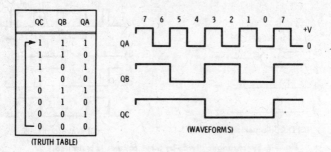

(B) Reverse, **subtract**, or down counter. Stages change on low-to-high transition of previous stage.

Fig. 5-8. Binary counter waveforms.

that invalid output counts will happen during the *settling times* caused by the stage-to-stage *propagation delays.*

We can control the count direction, depending on how we drive the clock of each stage. Fig. 5-8 gives the details. If we clock from $\overline{Q}$ of the previous stage, we get a normal, *add*, or binary-up sequence. On the other hand, if we use the Q output to drive a stage that is positive-edge clocked, we end up with a reverse, *subtract*, or down counter, as shown in Fig. 5-7B. If our logic blocks are negative-edge clocked (such as the 4024), the exact opposite is true—cascade from Q for a normal or up sequence and from $\overline{Q}$ for a reverse or down sequence. We will be looking at the longer binary and decimal MSI counters in detail in the next chapter.

Divide-by-Three Counter

Fig. 5-9 shows a *synchronous* divide-by-three counter using a 4027 or 74C73. Note that both stages are clocked at the same time from the input, so we don't have the propagation and ripple-delay effects of cascaded stages. The output of this counter is said to be *weighted* 1-2, meaning that one output counts for "1" if it is there and the other output counts for "2" if it is present. So, you can look directly at the

states and immediately tell what count is stored in the circuit. This
circuit is the shortest example of the odd-length walking-ring counter
of the next chapter. Two of the three counter states are self-decoding;
the third is picked up with the NOR gate.

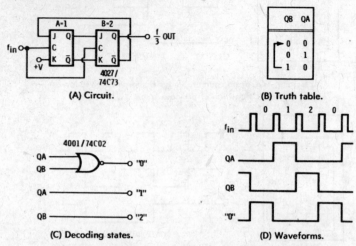

(A) Circuit.

(B) Truth table.

(C) Decoding states.

(D) Waveforms.

Fig. 5-9. Synchronous divide-by-three counter is weighted 1-2.

Divide-by-Four Counter

A synchronous alternate to the ripple divide-by-four counter is
shown in Fig. 5-10. We use the J and K low "do-nothing" state of a
4027 or 74C73 to inhibit the counting of the second stage for half of
the time. Weighting is also 1-2. Four two-input AND gates may be
used to decode the individual stages as shown.

This "do-nothing" inhibiting of a JK flip-flop is the key to building
longer synchronous counters. For a divide-by-eight, you let the third
stage count only one-fourth of the time and inhibit it three-fourths
of the time. A divide-by-sixteen counter can count only one-eighth
of the time, and so on. You can use either multiple-input gates or a
cascaded sequence of enabling two-input gates for longer synchronous counters.

Divide-by-Five Counter

Here is another example of our odd-length walking-ring counter
of the next chapter. As Fig. 5-11 shows, the circuit is synchronous
with all stages clocked directly from the input. We can decode this
particular circuit with five two-input AND gates as shown. The output is unweighted and has a 3:2 duty cycle.

Any of these counters can be reset to zero by using the direct-reset
inputs. You do have to be sure the direct input goes back to low

before the next clock pulse to be counted arrives. With combinations of direct-set and direct-reset inputs, you can load any desired count into your circuit anytime you want.

Shift Registers

A *shift register* is built as shown in Fig. 5-12. We cascade the Q output of a type-D flip-flop to the D input of the next stage. With JK flip-flops, we connect Q to J of the next stage and $\overline{Q}$ to K of the next stage, making sure the first stage always sees complementary data on the J and K inputs.

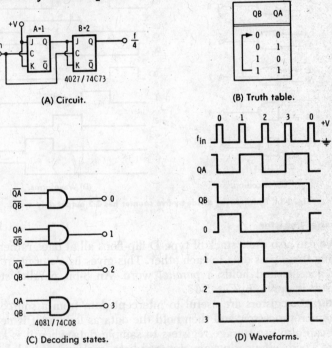

(A) Circuit.

(B) Truth table.

(C) Decoding states.

(D) Waveforms.

Fig. 5-10. Synchronous divide-by-four counter is weighted 1-2.

Each stage stores one *bit* of data, forming a *word* equal in length to the number of stages in the register. On clocking, each bit moves one stage to the right. The first stage picks up a new one or zero from the serial input. The last stage sends its output on to the outside world or loses it. We will be looking at shift registers extensively in the next chapter. The registers shown in Fig. 5-12 are usable as serial-in/serial-out (SISO) or serial-in/parallel-out (SIPO) registers. We can also build shift registers with parallel-loading direct inputs, and if we like, we can recirculate shift-register data from output to input with the circuit shown in Fig. 3-16A.

329

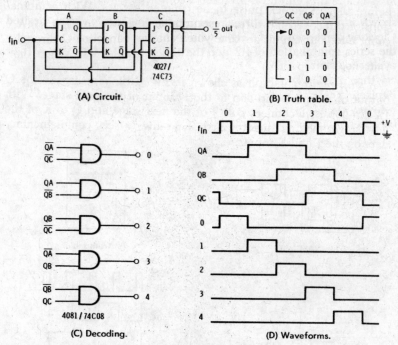

	QC	QB	QA
▶	0	0	0
	0	0	1
	0	1	1
	1	1	0
└	1	0	0

(A) Circuit.

(B) Truth table.

(C) Decoding.

(D) Waveforms.

Fig. 5-11. Synchronous divide-by-five counter has 3:2 output duty cycle.

Storage Register

We can also use a stack of type-D flip-flops all at once rather than having them pass data to each other. This gives us a *storage register* that accepts and holds a *parallel* word. An 8-bit parallel storage register is shown in Fig. 5-13.

Storage registers are useful to intercept data from a source such as a microprocessor, and then hold the data as long as it is needed. You can also use storage registers to sample data when it is known to be good, eliminating any intermediate garbage caused by settling times, propagation delays, and so on. A storage register on the output of an electronic-music digital keyboard will hold the note command after key release. This lets the note decay and lets fallback continue after the key is released, so that the rest of the circuit remembers what note it was working on.

Some MSI examples of storage registers include the 4175 quad, 4174 hex, and 4034 8-bit devices.

Monostable Multivibrators

We saw how to use the 4013/74C74 type-D flip-flop as a monostable multivibrator in the last chapter. More possibilities are

330

shown in Fig. 5-14. A normal monostable is shown in Fig. 5-14A. Clocking drives output Q high, which charges capacitor C through the series combination of R2 and the much smaller R1. When the cycle ends, C is rapidly discharged through R1 only. If a long recovery time is available, we can eliminate the diode and use R2 only. Replacing resistor R1 with a direct connection gives a very fast recovery time but distorts the Q output waveform. On time values for specific R and C values appeared in Chapter 4 in Fig. 4-34C.

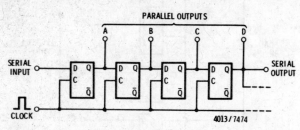

(A) Using type-D flip-flops.

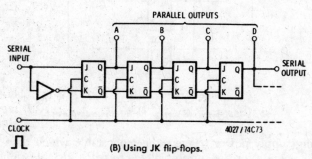

(B) Using JK flip-flops.

Fig. 5-12. Shift registers.

To obtain retrigger ability, use the arrangement shown in Fig. 5-14B. Here the capacitor discharges through R1 when the input clock is low. The positive clock edge drives Q high, and R2 charges C during the delay-until-reset time. The circuit may be triggered at any time and will time-out from the last triggering. Note that the monostable on cycle cannot *end* while the clock is low.

We can also use the alternate trigger method of Fig. 5-14C. Here we pulse the set input to start timing. This takes a resistor and a capacitor but gives us a second way to use positive-edge triggering. The time constant of the trigger circuit must be shorter than the on time for proper operation.

The system reset/power-on generator in Fig. 5-14D will give you a clean reset signal shortly after power is applied to the system.

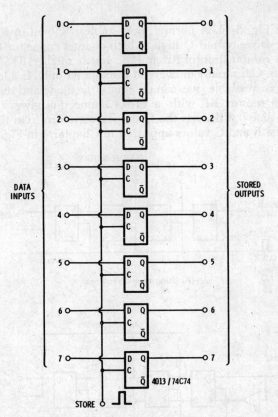

Fig. 5-13. 8-bit word storage latch for microprocessor.

Applying supply power triggers the monostable which then times-out long enough for the supply voltage to reach a stable value. The trailing edge of the monostable output can then be used for a system reset. This type of circuit is handy for initializing things like microprocessors, making sure everything comes up in a favorable state when first activated.

Touch and Proximity Circuits

The nearly infinite input impedance of CMOS makes it ideal for use in *touch* or *proximity* circuits. Usually, a touch-sensitive circuit needs physical contact, while a proximity circuit needs only the nearby presence of an object such as the human body.

There are several characteristics of the human body that may be used for touch sensors. The resistance of human skin is usually several hundred thousand ohms but varies with the individual, the contact spacing, and physiological factors. The capacitance of the hu-

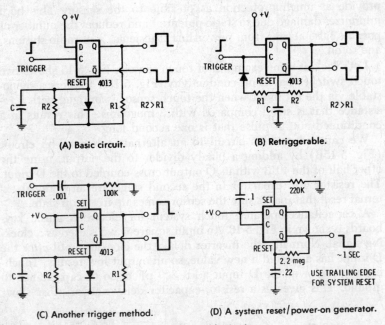

(A) Basic circuit.

(B) Retriggerable.

(C) Another trigger method.

(D) A system reset/power-on generator.

Fig. 5-14. Monostable circuits.

man body is usually around 300 picofarads referenced to ground. Touching a given point in a circuit usually has the effect of adding 300 picofarads of loading to that point. Just coming near a sensor provides a capacitive divider set by the 300-pF body capacitance and the capacitance between the sensor and the person. Usually, this capacitance is only a very few picofarads and drops dramatically as spacing is increased.

Finally, the human body acts as an antenna, picking up the 60-hertz power-line field in virtually all indoor and most outdoor situations.

There are several important things to watch for if you want a reliable touch-sensor circuit. The sensing circuit works best if it is solidly grounded. Portable or ungrounded touch circuits may not work at all. Hot-chassis techniques in which the sensor returns to one side of the power line should, of course, be avoided entirely. Even if ultrahigh series resistors are used, what may be an unnoticeable leakage current to one person can be a mild shock or at least a fuzziness sensation to another.

Most touch sensors should be debounced with a long time constant, preferably a second or so. This prevents any nervousness or hesitancy from being entered as multiple inputs. Always use only the absolute minimum sensitivity you need for any touch sensor and

333

provide as much protection as possible to the sensor. This both minimizes damage due to static potential and reduces the number of possible false alarms from power-line transients, a-m radio stations, and so on.

Let's look at some examples of touch sensors. Fig. 5-15 shows two touch switches based on conductivity. Fig. 5-15A triggers a monostable via the set input when the touch sensor is bridged with a resistance that is small compared with 3 megohms. This results in a conditioned output pulse that is one second long.

We can convert this circuit to an alternate-action off-on circuit (Fig. 5-15B) by adding a binary divider to the output, using the other half of the 4013 with its $\overline{Q}$ output cross-coupled to the D input. The resistor and capacitor in the second stage are an optional external reset that makes sure the sensor comes up in the off state.

A capacitance-operated touch system for electronic-music keyboards is shown in Fig. 5-16. An input square wave is used as a clocking signal. Normally, the inverter delays the clocking until *after* the D input has accepted a new value, so an output low results. Touching the sensor at the D input adds 300 pF or so of capacitance to ground. This gives us a resistor-capacitor delay network that slows

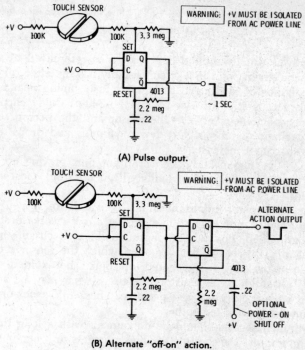

(A) Pulse output.

(B) Alternate "off-on" action.

Fig. 5-15. Touch switches based on conductivity.

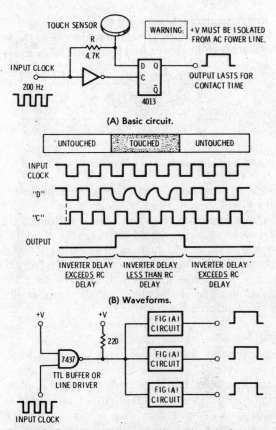

(A) Basic circuit.

(B) Waveforms.

(C) Input buffer minimizes interaction between multiple contacts.

Fig. 5-16. Touch switches based on capacitance have many electronic-music applications.

down the waveform reaching the D input. In fact, it slows things down so much that the clock gets there first, giving us an output-high condition. The input and output waveforms for this circuit are shown in Fig. 5-16B. The choice of clock frequency determines the debouncing you will get. Always use the lowest possible frequency for any particular application. For many electronic-music applications, 500 hertz or higher is a good choice.

Note that the series resistor is somewhat low in value. This resistor is used to set the sensitivity. Since the impedance is fairly low at this point, hum and interference pickup are unlikely, as is finger-to-finger conductivity when multiple keys are hit.

In multiple-key systems, such as that of Fig. 5-16C, some means of keeping the fingers from loading the clock-source rise and fall time must be provided. You can do this with an extra inverter, or

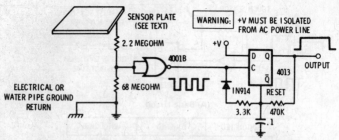

Fig. 5-17. Proximity switch based on power-line "hum" coupling.

you can provide an exceptionally stiff (low-impedance) driver for the
entire circuit. A TTL line buffer is a good choice.

Fig. 5-17 shows a proximity switch based on human coupling of
the 60-hertz power line. A hand held very near the plate induces hum
into the first gate. This hum is squared up and used to trip the retriggerable monostable as shown. A clean output results from the instant
of first proximity until a few milliseconds after release. The sensitivity depends on the size of the plate and the number of allowable
false alarms from induced noise sources.

Some Related Sensors

The object activating the resistance-style sensor in Fig. 5-15 does
not have to be a person; any high impedance that can't be heavily
loaded will work as well. Fig. 5-18 gives several examples. The rain
sensor in Fig. 5-18A uses the conductivity of rainwater to conduct
across a sensor grid. Two probes (Fig. 5-18B) in the bottom of a
hydroponic greenhouse bed will conduct when a nutrient solution
reaches them. A thermistor or thermostat is shown in Fig. 5-18C. As
the resistance drops, the output of the inverter will go high. A liquid-
level sensor system suitable for the booster tank on a fire engine is
shown in Fig. 5-18D. A photocell system is shown in Fig. 5-18E, followed by a splash alarm for swimming pool safety in Fig. 5-18F.

One thing is essential for all these applications to work. The difference between the "off" and "on" resistances has to be very large
and very stable. If the resistance drifts with time or if a sharply defined threshold is needed, you will have to add an input-conditioning
circuit, perhaps based on the 3130 op amp in Chapter 7. Fig. 7-6 is
one example of a photoconductive interface that automatically tracks
ambient light and resistance variations of the photocell.

Synchronizers

Synchronizers are a very important class of circuits. They are used
to lock outside-world signals (or other signals that are random or
somehow out of step) to system timing.

336

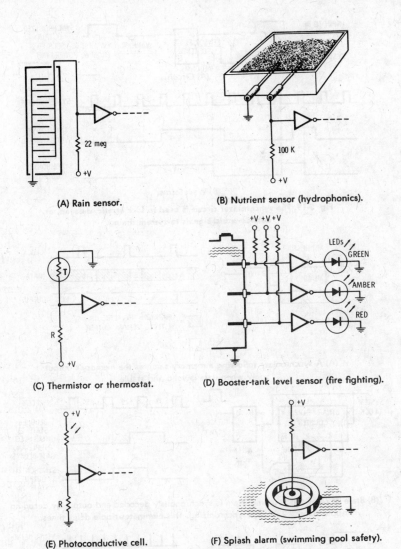

(A) Rain sensor.

(B) Nutrient sensor (hydrophonics).

(C) Thermistor or thermostat.

(D) Booster-tank level sensor (fire fighting).

(E) Photoconductive cell.

(F) Splash alarm (swimming pool safety).

Fig. 5-18. Other high-impedance conductivity sensors such as these
may be CMOS compatible.

The basic synchronizer connection for the 4013 or 74C74 appears
in Fig. 5-19. The signal to be synchronized is applied to the D input.
A system clocking signal goes to the clock. The locked signal output is
taken from Q. As the waveforms in Fig. 5-19B show, both the leading
and trailing edges of the input are delayed until the next positive
clock edge. The locked signal output will be a delayed replica of different width than the input signal, but the output will always be

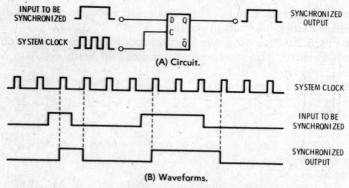

(A) Circuit.

(B) Waveforms.

Fig. 5-19. The synchronizer circuit is used to lock erratic, delayed, or outside-world signals to system timing.

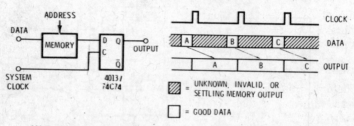

(A) A synchronizer following a memory samples the memory's output only when data is valid and stable.

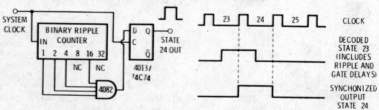

(B) State #24 of this counting chain is synchronously decoded and output by decoding ripple state #23 and resynchronizing. This eliminates ripple delay times.

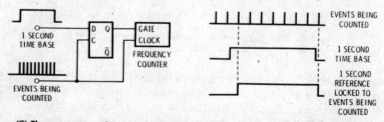

(C) The one-count ambiguity of a frequency counter may be eliminated by delaying the time reference so it starts and stops an event being counted.

Fig. 5-20. Using synchronizers.

338

locked to the clock timing. The output will also always be an exact number of clock intervals in width.

In Fig. 5-20A, we use a synchronizer following a memory or character generator. The synchronizer samples and holds the data only when it is valid, relocking the outputs to system timing. This eliminates the possible times when the memory is accessing, settling, or otherwise putting out garbage. What we have done is convert a short mixture of good and bad data into an always accurate answer that is delayed one clock-pulse interval in time. In Fig. 5-20B, we make a ripple counter look like a synchronous counter by decoding a state *one count early* and relocking it to the input clock. This eliminates any ripple delay and propagation times. In Fig. 5-20C, we have used a synchronizer to eliminate the "inherent" one-count bobble or ambiguity of a frequency counter. We do this by delaying the counter's time-reference gate until a pulse to be counted arrives. A whole number of input pulses are always counted this way. This eliminates the random starting and stopping points of the time gate and the apparent one-count bobble.

One-and-Only-One

The one-and-only-one is a very important synchronizer circuit that is detailed in Fig. 5-21. It will give you exactly one clock interval as an output in response to an outside-world command. Our outside-world command is shown as a positive edge. This sets the first flip-

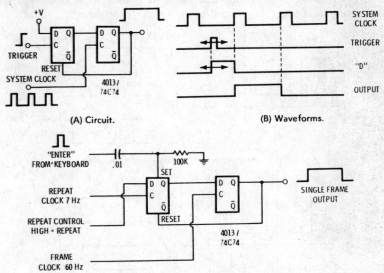

(A) Circuit.   (B) Waveforms.

(C) Single-frame update generator for a tv typewriter.

Fig. 5-21. The one-and-only-one.

flop. The first flip-flop absorbs the time difference between the arrival of the outside-world signal and the system clock edge. The second flip-flop generates a one-clock-wide output pulse and resets the first stage. Every time you trigger the circuit with an outside-world command, you get one and only one clock-interval pulse as an output. Waveforms are shown in Fig. 5-21B.

The one-and-only-one may be used as a single-frame update generator for a tv typewriter (Fig. 5-21C). The positive edge of an "enter" command from a keyboard is used to direct-set the synchronizer. An output lasting one frame interval (16.7 milliseconds) is provided. This circuit also gives us a repeat option. If we make D high and put a low-frequency (7.5-hertz) *blinker clock* on the first-stage clock input, we get repeated single-frame outputs recurring at the blinker rate. For normal operation, hold D low; for repeat, make D high.

You can convert a one-and-only-one circuit into an n-and-only-n circuit by adding a counter that delays resetting the first stage until n counts have gone by. A gate may be added to make this circuit into a generator of n clock pulses.

Two Sequential Circuits

Two sequential circuits that use 4013 or 74C74 flip-flops are shown in Fig. 5-22. The *bucket brigade* of Fig. 5-22A sequentially gives us self-decoded outputs as shown, first at A, then at B, and then at C. Only one start command must be provided per sequence if you are to

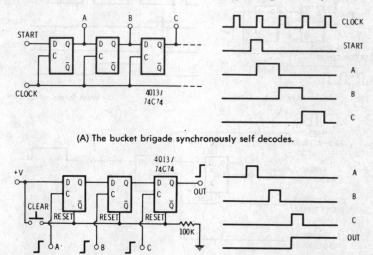

(A) The bucket brigade synchronously self decodes.

(B) Sequential pass-on provides an output only if C follows B and B follows A.

Fig. 5-22. Two sequential circuits.

avoid multiple outputs. You can close the circuit on itself for a continuous "electronic-stepper" operation, but you will have to add some method to make sure only one output is high at a time.

The *sequential pass-on* in Fig. 5-22B is handy for alarms, electronic locks, and other circuits where events must follow each other in a certain order. You get an output only if the leading edge of event A occurs and is followed by event B and then event C. Events that are in the wrong order or not present prevent an output. Since the event inputs all go to clocks, they must, of course, be properly conditioned, bounceless, and noise-free.

A Programmable Divider

An interesting if somewhat odd use of a 4013 flip-flop is the programmable divider in Fig. 5-23. The circuit will divide a clock either by one or by two. This is handy in digital data-recording circuits

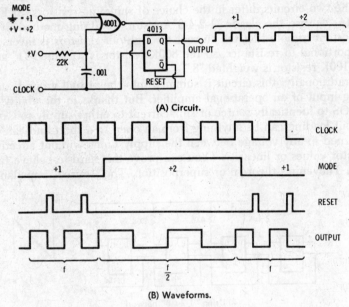

(A) Circuit.

(B) Waveforms.

Fig. 5-23. Programmable divider divides by one or two.

where a string of clock cycles related in frequency by 2:1 may be needed. With the *mode* input high, the 4001 NOR gate is disabled and the 4013 acts as a binary divider. We get the binary division through the $\overline{Q}$ to D external feedback connection. If you ground the mode input, the leading edge of the clock sets the 4013 and the trailing edge resets it via the direct-reset input, all in one clock cycle. For proper operation, the RC network has to have a short time constant

341

compared with twice the clock frequency. The clock input must have a symmetrical (50/50) duty cycle.

Digital-to-Analog Converters

Most digital-to-analog conversion schemes are based on digitally switching current sources, voltages, or resistors and then summing the resulting current as an analog output. CMOS output stages are especially attractive for this application because they swing through the full supply range and have no offsets.

Fig. 5-24 gives us two different digital-to-analog (d/a) conversion schemes. We have shown the digital portion as a binary divider using 4013s or 74C74s. The counter is advanced to the desired count and then is output as an analog voltage. As an alternate, we could use our type-D flip-flops as latches and apply a parallel digital word. The word is updated every time we want a new analog conversion.

The two circuits differ in the choice of summing resistors. In Fig. 5-24A, we use the classic "1-2-4-8" resistor weighting method. It is important to note that the current through our resistors is inversely proportional to resistance, so the 800K resistor is weighted "1" and the 100K resistor is weighted "8."

Traditionally, this circuit is summed into the virtual ground summing input of an operational amplifier. But thanks to the ability of CMOS to identically source or sink current to either supply rail, our output loading can be anything from an open to a short and can be returned to any voltage between the supply limits without affecting resistor values or linearity. (You can prove this hard-to-believe fact with Thevenin's theorem or superposition. The "source" impedance

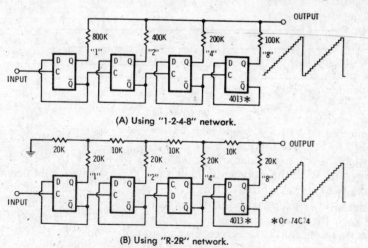

(A) Using "1-2-4-8" network.

(B) Using "R-2R" network.

Fig. 5-24. Digital-to-analog (d/a) converters.

of our multiple supply is constant; only the equivalent "single-source" voltage changes.)

For more resolution, the resistor values tend to get out of hand. We can combine the resistors with a current- or voltage-divider circuit and eliminate this restriction. This is called the "R-2R" method and is shown in Fig. 5-24B. Outputs of both circuits are the same, but the R-2R circuit easily expands to long word lengths. Another of its advantages is that the source and sink currents of each stage are identical, so any internal drops tend to compensate themselves rather than get progressively worse.

You can make these circuits into analog-to-digital (a/d) converters by feeding the analog output to a comparator that starts or stops the clock pulses, depending on an input analog signal. The digital output is then taken from the counter. Other examples of a/d converters were given in Chapter 4.

Phase Shifters

Digital *phase shifters* may be used to generate various delays in a signal. This is handy for color-tv processing, audio filtering, radar-signal correlation, and similar uses. The same idea is also used to generate multiphase ac power-source signals and the multiple clocks needed for some microprocessors. Fig. 5-25 gives us some details on digital phase shifters.

The binary divider of Fig. 5-25A is the simplest example. It starts with a double-frequency clock and gives us two clock phases that are spaced 180° from each other. If some "daylight" is needed between clock phases, a narrow negative-going clock pulse can be ANDed with the outputs to provide a zero state between the phase outputs.

In Fig. 5-25B, we generate two *quadrature*, or 90°, phase-shifted digital signals. We start with a 4f clock and use an even-length walking-ring counter (described in the next chapter) to generate the phases for us. Unlike audio networks, the 90° phase shift is independent of frequency and follows the input clock over any desired frequency range. By decoding both-outputs-high and both-outputs-low, we can generate the type of two-phase clock signal needed by a microprocessor in which one clock goes high, followed by "daylight," followed by the other clock high, followed by more "daylight."

Incidentally, it is extremely important in many electronic circuits to *exactly* obey the multiphase clocking requirements. Be sure to pay particularly close attention to the allowable clock overlap and underlap, minimum spacing allowable **width** variations, the states (if any) you are allowed to stop in, and so on.

A three-phase power-generator circuit is shown in Fig. 5-25C. It starts with a 6× clock (360 hertz in the case of a 60-hertz power

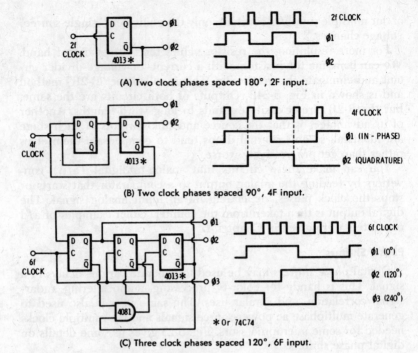

(A) Two clock phases spaced 180°, 2F input.

(B) Two clock phases spaced 90°, 4F input.

(C) Three clock phases spaced 120°, 6F input.

Fig. 5-25. Digital phase shifters form multiphase clock sources.

system) and gives us three square waves phase-shifted by 120°. The AND gate is a disallowed-state eliminator but otherwise doesn't enter into the operation of the circuit. Details on this walking-ring counter and ways to digitally generate output sine waves rather than square waves appear in the next chapter.

Phase Detectors

We can also use either a 4013 or a 74C74 and an RC filter network to produce an analog output proportional to the phase shift between two signals. We took a quick look at a phase detector in the last chapter. Two more possibilities are shown in Fig. 5-26.

We use phase two, or $\phi2$, as our reference. It sets the flip-flop with its positive edge. The positive edge of phase one ($\phi1$) resets the flip-flop. The time the Q output is high depends on the phase difference between $\phi1$ and $\phi2$. The output RC filter averages the high-time to low-time ratio into a continuous analog output voltage. As the response curve shows, the greater the phase lag of $\phi1$, the more output voltage you get. The best operating point is at the point halfway up, or 180°. Operating near 0° or 360° isn't recommended because of the discontinuity as the detector slips cycles.

344

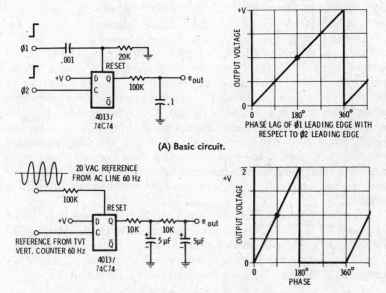

(A) Basic circuit.

(B) Line-lock phase detector for a tv typewriter simultaneously conditions power-line reference.

Fig. 5-26. Clocked-logic phase detectors.

Sometimes we can cheat a little and make a circuit do much more than we would at first suspect. Fig. 5-26B is an example of a phase detector that works only over the first 180° of phase shift and provides only one-half the output voltage of the first circuit. Its best operating point is at a 90° phase shift. This is not as desirable as the circuit of Fig. 5-25A but probably is still useful, particularly as the line lock for a tv typewriter.

But, now we can shove a plain old power-line sine-wave reference into the direct-reset input without any conditioning, and let the *same* flip-flop do our power-line conditioning and our phase-detecting simultaneously. This example of a do-more-with-less configuration shows how some extra thought time can often cut the amount of circuitry apparently needed in half.

Digital Mixer

For some uses, a 4013 or a 74C74 can act as a mixer that gives us the difference between two input frequencies. The circuit in Fig. 5-27 shows the basic idea for a digital mixer.

Suppose F1 and F2 were identical in frequency. The output would always be in the high or low state, depending on whether F1 was high or low at the positive edge of F2. Now, suppose that F1 is slightly lower in frequency than F2 and it starts "slipping cycles" with respect to F2. As it slips cycles, the Q output will also slip and give us a square wave nearly equal to the difference frequency between the two inputs.

There are several restrictions on mixers of this type. Unlike analog mixers, both F1 and F2 must be single-frequency, clean, digital waveforms. Our digital difference frequency is always *quantized* to be some exact submultiple of F2, so some apparent frequency-jitter noise will always appear on the output. This jitter becomes less pronounced when you are measuring the small difference between two nearly equal, high-frequency signals. (Actually, this frequency jitter represents the sum of the two frequencies. You normally can't separate sum from difference frequencies in a single mixer; it takes sine and cosine channels, single-sideband techniques, or something similar.)

The circuit is also harmonic-sensitive. For instance, if F2 is 10 kHz and F1 is 20.003 kHz, a 3-hertz beat note results. Mixers like this can be used in frequency synthesizers and phase-locked loops. They are pretty much limited to uses where a somewhat noisy, low-frequency difference between two faster signals is useful.

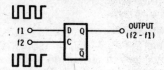

Fig. 5-27. Digital mixer provides difference frequency between F1 and F2.

A Tuning Indicator

We can wrap up or look at the clocked-logic flip-flops with another example of a do-more-with-less circuit. It uses a single 4013 or 74C74 to replace what at first glance would seem to be a bunch of MSI circuitry.

The problem and its solution appear in Fig. 5-28. The problem involves the bit-boffer digital-data cassette recording system discussed in Chapter 10. To minimize errors, this recording system must be properly tuned. This is easily done by adjusting a monostable, but this could require an oscilloscope or other special test setup. It would be far better to come up with a "turn the knob until the red light goes out" circuit.

The recording system involves two signals, A and B. The system is in tune if there are exactly eight A pulses for one B pulse. If there are seven or nine A pulses, it is out of tune. What we want to do is build a simple and cheap circuit to put a light *on* for seven or nine A pulses and *off* for eight A pulses.

We will need a divide-by-nine counter, a digital comparator, gating for the greater-than and less-than outputs, some control logic, a

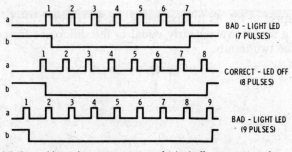

(A) The problem—show correct tuning of a bit-buffer cassette interface.

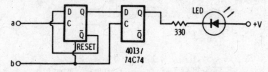

(B) Solution using a single 4013 or 74C74. First stage is even-odd detector, second stage stores result.

Fig. 5-28. Tuning indicator for a digital-data cassette storage system.

latch, and a lot of other parts. Can we do the whole job with a single 4013 or 74C74?

Instead of actually measuring the total number of A pulses, we observe that an *even* number (eight) is valid, while the *odd* numbers (seven) and (nine) are not. So, we will use half a 4013 or 74C74 as an even-odd detector and the other half as a result memory and LED driver, as shown in Fig. 5-28B.

Waveform B high holds the first stage in the low state. When waveform B goes low, the first-stage binary counter starts counting waveform A pulses, going high on odd counts and low on even counts. When waveform B returns to high again, the even-odd answer is stored in the second flip-flop and is output to light or not light the lamp. You can do the entire circuit with a 4013 or a 74C74, an LED, and an optional resistor.

Always watch for your chance to "do more with less" in any design problem. Very often, the simpler clocked flip-flops of this chapter can eliminate or simplify things enough to greatly reduce your circuit cost and complexity.

A CMOS Cameo

TOUCH-CONTROLLED LATCH

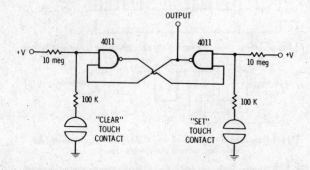

Touching "SET" with your finger drives output high.
Touching "CLEAR" with your finger drives output low.

Chapter 6

Counters and Shift Registers

We can build more complex clocked-logic blocks by taking groups of the clock flip-flops of Chapter 5 and suitably combining them with extra gates. A far better way is to adapt your problem to fit one or more of the dozens of available CMOS clocked-logic MSI circuits. Two of the most important types of these more complex CMOS blocks are *counters* and *shift registers*.

Strictly speaking, we can define a counter as a *sequential-state machine*. A counter has n different states into which it can go, always in a predetermined sequence. You can use counters to divide an input frequency by n or to provide one output pulse for n input pulses. Counters will also remember the number of events that have happened over a certain interval. By *decoding* the internal counter states, you can separately activate one-of-n output lines. Alternatively, you can combine your decoding and driving into other useful codes such as the seven-segment code used to drive LED readouts.

A shift register is a cascaded group of type-D flip-flops. Each flip-flop will pass its stored one or zero onto the next stage on clocking. You can use shift registers to store data, to delay or phase-shift signals, to *format* sequences from parallel to serial or vice versa, and as a *buffer* that absorbs timing and data differences between two different systems. Specialized shift-register connections will count, provide *pseudorandom* code sequences, or generate digital sine waves.

COUNTER QUALITIES

The features we may want in a particular counter are called its *qualities*. Some of the more important qualities that we are usually interested in include the following.

Modulo

The *modulo* of a counter is simply n or the number of different states the counter sequentially goes through. *Low-modulo* counters usually have nine or fewer states. We saw some examples of these counters in Chapter 5. *Decimal counters* usually count to ten. They are often the basis for frequency counters, timers, clocks, and other *digital-readout* displays. A *hexadecimal* counter normally consists of four binary dividers, and it has sixteen counts. A *variable modulo* counter is one whose count length usually is easy to change.

Synchronous versus Ripple

In a *ripple* counter, the output of one counter stage becomes the clocking input for the next stage. During the *ripple-through* or *propagation-delay* times, invalid counts can result and produce *glitches* on various decoded outputs.

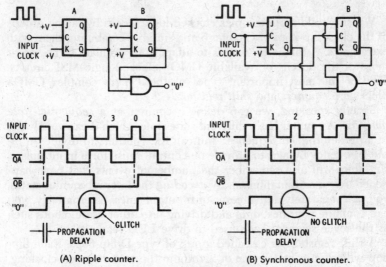

(A) Ripple counter. (B) Synchronous counter.

Fig. 6-1. Ripple counters can produce glitches and invalid states during propagation delays.

In a *synchronous* counter, all counter stages are directly driven from a single input, and all outputs change simultaneously. Most longer synchronous counters take extra gating, so they are usually more complex and more expensive than ripple counters.

Fig. 6-1 compares a divide-by-four ripple counter with its synchronous counterpart. If we try to decode the zero-count state, the ripple counter gives us an extra output glitch, but the synchronous counter does not.

350

For some uses, these short glitches may not matter. However, if we are driving clocked logic with an output that has glitches, all sorts of trouble will result. Generally, half the states of a ripple counter will decode glitch-free, while the other half of the states will provide at least one glitch. You can sometimes get rid of glitches by gating the input clock into your decoding process.

Besides being essentially glitch-free, another advantage of a synchronous counter is speed. No matter how long the synchronous counter is, you get only the propagation delay of a single stage from input to output. Ripple counters can involve as many as 14 sequential propagation delays between input clocking and final-stage output. With CMOS operated on lower voltage supplies, these composite delays can be one or two microseconds long and can be critical in high-speed circuits.

Coding

The *coding* of a counter tells us which states the counter is going to go through and in what sequence. If we are interested simply in dividing pulses or frequencies by n (this is called simple *scaling*), the coding usually won't matter. On the other hand, if we are driving a decoder or providing a particular set of outputs, the coding becomes all-important.

Popular codings include *straight binary*, or 1-2-4-8-16-32- . . . , and *binary-coded decimal* (bcd), or 1-2-4-8; 10-20-40-80; . . . . Binary-coded decimal is the same as straight binary up to count nine; it then repeats by decades, with 4 bits needed for each decade word. Another coding example is the *walking-ring counter* discussed later in this chapter. Walking-ring coding offers good high-frequency performance, easy decoding, and a simple way to generate digital sine waves.

Weighting

A counter is *weighted* if the code we select lets the output of each counter stage stand for so many counts if it is a one and for no counts if it is a zero. In a hexadecimal counter, the four outputs are weighted "1," "2," "4," and "8." When we are in state 1011, we are in state "8" + "0" + "2" + "1" = 11, or in state twelve of the count sequence.

Weighting is particularly handy for digital-to-analog conversion. Both straight binary and bcd are weighted codes; the walking-ring sequences are not.

Resettability and Presettability

A counter is *resettable* if we can force it into an all-zeros state. To do this, state zero must be part of the normal coding sequence, and some way has to be provided to directly put the counter into this state. Usually we do this with a separate direct-reset input pin.

A *presettable* counter can be directly forced into any desired state. This takes a group of new inputs called *preset* inputs and usually a controlling pin called load, jam, or something similar. Note that we can also sometimes reset a counter by presetting it to all zeros.

Decodability

We have a *decodable* counter if it is reasonably easy to use gates or MSI to separate the counter states into one-of-n lines or some other desired sequence such as a seven-segment readout coding. Two ways to get good decodability are to use a standard weighted code such as binary or bcd, or to use the walking-ring sequences that all decode with two-input gates.

The counter and decoder used must, of course, work with the same coding if they are going to communicate with each other properly.

Decade Cascadability

A variable modulo counter is *decade cascadable* if you can arrange the presetting on a "by-decades" basis. This is a very special feature and is available only on the CMOS 4522, the 4059, the 0320, and a very few others.

Decade cascadability takes a special gating called *state-zero feed-forward decoding*. Cascade a regular divide-by-300 counter, a divide-by-60 counter, and a divide-by-2 counter, and you get a counter that counts to 300 *times* 60 *times* 2 = 36,000. Cascade a decade cascadable counter set to 300 on the hundreds selector, 60 on the tens selector, and 2 on the units selector, and you get a count of 300 *plus* 60 *plus* 2 = 362.

Disallowed States

If a counter consists of n flip-flops, there are 2^n different states it can get into. Any leftovers are called *disallowed states*. If the counter ever gets into one of these states and can't get back out without help, it may not count at all, or it can count by the wrong modulo in the wrong sequence. Only straight binary counters do not have disallowed states; *all others do*.

In the case of many bcd counters, states 10 through 15 are disallowed. Most often, if the counter gets into one of these states (usually on power-up or through a noise pulse), it will correct itself after a few counts and jump back into the right sequence.

The walking-ring counters that are even length and modulo six or higher have disallowed states that must be eliminated with external gating. One way to do this is to detect two outside zeros and force all inside stages to zero with gating. Most CMOS counters using walking-ring coding have internal disallowed-state elimination. Typical examples are the 4017, 4018, and 4022.

Symmetry

A counter's output is *symmetric* if it has a 50/50 duty cycle. Symmetry is usually free on all binary counters. You can sometimes obtain symmetry by using a counter of one-half the normal length and then adding a divide-by-two binary counter to the output. Symmetrical, odd-modulo counters can be done with gating that forces the output high on a positive clock edge and low on a negative clock edge.

Symmetry is also free on walking-ring counters of even length. Note that the "8" output on a bcd output has 8:2 asymmetry or a 20-percent duty cycle.

Count Direction

Most counters are *up-only,* or *add,* counters that always add one to their sequence on every clocking until they overflow and then start over again. A very few counters, particularly the 4522 and 4526, are *down-only,* or *subtract,* counters. They remove one count each sequence until they underflow and jump to their *maximum* count.

Finally, we have the option of *add-subtract,* or *up-down,* counters. These counters will go in either direction. They may operate with a single clock and a separate up-down pin (4029, 4510, 4516) or with separate up and down clock lines (40192, 40193, 74C192, 74C193).

Clocking Edge

Most CMOS counters change one count on the positive, or rising, edge of the clock. The long binary ripple counters advance one count on the negative or falling edge of the clock.

A very few counters are available with an *enable* input that either may be used to *inhibit* or ignore count pulses, or may be used to give us a choice of positive or negative clocking edges. In the case of the 4518 and 4520, we gain negative-edge clocking simply by calling the enable input the clock and vice versa.

SOME CMOS COUNTERS

A sampling of some of the more popular and more available CMOS counters is shown in Table 6-1.

Long binary counters combine very large counts, low supply-power requirements, and low cost in a single package. They are among the oldest of CMOS designs. They are handy anytime you want to scale by a really large number. These counters trigger on

Chart 6-1. Some Typical CMOS Counters

Device	Type	Length	Internal Organization	Count Coding & Outputs
4020	Long Binary	$2^{14} = 16{,}384$	Ripple	Binary, Except for 2^2 & 2^3
4024	Long Binary	$2^7 = 128$	Ripple	Binary
4040	Long Binary	$2^{12} = 4096$	Ripple	Binary
4060	Long Binary	$2^{14} = 16{,}384$	Ripple	Binary, Internal Oscillator
4017	Decoded 1-of-n	10	Synchronous	Walking Ring, 1-of-10 Outputs
4022	Decoded 1-of-n	8	Synchronous	Walking Ring, 1-of-8 Outputs
4026	Readout Decoded	10	Synchronous	Walking Ring, 7-Segment Out
4033	Readout Decoded	10	Synchronous	Walking Ring, 7-Segment Out
4511	Readout Decoded	10	Synchronous	Walking Ring, 7-Segment Output with Driver
74C90	Decade	10	Synchronous	BCD Ripple, 0 or 9 Presettable
74C93	Binary	16	Synchronous	Binary Ripple
74C160	Decade	10	Synchronous	BCD, Asynchronously Presettable
74C161	Binary	16	Synchronous	Binary, Asynchronously Presettable
4518	Dual Decade	10	Synchronous	BCD, Clock Enable
4520	Dual Binary	16	Synchronous	Binary, Clock Enable
4029	Up-Down, Common Clock	10 or 16	Synchronous	BCD or Binary, Presettable
4510	Up-Down, Common Clock	10	Synchronous	BCD, Presettable
4516	Up-Down, Common Clock	16	Synchronous	Binary, Presettable
40192, 74C192	Up-Down, Dual Clock	10	Synchronous	BCD, Presettable
40193, 74C193	Up-Down, Dual Clock	16	Synchronous	Binary, Presettable
4018	Variable Modulo	2 Through 10	Synchronous	Walking-Ring Outputs
4522	Variable Down Counter	10	Synchronous	BCD Decade Cascadable
4526	Variable Down Counter	16	Synchronous	Hex, Unit Cascadable
0320	Programmable	3 Through 1024	Synchronous	Divide-by-n Output Only

the negative, or falling, edge. They will produce glitches on some counts, and the long ripple delays can present problems in decode-and-reset applications. Extra-long reset times are recommended.

The 4020 (14-stage), 4024 (7-stage), and 4040 (12-stage) are typical examples of long binary counters. Due to pin limitations, stages 2 and 3 have no output pins on the 4020. The 4060 is another 14-stage counter. This one has an optional internal oscillator which can be resistor-capacitor or crystal controlled, but has no output pins for stages 1, 2, 3, and 11.

Decoded one-of-n counters are useful in digital filters, electronic music, waveform generators, and other places where you may want an electronic stepper action that gives you one-of-n output lines active. The 4017 (ten output states) and 4022 (eight output states) both use walking-ring coding internally. This minimizes output overlap and count-to-count width variations, besides being glitch-free. Disallowed states are eliminated by internal gating.

Readout-decoded counters directly provide a seven-segment readout output. This output can provide a low to medium amount of drive to a seven-segment LED display. The 4026, 4033, and 4511 are typical readout-decoded counters. The 4026 has a display-enable pin that simplifies brightness control, while the 4033 offers ripple blanking of the leading edge of unwanted zeros.

Dual synchronous counters are often the first choice for timing-chain design in tv typewriters, video games, and most other general timing applications. The 4518 (dual-decade) and 4520 (dual-hexadecimal) counters both offer two-per-package counting systems with a choice of clocking edge and with optional clock gating. They are both resettable but not presettable. Note that the 4518 can be converted into a divide-by-eight counter simply by connecting the "8" output to the reset line.

Up-down counters are handy for applications where you want to both add and subtract from the basic stored count. The 4510 (decade), 4516 (hexadecimal), and 4029 (programmable—either decimal or hexadecimal) are typical versions that use a single clock and a separate pin for up-down control. The 40192 and 74C192 (decimal) and 40193 and 74C193 (hexadecimal) counters have two separate, normally positive, input clocks, one for up-counting and one for down-counting. All these counters are presettable.

Variable modulo counters include the 4018 which is handy for low modulo counts up to ten and doubles beautifully as a digital sine-wave generator. The 4522 (base 10) and 4526 (base 16) are decade cascadable counters with the count-zero, feed-forward decoding that is essential for "by-decades" entry of a number to be scaled. These counters may be combined with bcd thumbwheel switches to build a digital frequency synthesizer.

The 0320 frequency synthesizer and the unlisted 4059 offer single-package approaches to the same problem at higher cost. The 0320 also includes the ability to binarily offset your bcd count. This is a very handy trick that can be used in multiple-channel radio frequency synthesizers.

SOME COUNTER APPLICATIONS

Let's look at some practical examples of how these different types of CMOS counters can be used.

Polytonic Electronic Music Generator

Our top-octave generator circuit of Fig. 4-29 produced 12 equally tempered notes of any single octave. If we want all of the notes of all the octaves at once, we can add binary dividers to produce all of the lower-octave related notes. This is useful for either a polytonic synthesizer or an electronic organ.

A binary divider using the 4024 is shown in Fig. 6-2. The 4024 by itself produces high-level, symmetrical square waves. Normally, you would repeat this circuit 12 times. If you input note C8, you will get notes C7, C6, C5, C4, C3, C2, and C1 as outputs. If a five-octave range (including the input) is enough, you can get by with a 4520, and will need only six packages for the division, since there are two counters in each package.

The "raw" square waves usually have too much amplitude for electronic-music analog processing. They also consist only of odd harmonics. While useful for woodwind and stopped-pipe tones, they are limited in the kind of filtering that can be used with them to produce other voices. To overcome this problem, we can add the 1-2-4-8 resistor networks as shown in Fig. 6-2. This d/a converter converts the square wave into a 16-step approximation of a sawtooth that has all harmonics present except for the 16th, 32nd, and so on. Outputs 6, 7, and 8 will also have a few lower harmonics missing. You can replace these by using a top-octave generator whose output duty cycle is 30 percent, rather than 50/50. The Mostek MK5024-BA has this asymmetrical output.

The output amplitude is set with the load resistors. Often, this output amplitude is held under 100 millivolts maximum so that a CA3080 or similar voltage-controlled amplifier (vca) can be used for keying.

Note that this is one application where a stiff, properly regulated power supply is essential. This doesn't matter to the CMOS, but since the power supply directly feeds the stairstep output, any hum, variations, or interactions will distort or intermodulate the output tones.

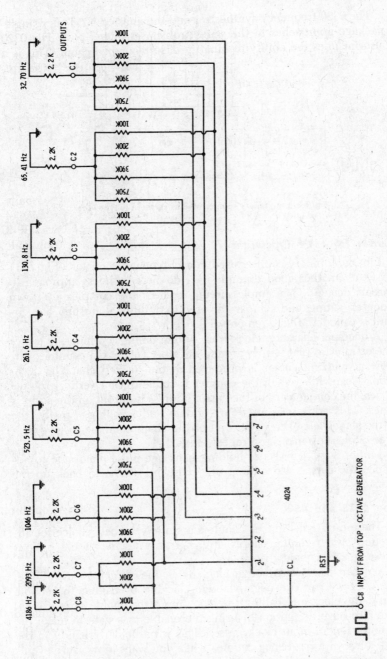

Fig. 6-2. Long binary counter used as polytonic note generator for electronic music.

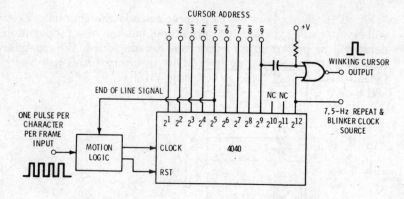

Fig. 6-3. Long binary counter used as cursor in tv typewriter.

Cursor for a TV Typewriter

One long binary counter used in a tv typewriter cursor (Fig. 6-3) gives us another good example of a do-more-with-less approach to circuit design. This simple circuit replaces the complex up-down counter, gating, and comparator used in traditional cursors. The circuit is called a *McFadden* cursor.

A once-per-character clock from system timing continuously cycles the counter so that it goes around once per frame. On counter overflow, a winking cursor is produced at the desired character location. Meanwhile, when the counter is stable during vertical retrace times, the *complement* of the cursor address is output on the address lines. This address is used for update and scrolling. Finally, the three stages left over at the end of the chain combine to give us a free 7.5-hertz blinker and repeat clock.

Motion logic consists of holding back one pulse per frame to advance the cursor and adding one pulse per frame to back up the cursor.

The 4518 and 4520

These dual synchronous counters are often the best choice for all-around system-timing chores. Fig. 6-4 shows some of the options these versatile chips give us.

We can use positive-edge clocking in the normal mode of Fig. 6-4A, or we can use negative-edge clocking by calling the enable input the clock and calling the clock input the enable (Fig. 6-4B). We also have the option of using the enable as a gate to make the counter ignore input clock pulses. This is particularly handy in the first stage of a frequency counter when you want to measure events per unit time.

Fig. 6-4C shows how we can ripple-cascade the counters. They will be *internally* synchronous per decade, but one new propagation delay will be picked up for every new decade added. We can gain fully synchronous operation by adding an external four-input gate for the 4520 or a two-input gate for the 4518 as shown in Figs. 6-4D and 6-4E.

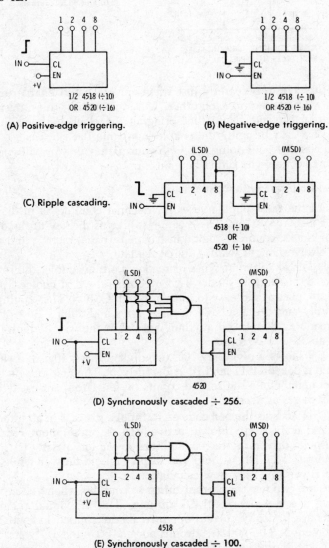

(A) Positive-edge triggering.

(B) Negative-edge triggering.

(C) Ripple cascading.

(D) Synchronously cascaded ÷ 256.

(E) Synchronously cascaded ÷ 100.

Fig. 6-4. The 4518 (decade) and 4520 (hexadecimal) are a versatile pair of dual synchronous counters.

CLOCK	ENABLE	COUNTS?
⎦	1	YES
⎦	0	NO
⌐	0 OR 1	NO
0	⌐	YES
1	⌐	NO
0 OR 1	⎦	NO

RESET INPUT MUST BE <u>GROUNDED</u> FOR
NORMAL COUNTING. BRINGING RESET
HIGH FORCES ALL ZEROS STATE.

Fig. 6-5. Input gating rules for the
4518 and 4520 counters.

We do have to be careful just when our clock and enable inputs change with respect to each other. The input gating rules are summarized in Fig. 6-5. Note that bringing the clock high while the enable is a 1 or bringing the enable low while the clock is a 0 will enter a count. Avoid doing this with your gating, or you will end up with extra counts you didn't bargain for.

Decimal Counters with Readouts

One of the most popular uses for decimal counters is in digital-readout display systems. These counters give us the key to the operation of clocks, counters, digital instruments, panel meters, and anywhere else you might want a digital readout.

While lots of different types of readouts are available, the seven-segment light-emitting diode (LED) is often the best choice. LEDs are bright, cheap, and easy to interface with CMOS. They multiplex beautifully and are available in many sizes and colors. Best of all, LED readouts don't need anything special in the way of high-voltage or ac drive waveforms.

Fig. 6-6 shows a divide-by-100, two-digit display. It uses a single 4518 and a pair of 4511 latch/decoder/drivers. The first decade handles the units, while the second counts by tens. You can easily add as many stages as needed.

The 4511 accepts the bcd code on its inputs, stores it in an internal latch, and converts it into an active-high, seven-segment readout code. Bipolar drivers on the readout lines give us lots of drive current. The internal latch lets us hold and display an old answer while the counter is working on a new calculation.

There are two very important things to consider when you are interfacing a seven-segment LED display. The individual segment elements are basically *current-operated* diodes, biased in their forward direction. The forward drop is normally around 1.7 volts. *External current limiting MUST be provided for all light-emitting-diode displays.* If we are using a low-impedance driver, such as the bipolar outputs of the 4511, a group of seven external current-limiting resistors will set the segment current for us. These typically range from 100 ohms to 1K, and vary with the supply voltage and desired brightness. If we are driving LEDs directly from conventional A- or B-Series CMOS output stages, we sometimes get an inherent current-limiting action inside the IC. This current may be too low for enough brightness, particularly with low supply voltages

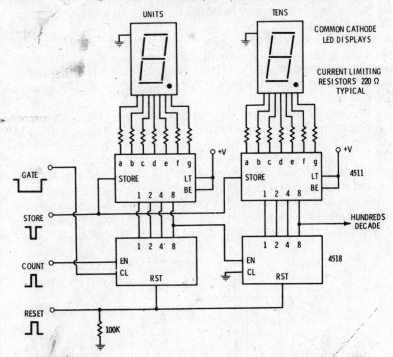

Fig. 6-6. Two-decade counter, latch, decoder, and readout system using a 4518.

and older, inefficient LED displays. At very high supply voltages, the dissipation limits on the IC must be watched. Note that the 4511 has bipolar outputs, and current limiting via resistors *must* be provided for this IC.

The second all-important detail is to watch what is connected to the common terminal of the readout. An LED lights only if the anode is more positive than the cathode. Common-cathode LED readouts must have their common terminal connected to ground. The segments are driven from a current-limited, positive = *on* source.

Common-anode readouts are exactly the opposite of this. Their common terminal goes to a positive supply, while the segments are driven by current-limited, ground = *on* sources.

Put an LED readout in the circuit with reversed polarity and it won't light. Mismatch it to the decoder/driver, and you will get a backwards display with the *out* segments generating the character. Forget to limit the current, and you will destroy the LED.

A one-package-per-decade approach to counters and displays appears in Fig. 6-7. Here we use a 4026 or a 4033 to directly drive the readout. Internal current limiting is provided by the CMOS output stage. No separate storage is available, so the display continuously follows the counter.

Either of these last two circuits may be gated at the enable input. Grounding the enable pin counts input pulses. A positive enable ignores count commands. To reset either circuit, bring reset high, being sure that it goes low before the next clock pulse to be counted arrives.

The decimal point (DP) in an LED display is simply one more light-emitting diode. The DP is lit in a common-cathode readout by applying a current-limited positive voltage to it. You can light the DP in a common-anode readout by connecting a resistor from it to ground. Usually the DP current is only a fraction of the segment current for apparently equal brightness.

We can add decades to either circuit. We can also put divide-by-six or divide-by-twelve counters in between for clock use. But, once you get past three digits, it is often better to use one of the *multi-*

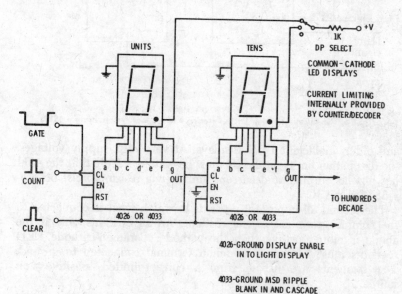

Fig. 6-7. Single-package-per-decade counting system.

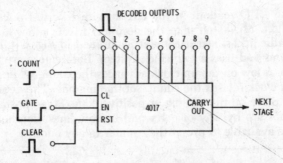

Fig. 6-8. The 4017 acts as a synchronous 10-point electronic stepper, self-decoding to 1 of 10 outputs.

plexed counter systems of Chapter 8. A multiplexed counter shares *one* decoder/driver among all the digits. Power is sequentially and rapidly applied one digit at a time at a fast *scan* rate. While the final appearance is the same, multiplexed displays are usually simpler and have fewer display-to-circuit interconnections.

Today, most of the routine counting applications (watches, stopwatches, timers, clocks, digital voltmeters, frequency counters, etc.) have LSI chips that are ready to use, giving you a one-package solution to a count-and-display problem. We will look at some of these chips in the final chapter.

An Electronic Stepper

Fig. 6-8 shows an electronic stepper using a 4017. An input clock is counted, decoded, and sequentially routed to one of ten output lines. Only the selected output goes high. The rest remain low. The internal circuitry is a self-clearing walking ring that is glitch-free and minimizes overlap between outputs. The 4022 is a similar eight-step unit. Either package can be shortened by routing one more than the desired length to the reset input.

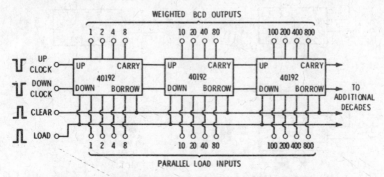

Fig. 6-9. Cascadable, presettable, synchronous up-down counting system.

Add-Subtract Counter

A versatile add-subtract counter is outlined in Fig. 6-9. It is fully synchronous and uses a two-clock system. Both clocks are normally held high. A low on the up clock advances the count, while a low on the down clock retards the count, subtracting one. The actual clocking takes place on the trailing or positive edge. Stage-to-stage interconnections are by way of carry and borrow lines. Parallel loading inputs are available to preset the counter to any number.

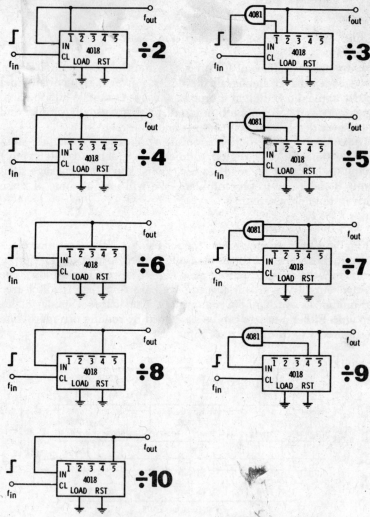

Fig. 6-10. Divide-by-2 through -10, using a synchronous 4018.

Some Low-Modulo Counters

We can divide by two through ten with the circuits shown in Fig. 6-10. Even counts use a single 4018, while the odd counts need an extra AND gate or a pair of Mickey-Mouse-logic diodes. The outputs are all the walking-ring sequences we will be looking at shortly. This type of coding is excellent for even-count symmetrical outputs, digital sine-wave generation, and anywhere else that you may need groups of phase-shifted outputs.

Decade Cascadable Counters

Figs. 6-11 and 6-12 both show decade-cascadable counting systems. The system in Fig. 6-11 uses a 4522 for each decade. Note that the decoded zero output of the tens stage is connected to the CF, or carry forward, input of the units stage. Only when both counters agree that they are in a zero state is a zero output provided. The zero output loads a new number into the counters via the parallel inputs.

Assume that we are in state zero. The zero output reaches around and loads the 2 and 6 from the thumbwheel switches into the counters. The circuit starts down-counting to zero from that point. Note that the units counter counts by *six* the first time around and

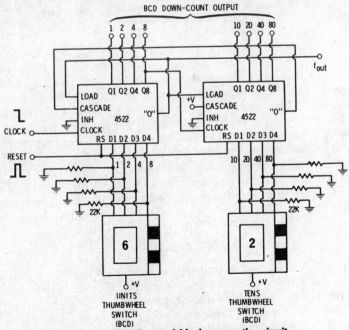

Fig. 6-11. Decade-cascadable down-counting circuit.

365

by ten the second two trips around. Only when both counters reach zero is a new count reloaded and the sequence repeated. Note also that we have shown the units thumbwheel on the *left*, working with the least-significant decade. Physically, of course, the thumbwheel goes to the right of the tens thumbwheel.

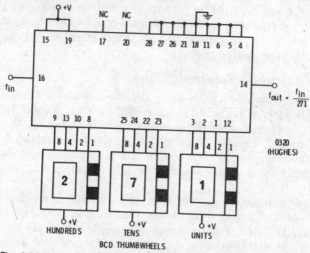

Fig. 6-12. Single IC decade-cascadable ÷ 3 through ÷ 999 is useful for frequency synthesizers.

The single-chip circuit of Fig. 6-12 has internal decade cascadability. Here we have three decades. Again, we work on a load-and-countdown basis with the zero-count state fed forward to reload the counter. This circuit offers only frequency or pulse scaling since there are no intermediate outputs. We can also add a binary-count *offset* into the count sequence to change the channel spacing. This is useful in alternating the channel spacing of a frequency synthesizer or in limiting the total number of channels.

Decade cascadable counters are useful anywhere you want a thumbwheel-switch entry of units, tens, hundreds, and so on. Important examples are predetermining process counters and frequency synthesizers.

SHIFT REGISTERS

We usually classify shift registers by their *length* and their *organization*. The length of a shift register is simply the number of serially connected stages available for data storage.

As shown in Fig. 6-13, there are four basic shift-register organizations. The simplest and cheapest is the serial-in/serial-out (SISO)

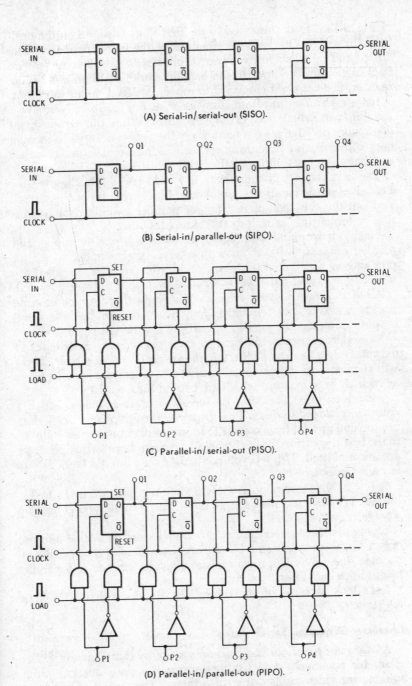

(A) Serial-in/serial-out (SISO).

(B) Serial-in/parallel-out (SIPO).

(C) Parallel-in/serial-out (PISO).

(D) Parallel-in/parallel-out (PIPO).

Fig. 6-13. Shift-register configurations.

367

of Fig. 6-13A. Data presented to the first stage is stored on the positive clock edge. On the same clock pulse, the contents of stage one are passed on to stage two, and so on, out to the final stage. The final-stage data is output to the outside world. With an eight-stage register, it takes eight clockings to move data from input to output.

If we take an output from every stage as in Fig. 6-13B, we have a serial-in/parallel-out (SIPO) shift register. While this arrangement takes more package pins, we now have the option of serial-to-parallel data conversion or *formatting*, as well as the simple storage, delay, and buffering capabilities of the SISO register.

The next step up in complexity is to add parallel loading inputs. This gives us the parallel-in/serial-out (PISO) register of Fig. 6-13C. You can use this type of register for parallel-to-serial data conversion or formatting, as well as the basic SISO uses.

Finally, if we build a shift register with both parallel inputs and parallel outputs, we end up with a very large number of circuit pins and we are limited to shorter register lengths. This is called the parallel-in/parallel-out (PIPO), or *universal*, shift register.

All of the shift registers have a serial input that lets you cascade packages end-on-end for longer lengths. Note also that the serial output doubles as the parallel-output pin for the last stage.

Most shift registers shift data to the right only, or from input to output. Only a few *right-shift/left-shift* registers are available. These will shift data in either direction. They take internal spdt switches at each stage to look ahead or look behind for new data.

Some long SISO registers also have spdt data switches, called a *recirculate*, added to them. This gives you the option of passing data around and around from output to input, or of entering new data into the first stage from the outside world. Often, both switch positions are uncommitted. This lets you recirculate data around two or more cascaded packages.

A few of the more popular CMOS shift registers are listed in Table 6-2. We have included simple 2-, 4-, and 6-stage type-D flip-flop *storage registers* as well.

Storage registers store an entire parallel word at a time and are only 1-bit deep. While an 8-bit shift register handles 8 serial bits on a single line, an 8-bit storage register provides storage of 8 parallel bits at once on 8 lines.

Let's look at some of the things we can do with CMOS shift registers.

Character-Generator Serial Video

A *character generator* is a read-only memory that converts ASCII character commands and timing information into dot groupings suitable for video display or strip printers. The output of a typical

Table 6-2. Some Typical CMOS Shift Registers

Part	Organization	Length	Shift	Clear	Features
4006	Variable	18 Maximum	Right	No	(4), (4), (4 + 1), (4 + 1) Stages
4013	Dual D	2 × 1	Right	Yes	Direct Set & Clear
4014	PISO	6, 7, or 8	Right	No	Clocked Load
4015	SIPO	2 × 4	Right	Yes	Cascadable for 8-Stage SIPO
4021	PISO	6, 7, or 8	Right	Yes	Immediate Load
4031	SISO	64	Right	No	Expandable Recirculate
4034	Universal	2 × 8	Right	No	Bus Transfer Right/Left
4035	PIPO	4	Right	Yes	J, $\bar{K}$ Inputs
4076	Quad D	4 × 1	Right	Yes	Tri-State Input & Outputs
40174	Hex D	6 × 1	Right	Yes	Six Common Clocked D Flip-Flops
40175	Quad D	4 × 1	Right	Yes	Four Common Clocked D Flip-Flops
74C164	SIPO or SISO	8	Right	Yes	No Parallel Loading
74C165	SISO or SIPO	8	Right	Yes	Parallel Loading

character generator is a parallel word. This word is valid and stable only after a processing or *access-time* delay. The problem is to convert this sometimes-there parallel information into continuous serial video.

A PISO shift register is ideal for this conversion process. One possible circuit appears in Fig. 6-14. The register is continuously clocked at the video-dot rate, typically 6 megahertz or so. The shift register is loaded by a brief load command. This load command is always timed so that the character generator is putting out valid data at the instant of loading. Usually, a one-character delay is provided to be absolutely sure of valid data. The load command also synchronizes the clocking astable to lock the loading to the video-dot positions.

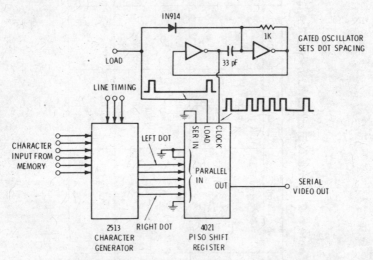

Fig. 6-14. PISO register used to convert dot output of character generator into serial video.

Grounds on the unused shift-register inputs give us blanks during the loading time and after the character is output. This gives us "undots," or the spaces between characters. The zeros fed to the serial input give us between-character line and retrace blanking.

This particular circuit approaches the maximum limit of what you can do with CMOS in fast circuits. It takes selected ICs and a higher supply voltage.

A Teletype® Transmitter

With the circuit of Fig. 6-15, you can convert the parallel output of a keyboard or a microprocessor into an ASCII-coded signal that a teletypewriter can use. Once again, we use a PISO shift register to format parallel data into a serial string of bits, but this time at a much slower speed.

Most computer teletypewriters (tty's) are of the ASR-33 variety. They respond to a serial 100-word-per-minute, ASCII-based code, as shown in Fig. 6-15A. This is an example of an *asynchronous* serial code because any amount of time can pass between characters.

There are 11 bits to the code. Each bit lasts 9.09 milliseconds for a total of 100 milliseconds per character sent. The between-character times are always a *mark* or a one, and can be any time from zero upwards. Our first bit is called a *start* bit and is always a zero. This is followed by the 7 ASCII-code bits, the least-significant bit first. A ninth bit is either a parity bit or an optional 1, 0, or flag. The two final bits are called *stop bits* and are 1's, identical to the between-character marking ones.

Our circuit needs a 110-hertz clock, perhaps from the baud-rate generator of Fig. 4-26. The frequency accuracy must be better than one percent. Preferably the frequency source should be adjustment-free.

A *send* command is derived from the key pressed on the keyboard and consists of a narrow positive pulse. This loads our PISO shift register. The register is 10 bits long. A dual type-D flip-flop in front preloads a 1 and a 0 ahead of the ASCII code and parity bit, which are fed to the parallel inputs on the 4021.

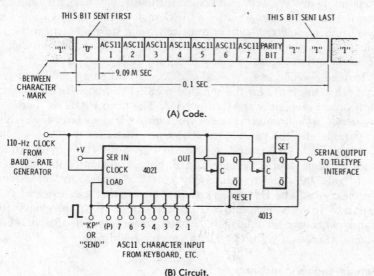

(A) Code.

(B) Circuit.

Fig. 6-15. 100-word-per-minute, parallel-to-serial teletypewriter interface uses a PISO shift register.

The load command should be delayed until the input data is valid. At that time, the rightmost stage immediately jumps to the "1" state. Since it was already in the "1" state due to previously clocked 1's that got there from the serial input, nothing happens to the output. This stage acts as a synchronizer that absorbs the time difference between the last clock pulse and the random arrival of a send or load command.

The clocking continues after loading. We first get the "short" 1 tacked onto a string of 1's. This is followed by a 0 start bit, which in turn is followed by the 8 code bits, and finally at least two extra 1's from the serial input. Any new load command must be delayed for at least 100 milliseconds, by way of either lockout or "handshaking" circuits. A conventional transistor or, preferably, an optoisolator is used to interface the serial code to the 20-mA current loop of the teletypewriter.

If a 74C165 is used in place of the 4021 and a 74C74 in place of the 4013 in the circuit of Fig. 6-15, then the "send" command must be a narrow *negative* pulse to both load the PISO shift register and preset the D-type flip-flops. In addition, the 74C165's "serial" input is grounded.

A CMOS UART

It would also be nice to be able to receive signals from our serially responding teletypewriter and convert these signals back into parallel form. This can be done with an SIPO register and some gating and synchronization circuitry that tests for a valid start bit, accepts serial bits, and then provides a parallel output. If we try this with shift registers and gates, quite a few parts are involved.

Instead, we can use a special CMOS chip called a UART and get by with a single IC for the serial-to-parallel-and-back-again conversion process which is so often needed. The term UART is short for *universal asynchronous receiver transmitter*. It does both the serial-to-parallel and the parallel-to-serial conversions for us.

A typical UART circuit appears in Fig. 6-16. This circuit works with a single supply voltage and internally generates its own 110 baud rate by starting with a low-cost 3.58-megahertz color television crystal. The IM6402 is a similar circuit that uses external and separate 16× receiver and transmitter clocks. Separate clocks are handy when you are interfacing cassette recorders and, also, in other places where you want to track a varying baud rate.

Pseudorandom Sequences

We can build counters out of shift registers by suitable feedback from various stages to the input. Two of the more interesting shift-

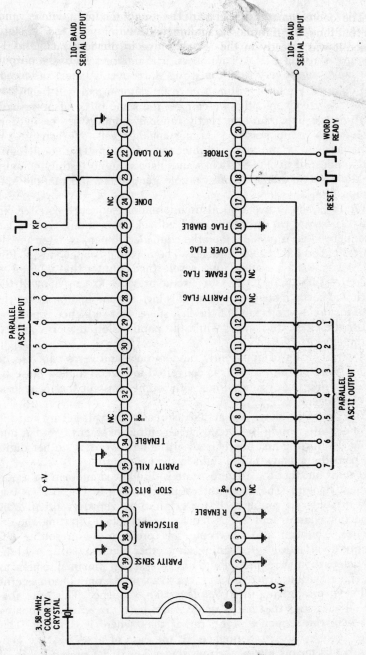

Fig. 6-16. Intersil IM6403 single-chip UART does two-way code conversion; includes internal baud-rate generator. Other formats and rates are easily programmed.

register counter configurations are the *pseudorandom sequence generator* and the *walking-ring counter* that follow.

A pseudorandom sequence is simply noise that repeats. The sequence is a long string of numbers. The numbers always repeat in the same order every time the string starts over. A short sample of the number sequence will *appear* to be random and will behave as noise. Very conveniently, we can get the same piece of noise back anytime we like. The apparently random numbers are easily converted into pseudonoise or varying analog levels. We can use the sequences in cryptography, protection of computer data, secure-communications channels, locks and alarms, audio testing, communication-system analysis, musical-note generation, and random-tone compositions.

Fig. 6-17 shows two typical pseudorandom-sequence counter circuits. If we are careful about our feedback connections, we can get a sequence that is *one less than* the equivalent binary counter length. This is called a *maximal* sequence. The maximal connections of Fig. 6-17A gives us a sequence for a seven-stage counter that is $2^7 - 1 = 128 - 1 = 127$ counts long. Our circuit provides EXCLUSIVE-OR of the sixth and seventh stages and uses this logical combination to set the data for the next clocking of the first stage. Our sequence repeats at 1/127 the clock frequency, while the parallel-output words appear at the clock frequency.

A serial string of apparently random ones and zeros can also be taken from the output. This is converted into good analog noise by low-pass-filtering or integrating with a time constant that is at least 20 clock cycles long.

You can write down the actual sequence you will get by starting in some state, noting the EXCLUSIVE-OR output of stages 6 and 7, and using this for the first one or zero of the next word. All other digits are passed one stage to the right in the new word.

It turns out that the all-zeros state is disallowed and can be a permanent hangup. The capacitor coupling shown in Fig. 6-17 is one way around this problem. If we get into all zeros, eventually the capacitor charges and a 1 is fed to the first stage, starting the sequence. You can also overcome this problem by presetting the counter to all ones or preloading some other nonzero word.

Each particular register length has at least four maximal sequences associated with it. If we use an EXCLUSIVE-NOR, the ones become zeros and vice versa, giving us a *complementary* sequence. This time, it is the all-ones state that hangs up. We also have the option of looking at where the sequence went, rather than where it is going. This gives us a backwards sequence. In the case of a seven-stage generator, this means feedback from stages 1 and 7. Combine both of these techniques and you will get a backwards and complementary sequence.

Picking the maximal feedback arrangement becomes tricky for longer sequences if we are after the $2^n - 1$ maximal length. Actually, almost any connection of EXCLUSIVE-OR gating will give you a sequence of some length, but practically all of the nonmaximal selec-

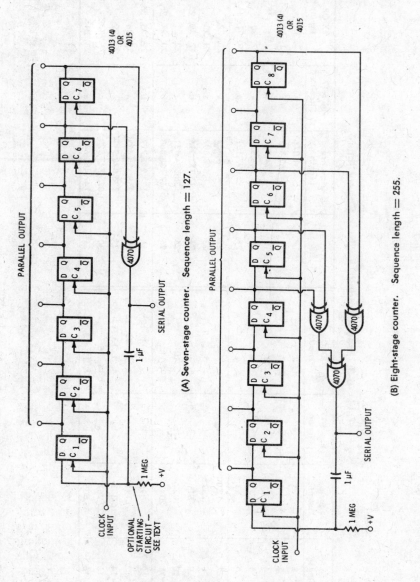

(A) Seven-stage counter. Sequence length = 127.

(B) Eight-stage counter. Sequence length = 255.

Fig. 6-17. Two pseudo-random sequence counters using SIPO shift registers.

375

Table 6-3. Pseudorandom Register Connections

Stages	Sequence Length	Fig. 6-17 Circuit	EX OR Inputs	Stages	Sequence Length	Fig. 6-17 Circuit	EX OR Inputs
2	3	a	1, 2	17	131,071	a	14, 17
3	7	a	2, 3	18	262,143	a	11, 18
4	15	a	3, 4	19	524,287	b	14, 17, 18, 19
5	31	a	3, 5	20	1,048,575	a	17, 20
6	63	a	5, 6	21	2,097,151	a	19, 21
7	127	a	6, 7	22	4,194,303	a	21, 22
8	255	b	4, 5, 6, 8	23	8,388,607	a	18, 23
9	511	a	5, 9	24	16,777,215	b	17, 22, 23, 24
10	1023	a	7, 10	25	33,554,431	a	22, 25
11	2047	a	9, 11	26	67,108,863	b	20, 24, 25, 26
12	4095	b	6, 8, 11, 12	27	134,217,727	b	22, 25, 26, 27
13	8191	b	9, 11, 12, 13	28	268,435,455	a	25, 28
14	16,383	b	4, 8, 13, 14	29	536,870,911	a	27, 29
15	32,767	a	14, 15	30	1,073,741,823	b	7, 28, 29, 30
16	65,535	b	4, 13, 15, 16	31	2,147,483,647	a	28, 31

tions will be short and uninteresting. One EXCLUSIVE-OR gate isn't enough for a maximal length eight-stage sequence. Instead, we have to use three gates as shown in Fig. 6-17B. Feedback comes from stages 4, 5, 6, and 8.

The maximal-length pseudorandom feedback connections for sequence lengths of 3 through 2,147,483,647 are shown in Table 6-3. As an example, a 20-stage register has a maximal sequence length of 1,048,575. It is built with a single EXCLUSIVE-OR gate and a 20-stage SIPO register. You take feedback from the 17th and 20th stages for the normal sequence. For a backwards sequence, use the "mirror" 20th and 3rd stage connections.

If you input a 1.048575-megahertz clock, the 20-stage sequence will repeat once each second. The same clock routed to a 31-stage sequence generator takes over 34 minutes to repeat, running through a million words a second.

Walking-Ring Counters

The *walking-ring sequences* are much shorter than the pseudorandom ones, but they do offer some special features all their own.

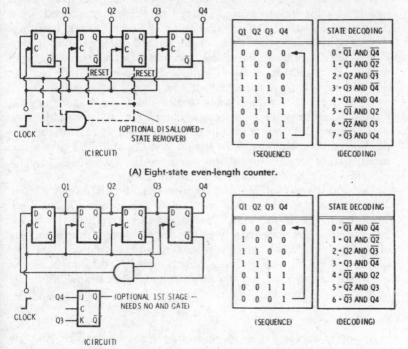

(A) Eight-state even-length counter.

(B) Seven-state odd-length counter.

Fig. 6-18. Two walking-ring counters using shift registers.

These sequences are also sometimes called *Johnson* sequences. Two walking-ring counters are shown in Fig. 6-18. One of these is the *even-length* walking ring (modulo 8, Fig. 6-18A); the other is the *odd-length* connection (modulo 7, Fig. 6-18B).

Our eight-state sequence starts with 0000. The output of the last stage is complemented and passed to the first stage, while everything else goes to the right. So, the next state will be 1000. This is followed by 1100, 1110, and 1111. This time, the end-stage 1 complements to a zero, so the next state is 0111, followed by 0011, 0001, and finally back to 0000. There are eight states to the sequence. In an even-length walking-ring counter, the sequence length equals *twice* the number of stages in use.

Walking-ring counters have several very interesting features. Their outputs are always symmetrical. Only one stage ever changes at a time, and all stages operate at 1/n the clock frequency. We get n phase-shifted outputs in every sequence. Regardless of the length of the counter, we can decode each and every state onto one-of-n lines with nothing but two-input gates. No glitches are produced in the decoding, and everything is fully synchronous.

A disadvantage of the even-length walking-ring sequence is its inefficiency for long counts. (Thirty-two stages are needed for a divide-by-64 counter, compared with six in a binary ripple counter).

A second limitation is that all of the *unused* states gang up to provide us with disallowed sequences. Modulo 6 and higher even-length walking rings must have their disallowed states eliminated. You can do this by resetting or presetting, or you can use extra gating.

There are lots of simple ways to gate-out the wrong sequences, jumping them into the main count. The circuit shown decodes two outside zeros and uses them to force all inside stages to zero. Internal disallowed-state elimination in the 4017, 4018, and 4022 uses a different gating scheme to get the same results. Two 4018s may be cascaded without any disallowed-state problems.

The odd-length walking-ring counter (Fig. 6-18B) counts to *one less than* the equivalent even-length circuit. Here we add feedback to eliminate the all-ones state. You can do this directly with J and K inputs on the first stage, or you can add the AND gate shown and use all D flip-flops. Most shorter odd-length sequences do not have any disallowed states. The output symmetry of the odd lengths is one count shy of 50/50. Thus, the seven-state version has a duty cycle of 4:3.

Electronic Dice

As an example of the simplification that the walking-ring sequences offer in some specialized circuits, look at the electronic-dice circuit shown in Fig. 6-19. Two modulo-6 walking-ring sequences are

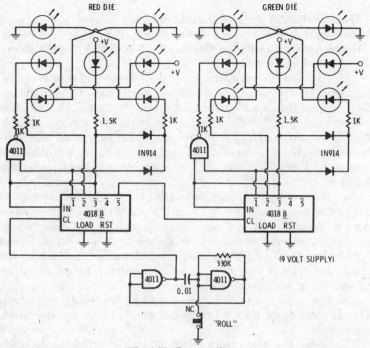

Fig. 6-19. Electronic dice.

used in this circuit. These are built with the self-clearing 4018B circuits. We call our sequence 1-3-5-6-4-2. This gives us a free "4, 5, and
6" and "even or odd" decodings. A 4011 gate decodes "6" for us, and
two diodes decode "not 1."

These decodings directly drive the light-emitting diodes, as shown,
to produce the familiar die patterns. The circuit has true dice odds.
Pressing the ROLL button starts the gated astable which cycles the
first die hundreds of times and the second die dozens of times. This
randomizes the result. The final state is held on release of the ROLL
button. If we like, we can use the rundown circuit of Fig. 4-38 for
more dramatic operation.

Be sure to use the B version of the 4018. Older A-version devices
may not have enough current drive on some outputs for uniform
brightness. A 9-volt power supply may be used.

DIGITAL SINE-WAVE GENERATORS

It would be nice if we could digitally generate sine waves. Actually, this turns out to be very simple with walking-ring circuits. Digital sine waves give precise control of amplitude and frequency.

Thanks to freedom from coils and capacitors, digital sine-wave generators have the ability to rapidly change frequency. All you do is change the input clock frequency, either manually or by digital or microprocessor control.

Actually *any* constant-frequency digital waveform consists of a fundamental sine wave and a bunch of harmonics. You can take any signal, run it through a low-pass filter, and produce a nice sine wave.

The hangup is that unless we are very particular about what our digital waveform looks like, there will be lots of strong, lower-order harmonics. These harmonics will be hard to filter, will distort the output, and will make operation over a wide frequency range rather tricky.

As a start, we will pick a symmetrical waveform. This at least gets rid of the even harmonics, particularly the strong and troublesome second harmonic. Now, if we are really careful, we can get rid of *all* the lower-order harmonics. In fact, we can get rid of as many harmonics as we like. The harmonics we choose to leave can be made small and high in frequency. They will be easy to get rid of with simple filters, and sometimes they can even be ignored.

The key to all this magic is the even-length walking-ring counter. One or more 4018s form a simple and nearly ideal digital sine-wave generator.

The basic idea is shown in Fig. 6-20. In Fig. 6-20A, we have built a modulo-ten, five-stage walking-ring counter. We get five phase-shifted outputs as shown. Next, we sum *four of the five* outputs with carefully selected oddball resistor values. This gives us the blocky stairstep waveform shown in Fig. 6-20B. Note that we do not use output E. This makes the dwell time longer at the peaks and valleys, and gives us a good fit to the sine-wave crests and troughs.

Our summed waveform may look pretty bad, but it is a simple matter to smooth it out with filters. It turns out that the first two harmonics present are the *ninth* and the *eleventh*. The next harmonic pair is the 19th and 21st, followed by the 29th and 31st, and so on. These are all small harmonics. The number of the harmonic also turns out to be the attenuation. So, the amplitude of the ninth harmonic is 1/9 the fundamental, the amplitude of the eleventh harmonic is 1/11 the fundamental, and so on.

The more stages we use, the smaller the harmonics and the higher their number. The harmonics are always one more than and one less than multiples of the sequence length. Digital sine-wave generators of three, four, and five stages are shown in Fig. 6-21, along with the appropriate resistor values. These generators run with clocks of six, eight, and ten times the output frequency. As with our earlier d/a converters, the output can be any loading from a short to an open and can return to any voltage within the power-supply rails. We have shown our output resistors as ratios (in parentheses) and as actual values. While one-percent resistor values are shown, ordinary five-percent resistors will work just about anywhere we want to use these digital sine-wave generators.

The resistor values and the leftover harmonics for any length are shown in Table 6-4. Once again, we have shown these values both as ratios and as one-percent resistor values.

Our output filtering can often be nothing but a capacitor. For wide-range operation, you can build a tracking filter, or several active low-pass sections with a sharp cutoff can be used. In digital

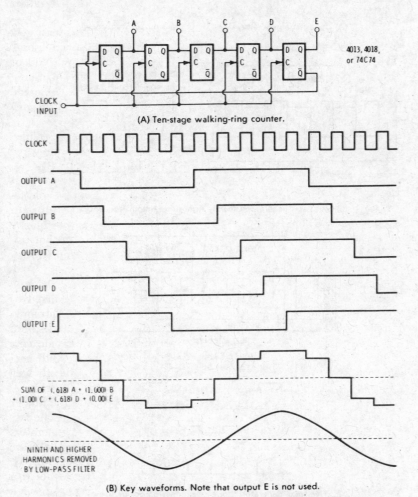

(A) Ten-stage walking-ring counter.

(B) Key waveforms. Note that output E is not used.

Fig. 6-20. Converting a walking-ring counter into a digital sine-wave generator.

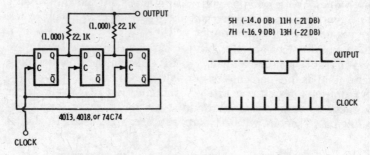

(A) Three-stage generator. Clock is six times output frequency.

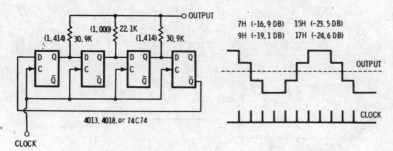

(B) Four-stage generator. Clock is eight times output frequency.

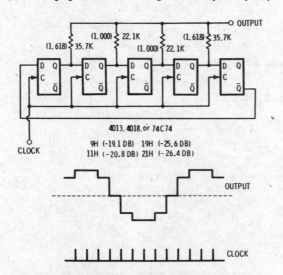

(C) Five-stage generator. Clock is ten times output frequency.

Fig. 6-21. Digital sine-wave generators.

Table 6-4. Design Information for Longer Sine-Wave Generators

Stages	Clock	Resistors	Harmonics	
6	12×	22.1K 38.3K 44.2K 38.3K 22.1K (1.000) (1.732) (2.000) (1.732) (1.000)	11H (−21 dB) 13H (−22 dB)	23H (−27 dB) 25H (−29 dB)
7	14×	22.1K 40.2K 49.9K 49.9K 40.2K 22.1K (1.000) (1.802) (2.247) (2.247) (1.802) (1.000)	13H (−22 dB) 15H (−24 dB)	27H (−29 dB) 29H (−29 dB)
8	16×	22.1K 41.2K 53.6K 57.6K 53.6K 41.2K 22.1K (1.000) (1.848) (2.414) (2.613) (2.414) (1.848) (1.000)	15H (−24 dB) 17H (−25 dB)	31H (−30 dB) 33H (−30 dB)
9	18×	22.1K 41.2K 56.2K 63.4K 63.4K 56.2K 41.2K 22.1K (1.000) (1.879) (2.532) (2.879) (2.879) (2.532) (1.879) (1.000)	17H (−25 dB) 19H (−26 dB)	35H (−31 dB) 37H (−31 dB)
10	20×	22.1K 42.2K 57.6K 68.1K 71.5K 68.1K 57.6K 42.2K (1.000) (1.902) (2.618) (3.078) (3.236) (3.078) (2.618) (1.902) 22.1K (1.000)	19H (−26 dB) 21H (−26 dB)	39H (−32 dB) 41H (−32 dB)
16	32×	22.1K 43.2K 63.4K 80.5K 95.3K 105K 110K 113K (1.000) (1.962) (2.848) (3.625) (4.262) (4.736) (5.027) (5.126) 110K 105K 95.3K 80.5K 63.4K 43.2K 22.1K (5.027) (4.736) (4.262) (3.625) (2.848) (1.962) (1.000)	31H (−30 dB) 33H (−30 dB)	63H (−36 dB) 65H (−36 dB)
n	2n×	$1; \dfrac{\sin\frac{2\pi}{n}}{\sin\frac{\pi}{n}}; \dfrac{\sin\frac{3\pi}{n}}{\sin\frac{\pi}{n}}; \ldots\ldots; \dfrac{\sin\frac{(n-1)\pi}{n}}{\sin\frac{\pi}{n}}; 1$	$(2n-1)H$ $(2n+1)H$	$(4n-1)H$ $(4n+1)H$

Resistor values in ( ) are exact ratios. 'K' values are rounded to stock 1% values.

cassette systems, modems, and other places where we are generating signals that are sensitive to group-delay distortion, the maximally flat-delay or Bessel active low-pass filter is often a good choice. More design details appear in the *Active-Filter Cookbook* published by Howard W. Sams & Company.

The digital sine-wave generator for a digital cassette recorder (Fig. 6-22) is a typical application. This circuit generates output sine waves of 1200 or 2400 hertz. The 4018 and the resistors generate the sequence, while the transistor and RC values give us a third-order Bessel filter. The output is a low-distortion sine wave that switches frequency with a minimum of group-delay distortion.

Other places where digital sine-wave generators are handy are in motor controls, modems, frequency synthesizers, electronic music, power system inverters, and computer-generated signal sources.

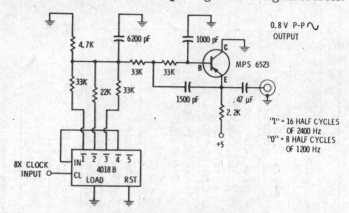

Fig. 6-22. Digital sine-wave generator and third-order Bessel active filter for 300-baud digital cassette recording.

Walking-ring counters are essential for longer digital sine-wave sequences, but we can use a binary approach for length eight, using nothing but a three-stage binary counter and a pair of EXCLUSIVE-OR gates as shown in Fig. 6-23. This circuit even has the advantage of needing only two resistors instead of three. Note that this circuit can be at the end of a long binary used to scale from a much higher frequency clock. Rather severe glitches may result when using ripple counters, particularly on states 4 and 8.

Unfortunately, there doesn't seem to be any ultrasimple binary approach to longer sequences without needing fancy gating or filtering. Go through the math, and you end up with too few equations in too many unknown resistor values.

Another interesting possibility with 8- or 16-state sine-wave generators is to use two resistor summing networks to produce *quadra-*

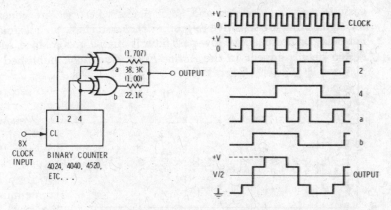

Fig. 6-23. Eight-step digital sine-wave generator using a divide-by-8 counter and complementary gating.

ture sine and cosine channels. This is useful in modulators, filters, and anywhere else that a pair of sine waves shifted 90° is needed.

For more information on digital sine waves, check out "Digital Generation of Low-Frequency Sine Waves" by A.C. Davies, *IEEE Transactions* IM-18, No. 2, June 1969.

SOME FINE PRINT

So far, we have assumed that there are no timing restrictions on our clocked logic. This isn't quite true. While CMOS is far more forgiving and far less restrictive on when and how you clock it, there are some definite limits you have to watch for. We have summed up these limits in Tables 6-5 and 6-6.

The *setup time* for a clocked-logic block is how long the input information must be stable before the clocking edge.

The *hold time* is how long the input information must be stable *after* clocking. Almost all CMOS clocked logic has a zero hold time, and data can be changed immediately after clocking. Three important exceptions are the 4076, 4174, and 4175 clocked storage registers. These have minimum hold times of 30 nanoseconds at 5 volts and 20 nanoseconds at 10 volts.

The *minimum reset time* is the narrowest pulse that will reliably reset the circuit. Note that long binary counters have long reset times to handle the ripple delays.

The *minimum clock-high time* is the narrowest allowable clocking pulse.

The *propagation delay* is the length of time after clocking that an output change can appear. This figure is also sensitive to load capacitance.

Table 6-5. Some Clocked-Logic "Fine Print"—5-V Supply

Type	Function	Minimum Setup Time Nanoseconds	Minimum Reset Time Nanoseconds	Minimum Clock High Time Nanoseconds	Propagation Delay Nanoseconds	Maximum Clock Frequency Megahertz
4013	Dual Type-D Flip-Flop	20	90	70	150	3.5
4014	8-Bit Shift Register, PISO	100	150*	75	200	3.0
4015	2 × 4-Bit Shift Register, SIPO	100	100	85	255	2.0
4018	Decade Divide-by-n	175	200*	200	350	2.5
4020	14-Stage Binary Counter	—	320	140	2800‡	3.5
4021	8-Bit Shift Register, PISO	150	150*	75	200	3.0
4024	7-Stage Binary Counter	—	375	200	1000‡	2.5
4026	÷10 Counter & 7-Segment Decoder	175**	200	200	350	2.5
4027	Dual JK Flip-Flop	70	125	165	150	3.0
4033	÷10 Counter & 7-Segment Decoder	175**	375	200	400	2.5
4034	8-Bit Universal Shift Register	35	240*	170	470	1.5
4040	12-Stage Binary Counter	—	320	140	2500‡	1.5
4060	14-Stage Binary Counter	—	500	200	2800‡	1.75
4076	Quad D Tri-State Latch	25	185	130	265	2.4
4174	Hex D Storage Register	30	35	25	60	9.0
4175	Quad D Storage Register	30	35	25	60	9.0
4192	Up-Down ÷ 10 Counter	175*	175	175	200	3.0
4193	Up-Down ÷ 16 Counter	175*	175	175	200	3.0
4518	Dual Synchronous ÷ 10	—	125	120	250	2.5
4520	Dual Synchronous ÷ 16	—	125	120	250	2.5

Typical values shown vary with supplier and series.
* Parallel Load Time
** Inhibit Clock Time
‡ To Final Output

Table 6-6. Some Clocked-Logic "Fine Print"—10-V Supply

Type	Function	Minimum Setup Time Nanoseconds	Minimum Reset Time Nanoseconds	Minimum Clock High Time Nanoseconds	Propagation Delay Nanoseconds	Maximum Clock Frequency Megahertz
4013	Dual Type-D Flip-Flop	15	40	30	65	8.0
4014	8-Bit Shift Register, PISO	50	75*	150	100	6.0
4015	2 × 4-Bit Shift Register, SIPO	50	50	185	95	6.0
4018	Decade Divide-by-n	75	100*	100	125	5.0
4020	14-Stage Binary Counter	–	120	55	1000‡	9.0
4021	8-Bit Shift Register, PISO	50	75*	150	100	7.0
4024	7-Stage Binary Counter	–	200	60	400‡	8.0
4026	÷10 Counter & 7-Segment Decoder	75**	100	100	125	5.0
4027	Dual JK Flip-Flop	25	50	55	60	9.0
4033	÷10 Counter & 7-Segment Decoder	175**	100	100	125	5.0
4034	8-Bit Universal Bus Shift Register	15	85*	70	175	3.0
4040	12-Stage Binary Counter	–	120	55	900‡	3.5
4060	14-Stage Binary Counter	–	250	75	1000‡	4.0
4076	Quad D Tri-State Latch	10	75	55	105	6.0
4174	Hex D Storage Register	15	20	10	25	16.0
4175	Quad D Storage Register	15	20	10	25	16.0
4192	Up-Down ÷ 10 Counter	70*	70	70	100	7.0
4193	Up-Down ÷ 16 Counter	70*	70	70	100	7.0
4518	Dual Synchronous ÷ 10	–	50	50	100	6.0
4520	Dual Synchronous ÷ 16	–	50	50	100	6.0

Typical values shown vary with supplier and series.

* Parallel Load Time
** Inhibit Clock Time
‡ To Final Output

The *maximum clock frequency* is the fastest allowable speed of operation.

These figures are important only in the fastest of CMOS circuits. If you stay under a megahertz and above one-microsecond pulse widths, you can completely ignore all of them.

All of the numbers in Tables 6-5 and 6-6 represent *typical* averages. They change with manufacturer and series and are NOT guaranteed minimums. Circuit loading may also affect them. If your system timing or frequency comes anywhere near these figures, get out the actual data sheets and double-check for any danger areas.

A CMOS Cameo

THREE-PHASE GENERATOR

Self-correcting circuit delivers three square waves phase-shifted by 120 degrees.

Op Amps, Analog Switches, and Phase-Locked Loops

CMOS can often open up the doors to genuinely new circuit capabilities—things that older logic families can't handle at all, or can only do expensively, complexly, or with lots of supply power. The digital sine-wave generators of the last chapter were good examples of things CMOS does better. Three other new and unique CMOS applications are the *operational amplifier*, the *analog switch*, and the *phase-locked loop*.

An *operational amplifier* is the two-input gate of the analog world. It usually is a high-gain dc amplifier with differential inputs. Most often, it uses feedback to force a desired response. Op amps are used for just about everything analog, including amplification, summing, differentiation, filtering, waveform generation, integration, comparing, latching, signal processing, and a lot of more-specialized things. The new CMOS operational amplifiers offer essentially infinite input impedance and the ability to swing the output over the full supply voltage. These two tricks simply weren't available in existing pre-CMOS, low-cost op amps.

An *analog switch* lets us turn signals on and off at will. CMOS analog switches offer zero-offset switching of bipolar signals at very low cost.

The *phase-locked loop* (PLL) is an extremely versatile way of tracking an input signal. This is done by comparing the phase and frequency of a voltage-controlled oscillator to that of an input frequency, deriving a correction voltage, and always forcing the vco to follow the input frequency. PLLs can filter a signal, demodulate fm, remove noise from a signal, multiply or divide frequencies, and perform many other signal-processing tasks. The CMOS phase-locked loop offers the ability to track over an extremely wide frequency range, gives a choice of phase-detection schemes, uses a single supply voltage, has direct CMOS logic compatibility, and needs very low supply current.

CMOS OPERATIONAL AMPLIFIERS

The 3130 is a CMOS operational amplifier made by RCA that costs less than a dollar. It is a very good choice for uses with lower supply voltages whenever you want the output to swing from rail to rail of the power supply. A simplified schematic of the 3130 is shown in Fig. 7-1.

The 3130 has two inputs, an *inverting* (−) input and a *noninverting* (+) input. These inputs go to a pair of p-channel field-effect MOS transistors set up as a differential amplifier.

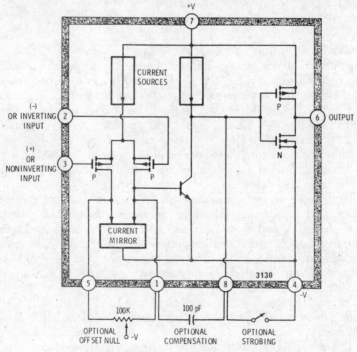

Fig. 7-1. Simplified schematic of a 3130 CMOS op amp.

Any difference in voltage between these two inputs proportionally splits the current from the upper current source, routing a fraction of the current through each transistor. With identical input voltages, half of the source current flows through each transistor. Any current unbalance between the two transistors is sensed by a special circuit called a *current mirror*. The sensed current unbalance is amplified with a gain of five and then routed to a second amplifier stage which is a bipolar transistor.

What we have done so far is to measure the input voltage difference and convert it into a proportional, single-ended current that is increased by a factor of five. Since the inputs go to the gates of the p-channel MOS transistors, we have an extremely high input impedance, and input loading is negligible for most uses.

The second amplifying stage is a single npn bipolar transistor. The gain of this stage is 6000. This very high gain results from the current-source load of the transistor, which behaves as a very high impedance and thus produces very high voltage gain.

The second stage in turn drives a push-pull output stage similar to a CMOS inverter. This inverter has a gain of roughly 30 and operates as a Class-A amplifier with a reasonable amount of output drive current.

Any small voltage difference on the (+) and (−) inputs is sensed, strongly amplified by the three gain stages, and output on pin 6. A rising signal on the noninverting, or (+), input drives the output positive, while a rising signal on the inverting, or (−), input drives the output negative.

There are several very interesting features of the 3130. We have already seen that both inputs work into the gates of p-channel MOS transistors, making the input impedance essentially infinite. The *common-mode* of the inputs ranges from slightly below the negative supply rail up to within 2.5 volts or so of the positive supply limit. If we don't heavily load our output, it can swing the full rail-to-rail range of the supply voltage. Maximum current in either direction is a short-circuit-proof 22 milliamperes.

Some Options

We also have some usage options for the basic 3130. These are shown along the bottom of Fig. 7-1.

All operational amplifiers have an inherent *offset voltage* at their inputs that cannot be distinguished from a legitimate input signal. Cheaper 3130s have an offset voltage that can be eight millivolts or more. We can essentially eliminate this offset voltage with the offset-null potentiometer shown in Fig. 7-1. This potentiometer unbalances the current mirror just enough to compensate for the input offset. One way to adjust the potentiometer is to short the two inputs together under in-circuit conditions and then adjust the potentiometer until you get an open-circuit, midrange output voltage.

We can *compensate* our op amp by adding a capacitor between pins 1 and 8. This reduces the high-frequency response of the op amp so that it can't oscillate or become unstable for certain combinations of feedback. As a general rule, the *lower* the circuit gain, the *more* signal fed back (details on this later) and the *more* likely the amplifier will oscillate. The tendency for the circuit to oscillate is due to its increased frequency response at lower gain. As the frequency response increases, there is additional phase shift of the feedback signal due to distributed capacitance of the op-amp circuit. The feedback signal normally has a phase shift of 180°. If the feedback signal is shifted an additional 180°, the circuit will become regenerative and oscillate. The effect of the compensation capacitor is to reduce the phase shift of the feedback signal. Compensation capacitor values from 50 to 200 pF are typical for lower-gain uses.

We also have the option of *strobing* the output to gain digital control over the presence or absence of a signal. Connecting pin 4 to pin 8 forces the output *high* and minimizes the 3130 operating current as well. Strobing can be done with a switch contact, an open-collector or open-drain transistor, a 4007, or a tri-state CMOS output.

Some Limitations

The 3130 sounds like a great device. But, there are several usage restrictions that will have to be observed very carefully if you are to use the 3130 successfully.

Compared with the popular 741 op amp, the 3130 offers about the same open-loop gain (several hundred thousand) and about the same input offset voltage (around eight millivolts). The input impedance at room temperature is a million times higher for the 3130 (2×10^{12} ohms versus 2×10^6), and the input bias and offset currents are proportionately lower. Slew rate, or the ability to handle fast, high-frequency, high-amplitude changes, is around 20 times better in the 3130, being ten volts per microsecond compared with one-half volt per microsecond for the 741. So, on all these counts, we either break even or end up way ahead using a 3130.

Now for the disadvantages. Our CMOS output stage on the 3130 swings very nicely from supply rail to supply rail if you don't load it very heavily. This is particularly handy for CMOS digital-output capability and is something the 741 can't touch. But the output of the 3130 is very sensitive to capacitance loading. It causes all sorts of problems if you try to drive a varying or a high-capacity load. The 3130 works nicely with a supply voltage as low as five volts. But, its upper supply-voltage limit is a *total* of 16 volts—or half that of a 741 with its typical +15-, −15-volt supplies. And the operating current of the 3130 tends to be rather high at the upper supply limits, particularly under heavy load.

The inputs of the 3130 can go as low as the negative supply. This lets you refer your inputs to ground for comparators in single-ended, single-supply applications. But don't let an input get more than two-tenths or three-tenths of a volt negative with respect to the pin-4 voltage. This starts conduction through a protection network and also can reverse the sense of the output, particularly in comparator applications.

The 3130 is also somewhat noisy. If very-low-level input signals are important, consider using a National LF156 or a similar, more-expensive, JFET-input op amp.

A final limit may be high-temperature operation. At room temperatures, the input impedance essentially acts like an open circuit. As the temperature rises, leakage current increases dramatically. At high temperatures, the input impedance drops to several hundred megohms or less. This is enough to load or otherwise introduce errors into very high impedance biasing circuits. For this reason, whenever you need all the input impedance you can possibly get, use a low supply voltage, heat sink the chip, and provide adequate air circulation for it.

The following rules summarize the usage hangups of the 3130:

- Keep the supply voltage under 15 volts total.

- Provide a compensating capacitor of 50 pF minimum for most closed-loop uses.

- Don't operate the inputs at the negative supply limit unless you are absolutely sure they will never go more than 0.2 volt negative with respect to the negative supply voltage, particularly in the presence of noise.

- Use a heat sink with the 3130 if you really need superhigh input impedances.

- Avoid any capacitance loading at the output, particularly any changing capacitive load.

- Use a different device if very low noise is important.

The 3140 is a newer CMOS operational amplifier that eliminates some of these problems—at a price. The 3140 has a bipolar output stage and operates up to the full op-amp voltages of +15 volts and −15 volts, or zero and +30 volts. Frequency compensation is internally provided, so there is no need for an external capacitor. The output stage easily drives capacitive or varying loads, and the prices of the two are about the same. The 3140 is a super 741 that does everything the 741 does without the poor slew rate and low input impedance.

Unfortunately, the 3140 won't swing within two volts of the positive supply rail, so it isn't nearly as suitable when you are trying to work with very small supply voltages (such as a single +5-volt supply) or, more importantly, when you are trying to directly interface CMOS digital circuits. Noise performance of the 3130 and 3140 are about the same.

The choice boils down to this: Use the 3130 anytime you must operate on a small supply voltage or anytime you need the entire rail-to-rail output swing. Use the 3140 as a general-purpose 741 replacement. Use a LF156 or something similar if low noise is very important.

Using the 3130 and 3140

With a very few exceptions, op amps are operated in a *closed-loop* mode where external feedback components are added to get a desired response. The very high amplifier gain inside the loop makes the circuit respond almost exactly to the *transfer function* of the feedback-network resistors and capacitors. Some basic op-amp circuits are shown in Fig. 7-2.

A gain-of-n inverting amplifier appears in Fig. 7-2A. You set the gain of the amplifier by the ratio of the feedback resistor to the input resistor. Use a 100K feedback resistor and a 10K input resistor, and you will get a gain of almost exactly −10. The minus sign means that the output is inverted with respect to the input signal. The input impedance of your circuit will be 10K.

Why only 10K if we are working into a 12-million-megohm input impedance? We get only 10K because the (−) input of the op amp appears as a *virtual ground* or a zero-ohms circuit point. It gets that way because of the feedback. Therefore, the input impedance is equal to the value of the input resistor.

Any voltage difference at all between the (−) and (+) inputs is strongly amplified and fed back to force the inverting input back to the same voltage that is on the (+) input, or ground, in the circuit shown. Since the (−) input always stays at a virtual ground, the gain of the amplifier is set by the stable ratio of two resistors. A 50 percent change in op-amp gain will have only a negligible effect on the −10 circuit gain.

The virtual ground on the (−) input is the key to the summing circuit of Fig. 7-2B. Since the inverting input is always at a virtual ground, there will be no interaction if you add extra inputs. This is useful in an electronic-music mixer or when you want to sum two or more analog signals. The gain of each signal path is independently set by the ratio of the common feedback resistor and the individual input resistors. As before, each separate input impedance equals the input resistor.

This virtual-ground concept exists only on the $(-)$ input and only when feedback is provided. Therefore, the unity-gain voltage follower in Fig. 7-2C does give a very high input impedance. You can think of this circuit as a super emitter follower with zero offset. It gives us a low-impedance, same-size, same-polarity replica of an input signal that we can't afford to load heavily. Since all of the output voltage is fed back to the $(-)$ input, the $(-)$ input will equal

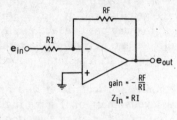

(A) Gain-of-n inverting amplifier.

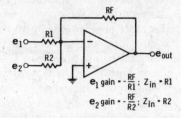

(B) Summing gain of n inverting amplifier.

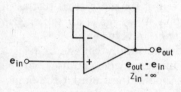

(C) Unity gain voltage follower.

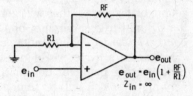

(D) Gain of (1 + n) noninverting amplifier.

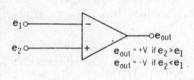

(E) Comparator.

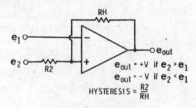

(F) Comparator with hysteresis or snap action.

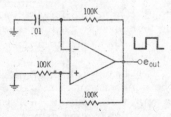

(G) Astable oscillator.

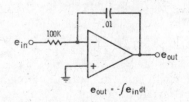

(H) Integrator or low-pass filter.

Fig. 7-2. Some basic op-amp circuits.

the output voltage. The high gain of the op amp will force the $(-)$ input to follow the $(+)$ input.

We can convert this circuit into a higher-gain, high-input-impedance, noninverting amplifier as shown in Fig. 7-2D. The feedback resistors form a voltage divider, and the output gain will be the inverse of their ratio as a voltage divider. Note that gains of less than one aren't available with this circuit.

Sometimes we can run our op amp open loop. If we do this, the output will be at one supply rail or the other practically all of the time. The comparator circuit of Fig. 7-2E is typical. If the $(-)$ input is more positive, the output rests on the negative rail. If the $(+)$ input is more positive, the output goes all the way positive. We can use this circuit to compare two analog voltages and to convert the result to a signal compatible with CMOS digital logic. The input impedance on both inputs is very high.

Comparators may be sensitive to noise, particularly if you are working with slowly changing or low-level signals, or if some high-level output works its way back into the input section of the circuit. As shown in Fig. 7-2E, we can add positive feedback to gain a snap action, or *hysteresis*, that will get us out of the noise area. Just as in the Schmitt-trigger circuits of Chapter 4, we gain an *upper trip point*, a *lower trip point*, and a *dead band* in between.

Normally, a hysteresis of one-tenth the supply voltage or so is a good starting point for design. The input impedance on e_1 will be very high; on e_2 it will be something around the value of R2. Sometimes hysteresis may not be desirable. If you are detecting a waveform at zero crossing and then later want to accurately measure the times between zero crossings, large amounts of hysteresis will make the crossing position amplitude-sensitive and can lead to jitter or other errors.

You will find an astable oscillator or signal source in Fig. 7-2G. The $(+)$ input acts like a comparator with lots of hysteresis, while the RC network on the $(-)$ input forms a triangle-wave generator. If the output jumps high, the RC network starts charging positive through the dead band until the upper trip point is reached. This jumps the output low, and the RC network starts discharging until the $(-)$ input hits the lower trip point, completing the cycle. You will get a square-wave output, along with an optional high-impedance and slightly nonlinear triangle wave that may be picked off at the inverting input.

An integrator or low-pass filter is shown in Fig. 7-2H. The capacitor will store the current through the input resistor. This circuit is extremely important in advanced op-amp work. You can use it as a ramp generator, a true analog integrator, an averaging device, or as a single-pole, low-pass active filter.

All of the circuits shown in Fig. 7-2 assume that you are using dual power supplies of equal value. For single-ended operation, the negative supply rail can be grounded and the inputs returned to some suitable voltage within the limits of the common-mode input. One-third to one-half the positive supply voltage is often a good choice.

Note that while you can run both inputs at the negative supply voltage, there are not many things you can do with them connected this way. This is good for an open-loop comparator provided you are sure you will never get more than a few tenths of a volt negative. For most op-amp uses, you need feedback from output to input, and at times the feedback must be provided with the output *lower*

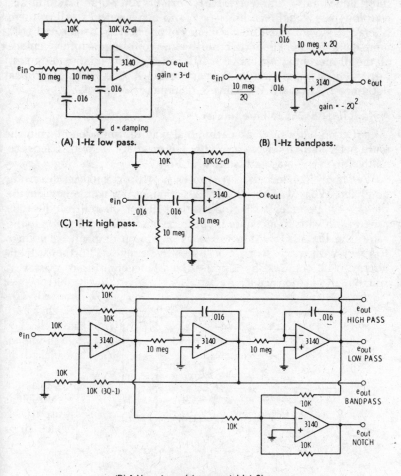

(A) 1-Hz low pass.

(B) 1-Hz bandpass.

(C) 1-Hz high pass.

(D) 1-Hz universal (state-variable) filter.

Fig. 7-3. Active filters using CMOS op amps.

or more negative than the input. There is no obvious way you can do this with an output that can't swing below the input level.

Active Filters

The 3140 is an excellent choice for active filters, particularly very low frequency ones such as those you find in brain-wave research, geophysical prospecting, and other subaudio uses. At these frequencies, you can use extremely high resistor values and still get by with cheap and sensible capacitor values.

As an example, consider the four second-order active filters of Fig. 7-3. We have shown a low-pass filter, a bandpass filter, a high-pass filter, and, finally, a universal *state-variable* filter that has simultaneous low-pass, bandpass, high-pass, and notch outputs. All of these filters operate with a cutoff frequency of one hertz. To scale up to the more common one-kilohertz cutoff values for design, simply replace all the 10-megohm resistors with 10K resistors. Much more design information on these circuits appears in the *Active-Filter Cookbook* published by Howard W. Sams & Company.

A High-Impedance Millivoltmeter

CMOS op amps offer a nearly ideal way to upgrade an ordinary multimeter into a high-sensitivity, 10-megohm, input-impedance millivoltmeter. Details are shown in Fig. 7-4.

We simply connect the op amp as a ×10 or ×100 noninverting amplifier. What was the 3-volt, 20,000-ohms-per-volt range on the meter now becomes a 300-millivolt (×10) or 30-millivolt (×100) range, both with 10-megohm input impedances. Opening the ground return on the 10K resistor will convert the op amp into a voltage follower with unity gain for less-sensitive ranging. Further attenuation can be obtained by tapping the 10-megohm input resistor at suitable attenuation values.

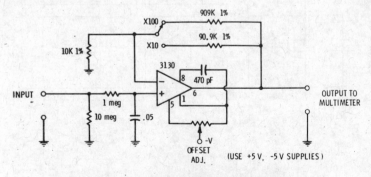

Fig. 7-4. High-impedance millivoltmeter using a 3130 op amp upgrades ordinary multimeter or VOM.

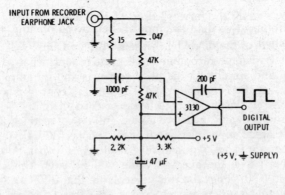

Fig. 7-5. Digital cassette signal-recovery and limiter circuit using a 3130 op amp.

Digital Cassette Signal Recovery

The circuit shown in Fig. 7-5 converts low-level audio recovered from a tape recorder into CMOS-compatible 5-volt square waves. It is an open-loop comparator with both inputs biased to +2 volts. The input-filter RC values minimize hum and bias interference. To prevent amplitude sensitivity and potential jitter, hysteresis is not used in this circuit.

A Precision Rectifier

The precision full-wave rectifier of Fig. 7-6 is used in digital voltmeters and anywhere else you want to convert an ac waveform into its dc full-wave-rectified equivalent, without worrying about diode threshold and offset problems. The first 3130 is used as a polarity separator. Positive-going signals appear across the lower 10K resistor, while negative-going signals appear across the upper 10K resistor. The pin-6 output of the op amp exceeds these positive and negative

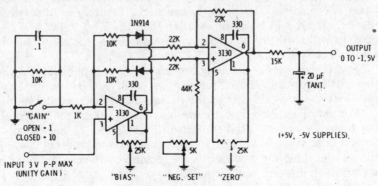

Fig. 7-6. Precision full-wave rectifier.

values by exactly the forward drop of the diode. Looking at things slightly differently, the net effect is to reduce the normal 0.6-volt offset voltage of the diode by the gain of the op amp.

The second stage recombines the positive and negative peaks by inverting and amplifying the positive peaks and by amplifying and not inverting the negative peaks. The NEG SET control is a trimming potentiometer that makes both peaks of equal height.

A low-impedance, negative-going, full-wave rectified replica of the input signal appears on pin 6 of the second 3130. If you are after an average dc value, you can take it off the low-pass filter output. For a sine wave, the average value will be 0.637 times the peak value, and 0.901 times the rms value. Note that different waveforms have different average values, and a true rms (root-mean-square) circuit has to be used for rms values of nonsinusoidal inputs.

A Tracking Photocell Pickoff

Getting a photoconductive cell to work properly in the real world can be a hassle, particularly for outdoor races and other sporting events. The resistance of the photocell may change with temperature, alignment, ambient light, and the cleanliness and length of the optical path. You can get spurious outputs, for instance, when a second pair of wheels on a racecar retrips the beam. And, unless your sensor circuit always slices the light/no-light conditions right down the middle, you can miss an event or lock up the output.

The simple 3140 op-amp circuit of Fig. 7-7 overcomes most of these hassles. It is a comparator circuit with a difference. In fact, it responds only to sudden changes due to interruption of the light, and ignores slow changes and the exact value of the photoconductor's *on* resistance.

In normal operation, the voltage at point v is set to about one half the supply voltage while the cell is receiving normal, unbroken beam illumination. The diode between the inputs guarantees that the (+) input will be more negative than the (−) input, and the op amp will normally be at output ground, since it is connected as a simple comparator.

Now, if the beam is *suddenly* broken, the (−) input drops below the (+) input, and the output swings positive. The output *stays* positive for a delay time set by the recharging of the capacitor on the (+) input. This locks out all spurious signals until the photocell has a chance to reset itself to normal illumination. A time-out of a second or so is a good choice for power boats and racecars. During the lockout time, any new signals are ignored by the circuit. You get a clean, conditioned, output pulse.

If the photocell resistance changes slowly, the (+) input tracks it with varying ambient light, interference, temperature, and so on,

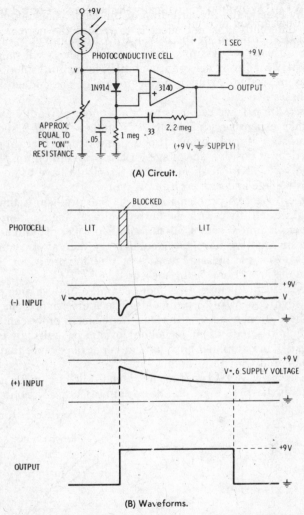

(A) Circuit.

BLOCKED

PHOTOCELL   LIT   LIT

(-) INPUT

(+) INPUT   V=.6 SUPPLY VOLTAGE

OUTPUT

(B) Waveforms.

Fig. 7-7. Photocell amplifier/detector offers snap action, high noise immunity, and automatic tracking of ambient light levels.

without tripping. Voltage v can move over a large range as long as it does so slowly. Any sudden drop in voltage at point v produces an output.

pH Meters

Sometimes you really need input impedance for a sensor that simply cannot be loaded. In the past, this meant an electrometer tube or an expensive modular FET op amp. One classic example is a pH meter. A CMOS op amp dramatically slashes the cost of pH measurement, either analog or digital.

The pH is an index of the chemical hydrogen activity of a solution. You usually express pH on a log scale. Values will range from pH1 to pH2 for strong acids, through pH7 for neutral solutions such as ultrapure water, on up to values of pH11 and higher for very strong bases like lye.

To measure pH, we can use any of a number of readily available, but rather expensive ($25–$100), pH probes. Fig. 7-8 shows the characteristics of a pH probe. It acts like a battery that proportionately generates positive dc voltage for low pH, nothing for pH7, and negative voltages for high pH values. So, all we have to do is measure this voltage and convert it to pH units.

But there are two problems involved. One problem is that pH is temperature sensitive, with the output voltage ranging from 54 millivolts per pH unit at zero°C up to 74 millivolts per pH unit at 100°C. This means that we have to manually vary the gain or conversion constant of our pH measurement to be able to correct for the temperature of the solution being measured.

The second problem is a bit more complex and explains the previously high cost of pH instruments. The source impedance of our pH probe is 15 megohms for the "low-impedance" probes and ranges upwards into hundreds of megohms for special units. In order to measure pH, our amplifier must have an input impedance that is very high compared with 15 megohms.

Once again, CMOS comes to the rescue. The circuit for a digital pH meter is shown in Fig. 7-9. The op amp is set up as a noninverting

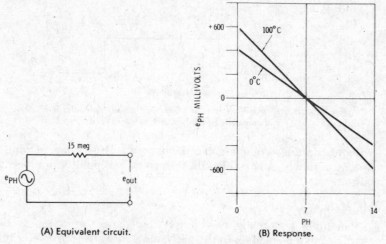

(A) Equivalent circuit.

(B) Response.

Fig. 7-8. Characteristics for a pH probe.

amplifier whose gain can be moderately changed with a potentiometer to compensate for solution temperature. If we only want an analog pH meter, all we have to do is route the output of the 3130 op amp to a center-scale milliammeter through a calibrating potentiometer.

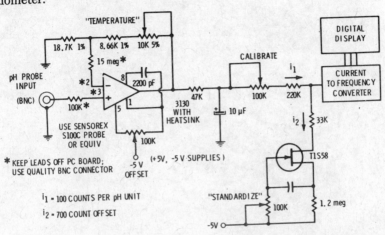

Fig. 7-9. Input circuitry for a digital pH meter.

On the other hand, for a digital display, we want a "7.00" display for zero output voltage. We can get this by summing a stable offset current equal to 700 counts and our pH output converted to a calibrated current, and in turn routing the sum of the two to a current-to-frequency converter followed by a digital display. The circuit of Fig. 4-34A is one possible approach.

Preserving an extremely high input impedance is of first importance for this circuit. This means that you should use very high quality cable and connectors, such as a BNC type, and the input resistors should go *directly* to the op-amp pins without being routed through a pc board or other material that may be slightly conductive. This is also one application where a heat sink is strongly recommended, along with a lower supply voltage and adequate air circulation.

A digital pH meter is shown in Fig. 7-10. The circuit is calibrated by using standard pH *buffer solutions*. These solutions are inexpensively available in liquid or tablet form at chemical supply houses.

CMOS ANALOG SWITCHES

Fig. 7-11 shows how we can connect a 4007 into an *analog switch*, either as a single-pole, single-throw switch (Fig. 7-11A) or as a

Fig. 7-10. Digital pH meter using a 3130 op amp.

single-pole, double-throw arrangement (Fig. 7-11B). To build one switch pole, we connect an n-channel and a p-channel transistor essentially in parallel. To turn the switch on, we make the n-channel gate positive and we ground the p-channel gate; to turn the switch off, we do the opposite. In the *on* state, point A is connected to point X. The applied signal voltage can be anything between the supply limits. For low input voltages, the p-channel transistor does most of the conducting. For high input voltages, the n-channel transistor does most of the work. For median input values, both transistors conduct together, each conducting about half of the total current.

The equivalent circuit turns out to be that of Fig. 7-12—an ordinary resistor connected in series with a switch. In the off state, the input is not connected except for a very small leakage current. This current is typically less than one nanoampere. In the on state, the equivalent circuit is simply a low-value resistor. Our switch can handle digital or analog signals of any value or polarity as long as the voltages are within the supply limits.

We call this type of analog switch a *simple* switch. It has some limitations in that the on resistance is rather high (300 to 1500 ohms typically), and the on resistance depends slightly on the polarity

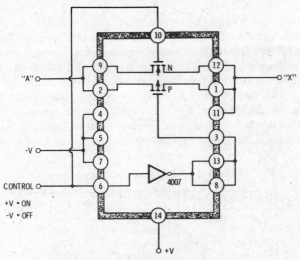

(A) Single-pole, single-throw (spst) switch.

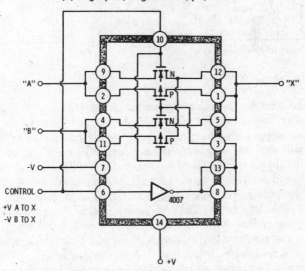

(B) Single-pole, double-throw (spdt) switch.

Fig. 7-11. Analog switches built with a 4007.

Fig. 7-12. Equivalent circuit of an analog switch. A or B may be input or output, and can handle analog or digital signals. All A and B signals and voltages must be between the supply limits. Inputs should not be "force fed" with the supply power off.

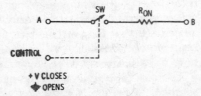

405

and size of the voltage being switched. You can get around these limitations by adding two new transistors to the basic configuration, giving an *improved* analog switch. The improved switch gives a better on resistance (80 to 250 ohms typically) and much less modulation of the on resistance by the input signals. Most of the ready-to-go CMOS analog switches are the improved type. The simple switches remain a good choice where you need absolute minimum leakage values, particularly in the upcoming sample-and-hold examples. Some popular CMOS analog switches are summarized in Table 7-1.

Table 7-1. CMOS Analog Switches

Type	Switching	Circuit	On Resistance	Package
4007	SPST or SPDT	Simple	600Ω	14-Pin
4016	4 SPST	Simple	300Ω	14-Pin
4066	4 SPST	Improved	80Ω	14-Pin
4051	1-of-8*	Improved	120Ω	16-Pin
4052	Dual 1-of-4*	Improved	120Ω	16-Pin
4053	Triple 1-of-2*	Improved	120Ω	16-Pin
4067	1-of-16	Improved	200Ω	24-Pin
4097	Dual 1-of-8	Improved	200Ω	24-Pin

* Translation Internally Available

Some Uses

We can run our analog switches in an *analog* or a *digital mode*. Either way, the switch basically does the same thing. In the digital mode, our supply might typically be ground and +10 volts, and our inputs will also run between ground and +10 volts. In the analog mode, our supplies could be +5 volts and −5 volts, and our switched signals could be bipolar ones referred to ground. Three of the available CMOS analog switches (the 4051, 4052, and 4053) can give us the best of both worlds. They include optional internal *translation* that lets you analog-switch with +5-volt, −5-volt supplies while digitally controlling the switching with 0-volt, +5-volt CMOS signals.

You will find some basic uses for analog switches in Fig. 7-13. Electronic-music preset or stop selection is a good analog switching example, as shown in Fig. 7-13A. Here we pick any of several voices or preset instruments and route them to a combiner. If the combiner is the inverting summing amplifier of Fig. 7-2B, we can sum as many voices together as we like. Another advantage of CMOS analog switches is that the switches can go where the signals are and then be remotely controlled by logic-level "dc" control lines. You don't have to route live signals all over the system using shielded cable.

You can digitally control the gain of an op amp by using analog switches to switch fixed resistor values in place (Fig. 7-13B). More switches can be added, and the switches can work with either input or feedback resistors. Eight analog switches connected to the parallel output bus of a microcomputer can digitally give us 256 different gain values.

In Fig. 7-13C, we have put an analog switch across the capacitor of our earlier op-amp integrator. Now, if we connect the input resistor to a fixed negative voltage, the virtual ground gives us a constant input current. This current also appears inverted at the output of the op amp and linearly charges the capacitor in the positive direction. Note that this gives us a linear ramp and not an exponential one. Closing the switch resets the ramp. When the switch is opened, the ramp starts and continues until the switch is once again closed.

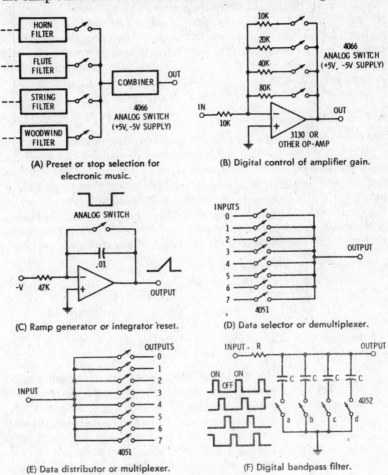

(A) Preset or stop selection for electronic music.

(B) Digital control of amplifier gain.

(C) Ramp generator or integrator reset.

(D) Data selector or demultiplexer.

(E) Data distributor or multiplexer.

(F) Digital bandpass filter.

Fig. 7-13. Analog-switch uses.

407

A 4051 lets us pick one of eight inputs for a common output (Fig. 7-13D) or lets us route one input to eight different outputs (Fig. 7-13E). Thus, the same circuit serves as a data selector or demultiplexer, or as a data distributor and multiplexer. The only difference between the two circuits is what we are calling inputs and outputs, and perhaps whether we are using single-ended or split power supplies.

The tracking digital bandpass filter of Fig. 7-13F is one way to build a digital filter. The capacitors are sequentially connected to ground one at a time so that each capacitor is grounded one-fourth of the time. The switching takes place at four times the intended center frequency so that each capacitor is sampled once during the period of the intended center frequency. The bandwidth is set by the RC values. This is an example of a *comb* filter that also responds to harmonics of the resonance frequency. Changing the switching frequency also changes the resonance frequency. Extra switches and capacitors can be added for better signal-to-noise performance. For instance, a 4051 can be paired with a divide-by-eight counter and driven at eight times the desired center frequency.

A Few Restrictions

As we can see, the no-offset bipolar resistive switching of the CMOS analog switch is a major advance in what we can economically do with electronics. For most circuit applications, you can use analog switches without any major problems cropping up. This is particularly true if your inputs come from the same power supply as the switch does and if you stick with higher values of load resistance.

However, there are some subtle problems with analog switches that you may have to watch for in certain applications:

- *If the inputs still provide positive voltage even with no voltage connected to the analog switch, you can have latching problems and excess current.* You can minimize these problems by making sure the main power supply can drive the load resistance by itself, being sure there is no resistance between the +V pin and the power-supply source, and adding series resistors to the outside-world inputs.

- *For very low values of load resistance, the improved analog switches can give you extra output current that is derived from the positive supply.* This is caused by forward biasing of a substrate in the transistors that are doing the "improving." There are two ways to get around this problem. One way is to keep the load resistance above 3K. The other way is to call the following pins *inputs* when delivering current from an outside source to the output load:

4051: pin 3 as input
4052: pins 3 or 13 as inputs
4053: pins 4, 14, or 15 as inputs
4066: pins 2, 3, 9, or 10 as inputs
4067: pin 1 as input
4097: pins 1 or 17 as inputs

Sample-and-Hold Circuits

A sample-and-hold circuit catches a brief part of an input analog waveform and stores it until the next sample or update command arrives. This is useful in electronic music for remembering the control voltage for a certain key even after the key is released, which in turn lets the decay cycle continue without forgetting which note is decaying. Sample-and-hold circuits are also useful in speech processing and analysis, in the stripping of various multiplexed data channels off a common carrier, in radar signal processing, and in similar applications.

The sample-and-hold circuit of Fig. 7-14 combines a simple analog switch with a CMOS high-input-impedance op amp. Closing the switch causes the capacitor to rapidly charge to its new value. Opening the switch causes the value to be held while the voltage follower gives a low-output-impedance replica of the input. With a good polystyrene or Mylar capacitor, the droop or charge time is very high, particularly when you use a simple analog switch rather than an improved version. The 50-picoampere leakage of the switch and the negligible op-amp input current will give you a droop or drift of about one millivolt per second with the circuit shown. Larger

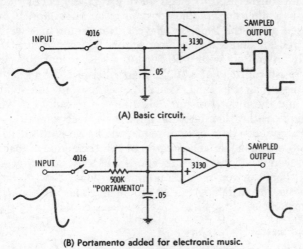

(A) Basic circuit.

(B) Portamento added for electronic music.

Fig. 7-14. Sample-and-hold circuits.

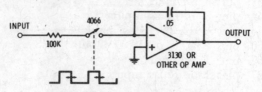

HIGH FREQUENCY, VARIABLE DUTY CYCLE CONTROL SIGNAL.
DUTY CYCLE SETS TIME CONSTANT OF INTEGRATOR.

Fig. 7-15. Electronically variable integrator or low-pass filter.

capacitor values can be used for even lower droop rates. If only a small charging time (called the *aperture time*) is available, the improved analog switches may be the better choice.

Sometimes we may intentionally want to take our time about charging the capacitor. In electronic music, this is called *portamento*, and it imitates slide trombone and other glide effects. Pressing a new key glides or sweeps the note to the new frequency, rather than jumping to it. As Fig. 7-14B shows, all we need do is add a variable resistor in series with the sample-and-hold to pick up the portamento. More elaborate current sources can also be used for linear rather than exponential changes of glide voltage.

Tracking Filters

Electronically varying the cutoff frequency of a filter can be a hassle, especially if lots of sections are involved or if we want to operate over a wide frequency range. Fig. 7-15 shows how analog switches can be used to electronically vary the cutoff frequency of a low-pass filter or the time constant of an integrator. If our switch is closed all of the time, we have an integrator that works with a 100K resistor and a 0.05-μF capacitor. But, if we turn the switch rapidly on and off with a 50/50 duty cycle, it looks like we have a 200K resistor instead of a 100K. This is because the actual resistor is only there half the time. Close the switch only one-tenth of the time, and the input resistor looks like one megohm. If we use this technique with a low-pass filter, we have electronically varied the cutoff frequency by a ratio of ten to one as we go from an apparent 100K to one megohm. The beauty of this technique is that when we have several sections, all the sections will track very well, since they are all being driven by the same proportional duty cycle. For best operation, the switching rate has to be much higher than the operating frequencies of the filter.

Microprocessor Data Entry

CMOS analog switches replace three-state gating in the manual data-entry system for a CMOS microprocessor as shown in Fig. 7-16.

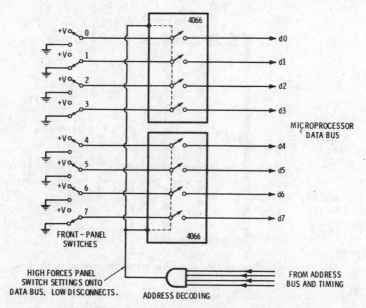

Fig. 7-16. Manual data entry for a CMOS microprocessor.

The inputs to the analog switches go to "1" and "0" selector switches on the front panel. The potential outputs go to the data bus on the microcomputer. The analog switches are activated only when a front-panel address is decoded. At that time, the front-panel settings are force-fed onto the data bus.

Video Combiner for a Game or TV Typewriter

The video combiner of Fig. 7-17 takes serial video, horizontal sync pulses, and vertical sync pulses and combines them into 100-ohm, sync-tip-grounded EIA video as shown. Both the sync switches are normally closed. Opening either sync switch unconditionally gives a grounded sync-tip output. Closing the video switch gives a 2-volt white level, while leaving it open gives us a 0.5-volt black level. The half-monostable RC networks on the sync inputs automatically shorten the delayed (for position) sync signals into pulses of the proper width. This is typically 5 microseconds for horizontal pulses and 180 microseconds or so for vertical pulses. Be sure to use the pin numbering sequence shown in Fig. 7-17 to avoid substrate current problems.

A 10-volt split power supply may be needed if you have to have a 100-ohm output impedance. The same circuit works with a 5-volt power supply if you can use a slightly higher output impedance. The remaining switch may be placed in parallel with the video input,

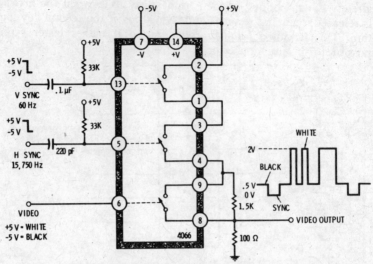

Fig. 7-17. Composite video output combiner for tv typewriter or video game.

either to increase the output swing or to provide a cursor or underline white output.

A Scanning ASCII Keyboard Encoder

You can convert the single-key closure on a keyboard into a composite parallel ASCII code by using the circuit in Fig. 7-18. Two-key rollover and debouncing are inherent in this circuit. Compared with existing single-chip encoders available at this writing, the circuit is cheaper, uses much less supply power, and needs only a single +5-volt supply.

The two NOR gates form a 50-kHz clock that is gated. If the clock is allowed to run, two cascaded binary counters are driven, and continuously cycle through all of their counts.

The slower counter is one-of-eight decoded and routed to a 4051 that sequentially connects columns of keyboard characters (usually *physically* arranged as *q w e r t y*, etc.) to the positive supply. Meanwhile, the faster counter is routed to a second one-of-eight 4051 selector that is monitoring sequential *rows* of characters. When a key is pressed, the output from +5V through both selectors stops the gated oscillator and holds the count. A delay is produced after the oscillator stops due to the EXCLUSIVE-OR gate that gives us a delayed keypressed output. Our "raw" ASCII output and keypressed command are then routed to the logic circuit shown in Fig. 9-15 for control and shift operations and for a choice of keypressed polarity and pulse width.

When the key is released, scanning resumes and continues until a new key is pressed. If a second key is pressed while one key is already down, nothing happens right away. But, when the first key is released, scanning resumes and then stops at the second key location. This is called *two-key rollover* and it makes faster typing possible with minimum error.

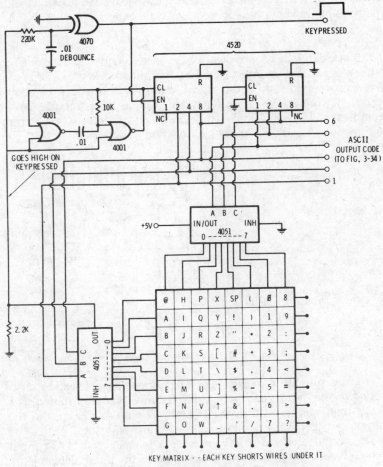

Fig. 7-18. Scanning ASCII keyboard encoder.

CMOS PHASE-LOCKED LOOPS

Moderate-cost analog phase-locked-loop ICs have been available for several years now and have proven extremely useful for solving all sorts of hard-to-do circuit problems easily and simply.

The 4046 is a CMOS phase-locked loop. It does many of the things the earlier PLLs did, over a frequency range from subaudio to 1

413

MHz. In addition, it is cheap and uses a single power supply at low operating current. But most important of all, the 4046 offers a choice of two phase-detection schemes. One of these arrangements lets the phase-locked loop track over an extremely wide frequency range—beyond 1000:1 if desired. This simply wasn't available before and opens up many new phase-locked-loop applications.

Loop Details

A block diagram for a typical phase-locked loop is shown in Fig. 7-19. We provide an input signal to the phase-locked loop circuit. We may want to lock to this signal, multiply or divide it by some factor, synchronize a sweep or system timing to it, remove noise from it, synthesize multiple channels with it, bandpass-filter it, strip information or audio off of it, or simply find out if it is there.

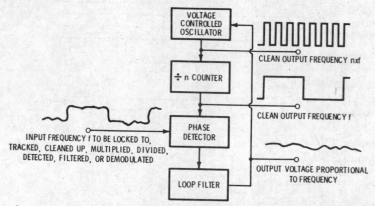

Fig. 7-19. Typical phase-locked loop.

We also provide a locally generated signal from a *voltage-controlled oscillator,* or vco. The output of the vco is then routed to a divide-by-n counter. Although n may be "1" in some cases, the vco frequency is generally higher than the input frequency.

The output of the divide-by-n counter is compared to the input frequency in a *phase detector*. The output of the phase detector is a voltage that represents the *error* between the input-signal phase and the phase of the divided-down vco signal. In turn, this error signal is filtered and used to control the frequency of the voltage-controlled oscillator. The final filter-to-vco connection *closes the loop,* forcing the vco to *track* or follow the input.

We can take the output from several points, depending on what we want the loop to do. The vco output will be a clean, logic-level square wave that is n times the input frequency. The divide-by-n output will be a logic-level square wave or pulse that equals the input frequency. Most often, it will be a cleaner, larger, less-noisy, filtered replica of the input signal.

We can also play games with our divide-by-n counter. If we add more n beyond the phase detector tap, we end up with a frequency *divider* whose output frequency is a clean submultiple of the input signal. By changing n, we can generate multiple channels separated by the input frequency. With two or more divide-by-n counters side by side, we can gain fractional or multiple-frequency references. If our vco drives separate divide-by-three, divide-by-four, and divide-by-five counters, a single electronic-music input frequency becomes a major triad chord output.

The error signal is normally a very high impedance. But we can unload this signal with a voltage follower and use the error signal itself as an output. This gives us a frequency-to-voltage converter, an input-tone-is-there indicator, or an audio fm (frequency modulation) or digital fsk (frequency-shift-keyed) demodulator.

The 4046

The 4046 has two separate internal circuits. One is a vco that runs from subaudio frequencies to beyond 1 MHz. The other is a dual-output phase detector. To this, you add a divide-by-n counter (or a jumper for n = 1) and a two-resistor, one-capacitor loop filter in order to build a complete phase-locked loop.

We have already learned about the vco portion of the 4046 back in Chapter 4. The center frequency range is set by a capacitor on pins 6 and 7. The maximum frequency is set by a 10K to 10-megohm resistor connected from pin 11 to ground; any frequency offset is provided by a usually larger resistor connected from pin 12 to ground. A voltage input to pin 9 will sweep the frequency between the limits set by the two resistors. Ground gives you a minimum frequency, and +V gives you the maximum frequency. This is a typical high-impedance CMOS input that won't load the loop filter.

Rather than go into the complicated internal details of our dual phase detector, we will just summarize what each detector option does for you in Table 7-2. There is an input for an external reference on pin 14, and an input for the divide-by-n on pin 3. Note that these two pins cannot be interchanged because it will reverse the sense of the loop. If n is going to be 1, you simply jump the pin 4 vco output to pin 3; if n comes from a counter, the vco output clocks the counter, and the counter output drives pin 3.

There is a choice of two phase-detector outputs. One is a conventional (EXCLUSIVE-OR) or *low-noise* phase detector the other is the new-to-CMOS *wideband* phase detector.

The low-noise phase-detector output on pin 2 has a limited tracking range that is hard to get beyond ±30 percent, but it provides

Table 7-2. Comparison of Wideband Phase Detector and Low-Noise Phase Detector, Both Using a 4046 PLL

Wideband Phase Detector	Low-Noise Phase Detector
Outputs on pin 13	Outputs on pin 2
Tracks over very wide frequency range, to 2000:1 if needed	Maximum tracking range is ±30%
Noise immunity is limited	Excellent noise immunity
Input signal can be narrow pulse or other duty cycle	Input signal must be a square wave
With no input, output frequency goes to lowest designed value	With no input, output frequency goes halfway between high and low limits
Output phase is 0° with respect to input	Output phase is 90° for midfrequency input, varies with frequency
Insensitive to harmonics	Harmonic sensitive
Loop filter acts as sample-and-hold	Loop filter acts as integrator

very good noise rejection. *For this phase detector to work, both input signals must be 50/50 duty-cycle square waves.* You use this phase detector when replacing traditional analog PLLs with the 4046. The output phase of your vco will be roughly 90° for midband input frequencies and will increase or decrease as your input reference frequency changes.

The wideband phase detector is actually a digital-logic phase/frequency detector. It provides a tri-state sample-and-hold output on pin 13 for the loop filter. If the input *frequency* is higher than the vco, a steady high output results. If the input frequency is lower than the vco frequency, a steady low output results. If the two frequencies are identical, the phase detector outputs a pulse proportional to the phase difference. This pulse is positive going for lagging vco phase and is negative going for leading vco phase.

The input frequency can range over an extremely wide area, even beyond 1000:1 if desired, and the vco phase is always 0° with respect to the input frequency when locked. The input and reference frequencies can be any duty cycle from a square wave down to narrow pulses of either polarity. This detector is the best choice for most newer PLL designs, particularly those that have to operate over a wider frequency range. The noise-rejection characteristics of this wideband detector are generally poorer than those of the low-noise phase detector; this is about its only real disadvantage.

A separate "test-phase" output that may be used for lock detection with either detector is available on pin 1. The input to either phase detector is pin 14. This input will, in theory, accept capacitively coupled low-level signals and amplify them to digital logic levels, or it can be driven directly by CMOS logic that swamps the internal amplifier.

The linear amplifier operation of pin 14 is an unmitigated disaster when the wideband phase detector is being driven. Don't use it this way! Linear operation causes extra amplitude-variation sensitivity, jitter, tearing, and generally poor noise immunity. Instead, drive your input with a full logic signal or use a resistor to pull it over to one extreme supply limit and capacitively couple an input signal. If your input signal is very small, amplify and limit it externally, particularly for low-frequency references such as the power line.

Loop Design

To get your phase-locked loop to work, you have to add a *loop filter* to it. All this takes is two resistors and a capacitor. While deceptively simple, the loop filter is the single most important part of any PLL. Fig. 7-20 gives some details on loop-filter design.

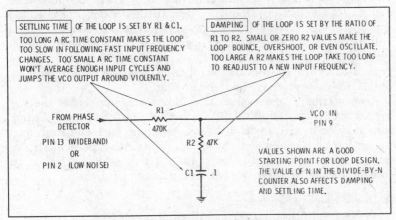

Fig. 7-20. The loop filter for a 4046 is a crucial design detail.

The loop filter decides two things for us—the *settling time,* or how many cycles the loop is going to average, and the *damping,* or the ability of the loop to accept new changes without excessive overshoot or possible oscillation. If the PLL was a pendulum, the settling time would be decided by the mass and length of the arm; the damping would be determined by air resistance and the friction of the pivot.

Generally, the low-noise phase detector takes a much longer settling time than the wideband detector since the capacitor is a continuously driven integrator or averaging device, while the wideband detector uses the capacitor as a sample-and-hold that is difference driven. Also, you will probably want long settling times when n in the divide-by-n counter is large. This prevents the early vco output from being different than the later vco output *for each cycle of phase-detector operation.*

It may seem that the "smoothest" loop operation would be gained by omitting R2 so that the capacitor directly "filters" the vco input. However, this is not so! Omitting R2 will almost always drive the loop into near oscillation, causing extreme underdamping and taking a long time rebounding in a damped oscillatory manner. The math behind this is detailed in *Phaselock Techniques* by Floyd M. Gardner, but the important point for us is that we will get the best loop operation with a series resistor, typically one-tenth to one-third the input resistance. Another handy book is the *Design of Phase-Locked Loop Circuits, with Experiments,* published by Howard W. Sams & Company.

The actual loop design isn't really difficult. You should use the longest possible settling time you can. Generally, you start with the values given in Fig. 7-20 and close the loop. Then you produce a sudden change in input frequency near the lower limits of intended vco operation, and watch the output frequency or the vco control voltage on a scope. The quickest operation is obtained when you let the vco overshoot around 40 percent on a step error. This is near *critical damping.* If you don't want overshoot, use correspondingly higher R2 values to increase the damping.

After you have some values that look reasonable, check the vco output for steady-state jitter. With high values of n, more settling time may be needed to keep the vco sweeping back and forth during each cycle of phase-detector comparison. This is particularly important if you are using the low-noise phase detector. Let's clear up some of these PLL mysteries with some design examples.

A Low-Frequency Counter

Suppose you were doing some brain-wave research or other low-frequency work. How would you accurately and quickly measure, for example, a 9-Hz signal? Well, you could count for a minute or so to get enough cycles for reasonable accuracy. Or, you could start and stop a high-frequency reference signal on the zero crossings of a single cycle of the 9-Hz signal. The number of high-frequency counts would tell you the time period, and you could then calculate the frequency from its inverse. This technique takes a very good comparator and requires excellent signal-to-noise ratio, but it works.

These are the two traditional ways of doing the job. A better way is to simply use a phase-locked loop to ×100 multiply the signal. You then measure the faster signal with an ordinary counter and a one-second time gate. Fig. 7-21 shows one possibility.

For a 1-Hz to 100-Hz input range, we design a vco to cover 100-Hz to 10-kHz, with some extra range on each end. The vco output is divided by 100 and then compared to the input signal using the wideband phase detector. The phase-detector output goes to the loop filter and then reaches around and closes the loop by running the vco.

This circuit brings out several subtle points about the 4046. Don't forget to ground the enable input on pin 5. Your loop capacitor has to be a quality, low-leakage unit. Since this loop works as a sample-and-hold, even a 10-megohm scope probe will shift the loop's operation. If you want to use or look at the control voltage (except for a quick check on limits), use a 3130 as a voltage follower to isolate the loop.

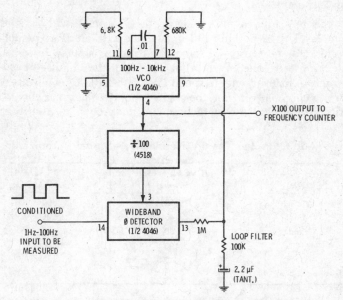

Fig. 7-21. ✕100 frequency multiplier used for fast measurement of subaudio frequencies.

Another way to observe the circuit operation is to monitor pin 1. When the loop is locked and working on a stable input frequency, you should get a normally positive waveform that goes to ground for 20 *microseconds* or so once each cycle and then returns to a steady high state. These 20-microsecond pulses represent the minimum time that the loop-filter sample-and-hold capacitor is being charged or discharged to a new value as a result of a phase or frequency error.

If the pulses are steady but much wider, you have a leaky capacitor, another load on the loop, a poorly conditioned input, input frequency jitter, or something else wrong.

419

If the pulses tear and vary all over the place, you are out of lock. This is caused by excessive input noise, an unconditioned input, or a vco that is out of range. If your scope shows pin 9 at ground, you are out of range on the low-frequency end. If pin 9 is at +V, you are out on the high end.

If you like, you can also put a 10K resistor to ground on pin 1 and use this as an output to a lock detector.

Note that your input signal has to be there for the lock time and the measurement time. If you are working with something shorter, like a piano note or a magnetometer's output pulse, you will have to think of something else.

Digital Tachometer

We can measure engine or shaft speed (rpms) with a digital tachometer. This is nothing but a modified version of a frequency multiplier. For tach use, we have to introduce a *scale factor* that changes rpm to frequency and that allows for the number of cylinders when measuring engine speed. We also have to condition our input for a noisy pickoff from the engine's points. These are about the only two differences between tachometers and frequency multipliers.

Table 7-3. Multiplier Factors Needed for Digital Tachometer

Range	N for 0.5-Second Interval	N for 0.1-Second Interval
4-Cylinder	60	300
6-Cylinder	45	225
8-Cylinder	30	150
RPM	120	600
RPM/10	12	60
PPS	2	10

The scale factors we will need for 0.1- and 0.5-second frequency measurement times are shown in Table 7-3. These values are the "n" that you program into your divide-by-n in order to get the frequency display to read actual rpms. Much of the digital-tachometer circuit is shown in Fig. 7-22. After input filtering and limiting, we use a 3130 as a down-the-middle comparator to complete the conditioning. You set n with the selector switches and Mickey Mouse logic around a 4520 or something similar, or else use the newer 4568. The output is routed to a suitable counter and display unit that measures for 0.5 second.

A car's idling engine varies its speed wildly. This can cause annoying jitter on the display, particularly if the engine is missing or otherwise sick. You can minimize this jitter by using a long loop

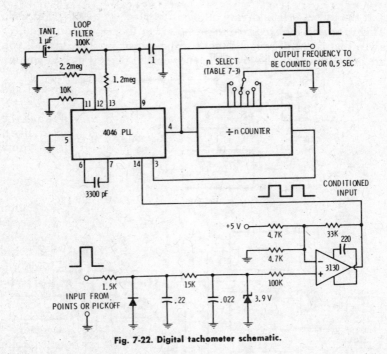

Fig. 7-22. Digital tachometer schematic.

settling time. While compromise values are shown, you can also switch two loop values, one for idle and one for measuring fast changes in acceleration.

A Frequency Synthesizer

We can generate any frequency from 1 kHz to 1 MHz with the frequency synthesizer shown in block-diagram form in Fig. 7-23A. Here we start with an accurate 1-kHz reference, perhaps from a crystal and divider. We then build a 1-kHz to 1-MHz vco and run the output through a thumbwheel-switch-controlled divide-by-1-through-1000 divider. The output goes around the loop filter and forces the vco frequency to n times 1 kHz. For a switch setting of 362, you get a 362-kHz output. The same idea works well for channel synthesis of radio and communications channels.

There is a not-so-obvious improvement we can make in this basic circuit. It would be much better to add extra switching and restrict the range of our vco one way or another. For instance, with a 10-volt supply and a 1-MHz vco, a 1-mV noise or error pulse in the loop means a 100-Hz shift in frequency. This may not be bad at 300 kHz, but it is intolerable on a 2-kHz setting.

The way to overcome this noise sensitivity is to restrict your vco so that n is always in the hundreds. Fig. 7-23B shows how we always

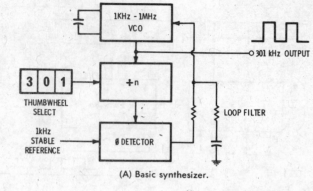

(A) Basic synthesizer.

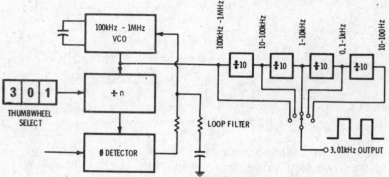

(B) More-stable, wider-range circuit.

Fig. 7-23. Frequency synthesizers.

synthesize a 100- to 1000-kHz signal and then use decade dividers to pick the lower ranges. This way, the error goes down as the frequency does. On a 2-kHz setting, we now get 1-Hz-per-millivolt error rather than the 100 Hz we had before, a 100:1 stability improvement.

Always use the minimum feasible vco range on any PLL to minimize noise sensitivity. Higher supply voltages will often help as well.

Line Lock for a TV Typewriter or Video Game

Fig. 7-24 gives yet another use for a multiplying phase-locked loop. In this case, the multiplying is more or less incidental to our main objective. We end up locking the system timing of a tv typewriter or video game to the power line. This locking is one very good way to prevent weaving or dancing of the display for vertical frequencies that are almost, but not quite, equal to the power-line frequency. Such weaving is caused by poor shielding and filtering in a tv set as the sweep "slips cycles" with respect to the power line.

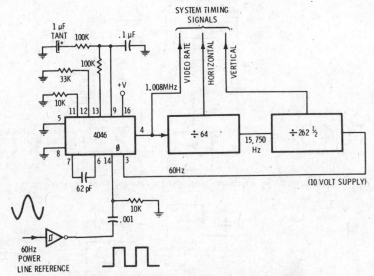

Fig. 7-24. Line lock for a video game or tv typewriter.

Our vco typically runs at 1.008 MHz, and is divided horizontally by 64 and vertically by 262.5 to get a 60-Hz output. This is the vertical-sync frequency and field rate. It is compared with the 60-Hz power-line reference by the wideband phase detector in the 4046 and locked to the timing of the power line. A similar technique can lock the timing to an external video program for superposition.

This circuit requires good power-line conditioning. Note that we have also disabled the Class-A operation of the phase-detector input to minimize any noise tearing. We have also added a small capacitor directly to the vco input on pin 9. This point normally has a very high impedance, and any nearby timing signals can produce a disconcerting brief shift of a character or character line. The capacitor eliminates this problem. The capacitor has to be small enough that it has a negligible effect on loop damping; values of one-tenth the loop capacitor or less are recommended.

Electronic-Music Tracker

Pair a top-octave generator with a 4046 and you get a very useful electronic-music *tracker*. It accepts a single wide-range input frequency and generates all possible chords off it. Fig. 7-25A shows the general idea. Our top-octave generator (see Chapter 4) is clocked by the vco in the 4046. It produces 12 or 13 outputs corresponding to the equally tempered scale over a single octave. One of these notes, usually the lowest one, gets compared to the input frequency by using the wideband phase detector in the PLL. The top-octave

423

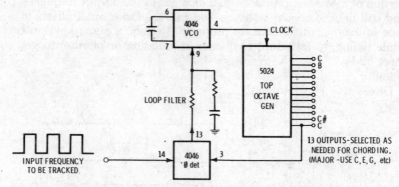

(A) Basic tracker produces accurate chords for any input frequency.

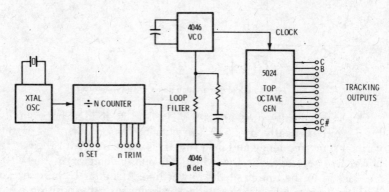

(B) Adding a crystal and programmable counter for accurate
vibrato, tuning, or short glides.

Fig. 7-25. PLL trackers and chord generators for electronic music.

generator then in turn outputs this note along with all the others.
They can be selected three at a time for chords. A divide-by-two
counter on each output gives you another whole octave of notes for
additional chording options.

The input frequency can vary over the entire audio range, but the
selected chords stay locked to the selected input note. Your notes can
sweep or glide and still produce accurate chording, a trick that is
hard to handle with fixed-frequency top-octave generators. A purposely underdamped loop lets the notes all "bounce" for interesting
special effects; otherwise, you use large values of settling time and
damping. Extreme values will give you portamento and glides.

Another variation on the tracker is shown in Fig. 7-25B. This circuit solves the dilemma of having both stable *and* changeable notes.
We start with a crystal-controlled reference frequency and then
divide it down by a programmable counter. By changing a *small*

portion of n, we can produce small changes in the output frequency and still hold to crystal stability. You can use these small offsets to tune to another instrument, to properly "stretch" a piano keyboard while tuning, to introduce frequency modulation or *vibrato*, to get short glides, or simply to introduce a randomness or noise modulation.

Direct microprocessor control of this circuit is an obvious possibility.

+1, −1 AMPLIFIER

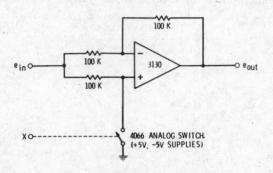

When X = +5, amplifier gain is −1.
When X = −5, amplifier gain is (+2 − 1) = +1.

Chapter 8

Digital Displays

The design and operation of CMOS circuitry may be enhanced by using either colored LEDs, 7-segment LEDs, or liquid-crystal numerical displays. LEDs are bright, cheap, and easy to interface with CMOS. They multiplex beautifully and are available in many sizes and colors. Because of the limited output drive associated with most CMOS gates and counters, the majority of LED displays are not able to be directly connected to the outputs of CMOS devices without some additional circuitry.

INTERFACING LAMPS AND LEDS

Fig. 8-1 shows how to drive a lamp using a CMOS gate and either pnp or npn transistors.

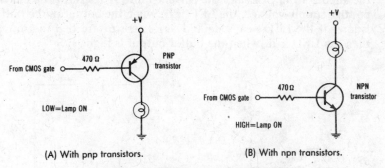

(A) With pnp transistors.

(B) With npn transistors.

Fig. 8-1. CMOS to lamp transistor drivers.

As shown in Fig 8-2, the typical LED has a voltage drop of approximately 1.7 volts when forward biased. In addition, the amount of light emitted, which may be red, orange, yellow, or green in color, is proportional to the forward current I_f. Typically, the LED forward current ranges from 10 to 20 mA (50% to 100% relative brightness), which is controlled by a series current-limiting resistor R.

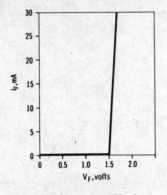

(A) Voltage drop across LED when forward biased.

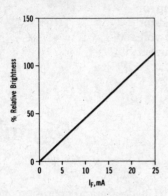

(B) Amount of light emitted is proportional to the forward current.

(C) Schematic diagram.

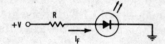

Fig. 8-2. LED characteristics.

Fig. 8-3 is used to indicate the output state of a 4049 buffer. When the anode of the LED and the current-limiting resistor is tied to the positive supply voltage, the LED is lit when the output of the buffer is low. If the LED's cathode and series resistor are tied to ground, then the LED is lit when the buffer output is high.

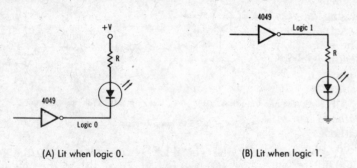

(A) Lit when logic 0.   (B) Lit when logic 1.

Fig. 8-3. LED logic indicators.

Because of their limited output current at low supply voltages, most CMOS devices are unable to drive LEDs directly. Fig. 8-4 shows several practical connections for driving LEDs. Using either a pnp or npn transistor and a 5-volt supply, the current-limiting resistor is typically 5.6K for a LED current of 15 milliamperes. For a 10-volt supply, use 12K. If a 4049/4050 buffer is used at 5 volts, the current-limiting resistor should be 330 Ω.

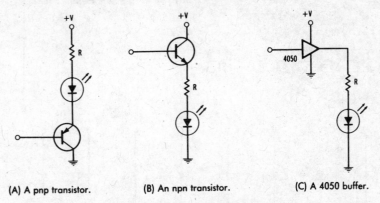

(A) A pnp transistor.   (B) An npn transistor.   (C) A 4050 buffer.

Fig. 8-4. LED drivers for CMOS devices.

7-SEGMENT LED DISPLAYS

Many pocket calculators, frequency counters, synthesized CB transceivers, microwave ovens, and television sets all have something in common. They all use 7-segment LED displays on their front panel to indicate numbers.

Each display contains seven separate LEDs arranged in a pattern to form any possible number from 0 to 9. As shown in Fig. 8-5, each segment location is given a specific letter, from a to g. By forward biasing one or more segments, you can display the numbers 0 through 9. Actually, most 7-segment LED displays have *eight* internal LEDs. The extra LED is for the decimal point, usually left-handed. Depending on the intended application, 7-segment display packages come as either a single digit or as a multidigit LED display, which can contain up to as many as ten digits.

All 7-segment LED displays are catagorized as either *common cathode* or *common anode*. A common-cathode LED display has the cathodes of all the internal LEDs tied together, as shown in Fig. 8-6A. In operation, the common cathode is then grounded while the individual segment anodes are each connected to separate current-limiting resistors or LED drivers, as shown in Fig. 8-6B, to form the number "1."

429

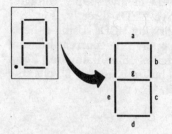

Fig. 8-5. A 7-segment display with
segment identification.

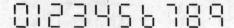

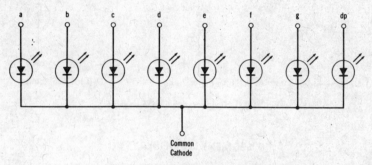

(A) Equivalent circuit, with decimal point.

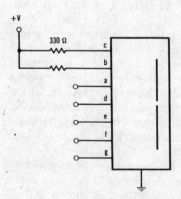

(B) Forward biasing LED segments *b* and *c* to
display the number 1.

Fig. 8-6. A 7-segment common-cathode LED display.

On the other hand, as shown in Fig. 8-7, a common-anode display
has the anodes of all the individual LED segments tied together. In
operation, the common anode is connected to the positive supply

430

voltage, while the individual cathode segments are each connected to ground via either resistors or suitable LED drivers.

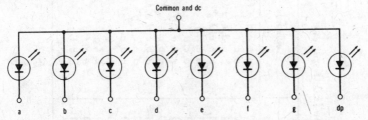

Common and dc

a   b   c   d   e   f   g   dp

Fig. 8-7. Equivalent circuit for a 7-segment common-anode LED display, with decimal point.

DISPLAY DRIVERS

One of the easiest 7-segment display drivers to use is the 74C48, which is used with *common-cathode* displays. Internally, it has a BCD-to-7-segment decoder that provides high outputs for the segments to be lit, while all other outputs are low. For each of the seven segments, a current-limiting resistor is required, as shown in Fig. 8-8. Besides the numbers 0 through 9, the 74C48 produces non-numerical display patterns for the equivalent BCD inputs, 10 through 14, while the display is blanked (no segments are lit) for the equivalent BCD input 15. The truth table for the 74C48 is given in Table 8-1.

Besides the BCD inputs and 7-segment outputs, the 74C48 has *zero-suppression* logic when used with two or more displays. This is used when we want to blank out unwanted zeros. For example, if a seven-digit display without zero suppression reads 00345.80, the first

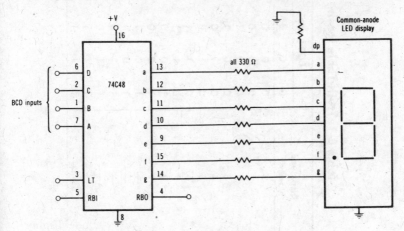

Fig. 8-8. The 74C48 common-cathode display driver.

Table 8-1. Truth Table for the 74C48 Decoder/Driver (for Common-Cathode LED Displays)

Displayed Number	BCD Input				Seven-Segment Output							Segments Lit
	D	C	B	A	a	b	c	d	e	f	g	
0	0	0	0	0	1	1	1	1	1	1	0	a, b, c, d, e, f
1	0	0	0	1	0	1	1	0	0	0	0	b, c
2	0	0	1	0	1	1	0	1	1	0	1	a, b, d, e, g
3	0	0	1	1	1	1	1	1	0	0	1	a, b, c, d, g
4	0	1	0	0	0	1	1	0	0	1	1	b, c, f, g
5	0	1	0	1	1	0	1	1	0	1	1	a, c, d, f, g
6	0	1	1	0	0	0	1	1	1	1	1	c, d, e, f, g
7	0	1	1	1	1	1	1	0	0	0	0	a, b, c
8	1	0	0	0	1	1	1	1	1	1	1	a, b, c, d, e, f, g
9	1	0	0	1	1	1	1	0	0	1	1	a, b, c, f, g
—	1	0	1	0	0	0	0	1	1	0	1	d, e, g
—	1	0	1	1	0	0	1	1	0	0	1	c, d, g
—	1	1	0	0	0	1	0	0	0	1	1	b, f, g
—	1	1	0	1	1	0	0	1	0	1	1	a, d, f, g
—	1	1	1	0	0	0	0	1	1	1	1	d, e, f, g
Blank	1	1	1	1	0	0	0	0	0	0	0	None

two zeros and the trailing zero can be blanked out by using the zero suppression feature, and we will then have 345.8. Blanking of zeros at the beginning of the significant numbers is called *leading-zero suppression*, while the blanking of the trailing zeros is called *trailing-zero suppre sion*.

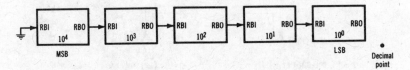

Fig. 8-9. Leading-zero suppression for a 74C48 decoder/driver display.

Ripple blanking input (RBI) and ripple blanking output (RBO) signals are used to blank the display. Fig. 8-9 shows how to connect the 74C48 for leading-zero suppression for a five-digit display. Note that the RBI pin of the most-significant digit is tied at logic 0, and the RBO pin is connected to the RBI of the next lower-order digit, etc. Fig. 8-10 shows the connections for trailing-zero suppression.

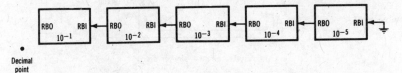

Fig. 8-10. Trailing-zero suppression for a 74C48 decoder/driver display.

In addition, the 74C48 has a *lamp test* (LT) input. If this input is low, all seven segments of the display are then lit, displaying the number "8." This provides us with a simple way of testing for burned-out LED segments.

For both the 4026 and 4033 decade counters, the decoded output for the numbers 6 and 9 will differ from that produced by the 74C48 decoder/driver. As shown in Fig. 8-11, the decoded output will produce a "tail" on numbers 6 and 9. For the number 6, the *a* segment is added, while the *d* segment is added for number 9.

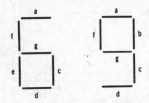

Fig. 8-11. Decoded numbers 6 and 9, showing their "tails."

433

Another way of connecting the output of most CMOS devices, in order to drive a wide variety of 7-segment LED displays, is to use an integrated-circuit transistor array like the 3081 or the 3082, such as shown in Fig. 8-12. The 3081 consists of seven high-current tran-

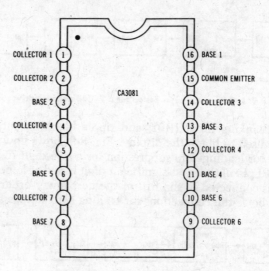

(A) CA3081 common-emitter array.

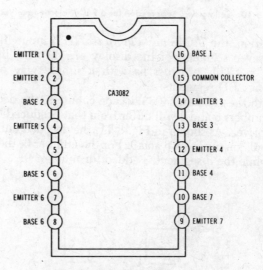

(B) CA3082 common-collector array.

Fig. 8-12. Transistor driver array pin diagrams for 7-segment LED displays.

sistors in a common-emitter arrangement for use with *common-anode* displays, while the 3082 is a common-collector array used with *common-cathode* displays.

For supply voltages greater than 3.5 volts, Fig. 8-13 shows how we can use the 3082 to connect a common-cathode display to either a 4026 or 4033. Each anode is connected to a current-limiting resistor (RE) placed in series with each emitter lead.

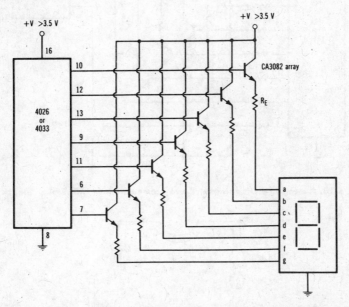

Fig. 8-13. A common-collector transistor array/common-anode LED display interface for common-cathode displays.

Fig. 8-14 shows how we can use the 3081 with a common-anode display. For both arrays, the minimum transistor beta (ß) is typically 30.

OBTAINING BCD INFORMATION
FROM 7-SEGMENT DISPLAYS

For a number of applications, particularly with microcomputers, it may be useful to obtain the equivalent 4-bit BCD code that corresponds to the number on the display. Few of the CMOS-type digital clocks, panel meters, pocket calculators, and counters, such as the 4026 and 4033, allow us to get at the BCD data.

Using a 74C915 decoder, as shown in Fig. 8-15, we are able to convert the 7-segment display information into BCD code; this is sum-

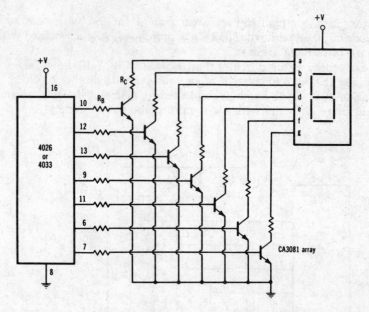

Fig. 8-14. A common-emitter transistor array/common-anode LED display interface.

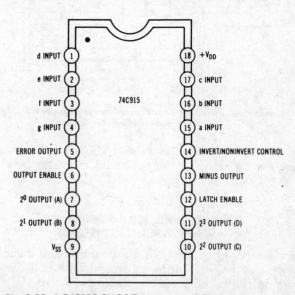

Fig. 8-15. A 74C915 CMOS 7-segment-to-BCD decoder.

marized by the truth table of Table 8-2. The states of the 7-segment inputs can be selected by the Invert/noninvert control input (pin 14)

Table 8-2. Truth Table for the 74C915 7-Segment-to-BCD Decoder

Displayed Number	7-Segment Input a	b	c	d	e	f	g	BCD Output A	B	C	D	ERROR Output	MINUS Output
0	1	1	1	1	1	1	0	0	0	0	0	0	0
1	0	1	1	0	0	0	0	0	0	0	1	0	0
2	1	1	0	1	1	0	1	0	0	1	0	0	0
3	1	1	1	1	0	0	1	0	0	1	1	0	0
4	0	1	1	0	0	1	1	0	1	0	0	0	0
5	1	0	1	1	0	1	1	0	1	0	1	0	0
6	0	0	1	1	1	1	1	0	1	1	0	0	0
6ᵃ	1	0	1	1	1	1	1	0	1	1	0	0	0
7	1	1	1	0	0	0	0	0	1	1	1	0	0
8	1	1	1	1	1	1	1	1	0	0	0	0	0
9	1	1	1	0	0	1	1	1	0	0	1	0	0
9ᵃ	1	1	1	1	0	1	1	1	0	0	1	0	0
Blank	0	0	0	0	0	0	0	1	1	1	1	0	0
Minus(—)	0	0	0	0	0	0	1	X	X	X	X	1	1

ᵃ6 or 9 with "tail."
X represents three-state condition.

in Fig. 8-15. A logic 0 selects active-high true decoding at the 7-segment inputs. The Error output (pin 5) is high whenever a nonstandard 7-segment code is detected. When the Latch enable input (pin 12) is 1, the BCD output is latched via three-state outputs.

LIQUID-CRYSTAL DISPLAYS

Because of their low-power consumption, many CMOS circuits requiring digital displays now use liquid-crystal displays (LCDs). Unlike LEDs, LCDs are not available as a single digit and must be selected for a particular application. Fig. 8-16 shows how the 74C48 can also be used for LCDs. Unlike LEDs, external excitation for the LCD's backplane is required in the form of a square-wave signal with a frequency between 30 and 200 Hz.

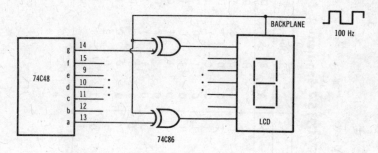

Fig. 8-16. A CMOS LCD decoder/driver, using a 74C48.

Besides the 74C48 driver, the 4055, shown in Fig. 8-17, can also be used, but without the EXCLUSIVE-OR gates. The square-wave signal for

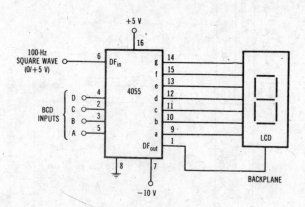

Fig. 8-17. A CMOS LCD decoder/driver, using a 4055.

the backplane is applied to pin 6. However, the 4055 also requires a negative supply voltage.

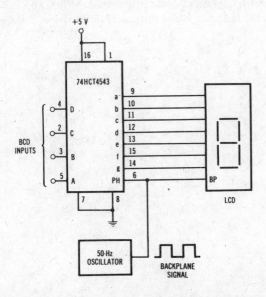

Fig. 8-18. Using a 74HCT4543 decoder/driver for liquid-crystal displays.

A 74HCT4543 decoder/driver can also be used to drive LCDs, but without the negative supply voltage. Fig. 8-18 shows the connections, when a 50-Hz square-wave oscillator is connected to the Phase Input (PH) of the 74HCT4543 and the backplane of the LCD.

Chapter 9

IC Memories
and Programmable Logic Arrays

This chapter is intended to present only a basic discussion of integrated-circuit memories and programmable logic arrays (PLAs), for this area is so extensive that an entire book could be devoted to it. Taking this into account, this chapter thus presents only the basic concepts of the major types of CMOS integrated-circuit memories that are available.

Up to this point, we have shown several examples of how to combine various logic functions to perform specific tasks. However, if we wanted to change from one task to another, we may have had to physically rewire the circuit. On the other hand, today's computers are built around a microprocessor, and they have the ability of being reprogrammed so that the user can easily change from one task to another.

In order for a computer to function properly, however, it must have some form of memory. For example, a microprocessor-controlled microwave oven may at one time be set, or "programmed," to defrost meat for 10 minutes and then cook it for 30 minutes at 350 degrees. On another occasion, the oven may be used to cook a slice of pizza or a hot dog (without prior defrosting) for only two minutes. At the completion of both tasks, a bell is sounded to alert you to the fact that the total cooking cycle has been completed. In order for the microwave oven to "remember" any of the sequence of commands being entered on the oven's front panel for a given cooking task, it must have some form of programmable memory to remind and tell the microprocessor "brain" what to do next.

WHAT IS A MEMORY?

Integrated-circuit (IC) memories are digital devices that are capable of storing logic 0s and logic 1s in a predefined arrangement. In a very loose sense, flip-flops and shift registers can be thought of as memories since their outputs will remain in a given state even after the input data is removed, and they will retain these states until changed (reset, application of a clock pulse, etc.).

Integrated-circuit memories can be divided into two types: read-only and random access, each of which can be subdivided further, as will be discussed in the following sections.

BASIC MEMORY ORGANIZATIONS

Before discussing the characteristics of the various memory types, let's first discuss those basic characteristics that are common to all. As shown in Fig. 9-1, a typical memory device can be thought of as being organized in terms of a matrix arrangement. Each matrix location, or *cell*, is designated by a row and a column location. This cell location is selected by an *address decoder*. As a general scheme, an internal decoder selects the *row* of the particular memory cell while another decoder selects the corresponding *column* of the cell. The binary inputs to the decoders are referred to as the *memory address*.

As far as the internal memory is organized, there are two conventions used: bit organized and word organized. A *bit-organized* memory device has the capability of storing only a single bit for each address location. Consequently, this arrangement has a single data input line, and a single, but usually separate, data output line, as

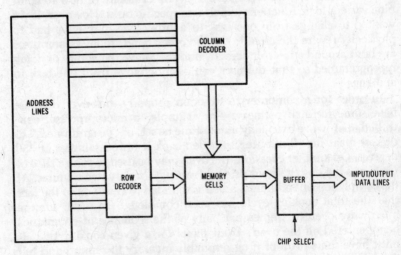

Fig. 9-1. Typical memory block diagram.

shown in Fig. 9-2A. However, some bit-organized memory devices have a combined input/output (I/O) data line, with a separate I/O control line that selects whether this combined data line will function as the input or output line at any given time (Fig. 9-2B).

In a *word-organized* memory device, a memory address selects several memory cells at one time instead of just a single cell. The number of cells selected simultaneously is called the *word length* and is usually 4 or 8 bits in length, although with today's home computers, 16-bit words are possible. Also, word-organized memories have the same number of I/O lines as there are bits in the word.

THE 1K CONVENTION

Whether bit organized or word organized, integrated-circuit memories are further characterized by their storage capacity. For large memory capacities, the suffix "K" is generally used as a shorthand notation for a value of 1024 bits. The following list illustrates use of this notation for several CMOS memory capacities (in bits).

Bits	Equivalent Notation
1,024	1K
2,048	2K
4,096	4K
8,192	8K
16,384	16K
65,536	64K
262,144	256K
524,288	512K
1,048,576	1 Meg

Since 1K equals a capacity of 1024 bits, a 4K memory has 4 × 1024, or 4096 bits. Note in the listing that 65,536 bits is abbreviated as 64K, although some sources incorrectly use 65K. For capacities higher than 64K, the difference between the actual number of bits and their shorthand notation becomes increasingly larger.

For memory segments smaller than 1K (1024 bits), the term *page* is often used to indicate an equivalency of 256 bits, so that 4 pages equals 1K. For a 64K-memory system, there are 4 × 64, or 256 pages.

INPUT/OUTPUT CONNECTIONS

Besides the necessary power-supply requirements, connections to and from integrated-circuit memories can be divided into the follow-

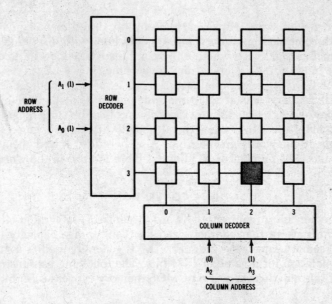

(A) Address 1011_2 of a 16K × 1 bit-organized memory.

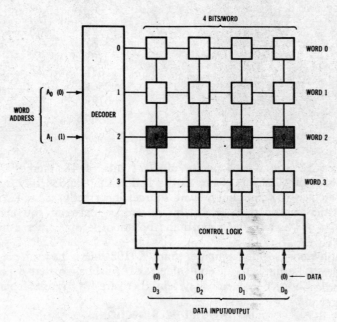

(B) Address 10_2 of a 4 × 4-bit word-organized memory.

Fig. 9-2. Decoding of memory addresses.

ing three categories: *address inputs, data input/output,* and *control inputs.*

Address Inputs

All memories need a series of address inputs to select the proper memory cell, or cells, in the case of a word-organized device. The total number of address inputs, or lines, depends upon the total number of addressable words. As an example, a 16K × 1 memory requires 14 address lines, while a 2K × 8 device requires 11 address lines. Even though both devices each have a total capacity of 16K bits, the first example has 16K words while the second example has only 2K words.

Data I/O Lines

The number of equal data I/O lines depends on the word length of the memory device. A 4K × 1 device will have one data input and one data output line, while a 1K × 8 device will have 8 data I/O line pairs. There may exist separate sets of lines for input and output, or there may be only one set of lines that handle both the input and the output data. Such a dual-purpose arrangement is generally referred to as "bidirectional," and is generally used in computer systems to reduce the number of required connections.

Chip Select Input

The Chip Select ($\overline{CS}$), or Chip Enable ($\overline{CE}$), input is one of several available control inputs. This input is used to select, activate, or enable the memory device before data can be written to or read from it. As a general rule, integrated-circuit memories have three-state data outputs and, when used with other memory devices to increase the available memory capacity, the $\overline{CS}$ (or $\overline{CE}$) control input is used to effectively isolate that device from the rest of the circuit. Usually a logic 0 is required to enable, or select, a given memory device; a logic 1 disables it.

READ-ONLY MEMORIES

The *read-only memory,* or ROM, is an extremely powerful and simple one-package way to solve truth tables, particularly those with multiple outputs. Even the smallest, under-$2, commercial ROMs handle eight outputs of five variables in one package, and much larger units are available.

A ROM is nothing but a hardware realization of your truth table. All inputs are exhaustively one-of-n decoded into individual words, and these words are then re-encoded onto output lines using OR logic.

The best way to understand how a ROM works is to build one yourself. Fig. 9-3 shows the truth table for a hexadecimal-to-ASCII converter that we will be using in a later example. Suppose we wanted to build only the top half of this table. We could, in theory, use nothing but lots and lots of gates, and we could also do the job with seven data selectors.

INPUTS	OUTPUTS	
S 8 4 2 1	8 7 6 5 4 3 2 1	SYMBOL
00000	00110000	(0)
00001	00110001	(1)
00010	00110010	(2)
00011	00110011	(3)
00100	00110100	(4)
00101	00110101	(5)
00110	00110110	(6)
00111	00110111	(7)
01000	00111000	(8)
01001	00111001	(9)
01010	01000001	(A)
01011	01000010	(B)
01100	01000011	(C)
01101	01000100	(D)
01110	01000101	(E)
01111	01000110	(F)
10000	00100000	(SP)
10001	00100000	
10010	00100000	
10011	00100000	
10100	00100000	
10101	00100000	
10110	00100000	
10111	00100000	
11000	00100000	
11001	00100000	
11010	00100000	
11011	00100000	
11100	00100000	
11101	00100000	
11110	00100000	
11111	00100000	(SP)

Fig. 9-3. Truth table for hexadecimal-to-ASCII converter.

Suppose, instead, that we take the approach of Fig. 9-4. We route our four inputs to a one-of-sixteen decoder, the 4514 or 74C154. For each and every possible input word, one, and only one, of these intermediate outputs goes high. We then arrange a group of eight final output lines and put a diode in every place where we want a one in our output word. Output buffers are then used to regain a low output impedance.

Admittedly, this takes a lot of diodes, but we have done the logic with only two packages. Better yet, regardless of the truth table we select, we can build it simply by adding or removing diodes as needed. Unlike our data selector that could handle only a paltry four billion truth tables, this one can handle 2^{128} different truth tables, a number bigger than 10 followed by 38 zeros. But even this is tiny by ROM standards, and most commercial ROMs offer far more possible combinations.

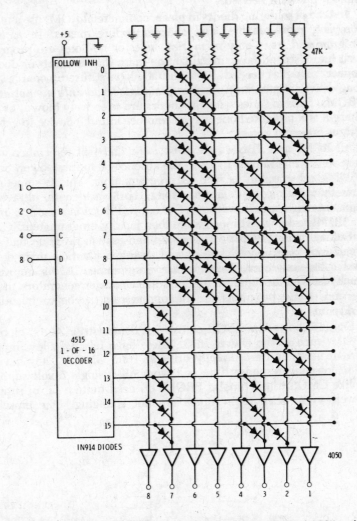

Fig. 9-4. "Do-it-yourself" ROM circuit builds top half of hexadecimal-ASCII truth table. Rearranging diodes yields any of 2^{128} different truth tables.

Out of this mind-boggling redundancy, we have forced our circuit into one table lookup response that directly realizes the truth table we are after. This particular ROM is called a 128-bit ROM,

447

since there are 128 possible diode locations. Its *organization* is 16 × 8, meaning 16 words of 8 bits each. Our 16 words correspond to the exhaustive decoding of four binary-address input lines. For 2^n words, n input lines are needed.

Instead of soldering diodes in place, commercial ROMs use a more-compact, more-reliable, and cheaper coding means. In *factory-programmed* read-only *memories,* holes or no-holes are programmed into a second-layer metallization mask that connects or doesn't connect diodes as needed. ROMs usually have stiff setup charges and long delivery times. In *field-programmable read-only memories,* or PROMs, fuses in series with the diodes are selectively blown, or else charges are implanted one way or another into the device to get the output ones and zeros.

All ROMs and PROMs are *nonvolatile* and will remember what they were programmed to do in the absence of supply power. Some PROMs are erasable as well as programmable. This is sometimes done by exposing the device to strong, short-wavelength, ultraviolet light through a quartz window. The entire PROM is erased at once.

PROMs used to cost much more than factory-programmed ROMs, but today many PROMs are so cheap and easy to program that the break-even point is well into several thousand identical units delivered at the same time. Today, factory-programmed ROMs are pretty much reserved for standard products like character generators, monitor and debug programs for microprocessors, keyboard encoders, and so on.

Fig. 9-5 shows Hex-ASCII conversion done with a 32 × 8 bipolar PROM, such as the Harris 7603. These have relatively low input-current needs and are directly drivable from CMOS. They have a 50-nanosecond response time and work off a single 5-volt supply. Unlike CMOS, these bipolar PROMs do need quite a bit of supply power, even when their inputs are not changing. One erasable

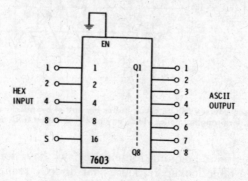

Fig. 9-5. Hexadecimal-ASCII converter programmed into a 32 × 8 PROM.

CMOS-compatible PROM is the Intel 1702 (256 × 8). Response time is normally one microsecond, and present units need a split +5-volt, −12-volt supply.

While you can program PROMs yourself with nothing but some switches, a meter, and a power supply, it is usually better to use one of the programming services available at low cost from many electronics distributors and hobbyist supply houses. When using these services, be absolutely certain the truth table you send in is the one you want, for they have no way of second guessing what you really had in mind.

If you find your initial PROM design takes too many inputs or needs more than one package, it can almost certainly be cut down in size to something reasonable by using one or more minimization techniques. These tricks include: using folding or symmetry (such as doing only one quarter of a sine wave); leaving simple, easy, or obvious logic functions outside the PROM and doing them with gates (such as the 4pdt switch in the Hex-ASCII example shown later in Fig. 9-14); pre-encoding two or more variables to use up some "don't-care" spaces in the truth table (similar to the residue technique with the data selector); and factoring so that one large ROM is split into two or four much smaller ones (such as can be done with digital multipliers and other math functions).

ROMs and PROMs are by no means limited to table lookup applications. We can sequentially address these devices from a counter to generate test waveforms, musical notes, and elaborate digital sequences. We can also use a PROM output, through an external latch, to select its *own* next input address or at least influence it. This is an example of *microprogramming*. Branches and loops are easily introduced with simple external gating.

ROM STORAGE

Computers frequently use ROMs to store a sequence of instructions for a computer program. Many microcomputers were originally 8-bit types, although 16-bit systems are now becoming commonplace. Thus, each basic computer instruction is either 8 or 16 bits in length. With an 8-bit computer system, it is possible to address 65,536 different memory locations of 8 bits each. The total possible storage capacity then is 64K × 8, or 524,288 bits.

In an 8-bit computer system, ROM usually occupies a small portion of the total available memory. Suppose we wish to have a 1K × 8 ROM system, containing one or more permanently stored programs. With all the different memory types available, both bit- and word-organized, there are many possible ways to accomplish this. The easiest way is to simply use a 1K × 8 ROM, such as the 1833.

On the other hand, the same 1K × 8 organization can be accomplished by using two 512 × 8 ROMs, as shown by the circuit of Fig. 9-6. In this arrangement, the basic 512 × 8 configuration has been expanded to a 1K × 8 system by connecting the eight three-state output data lines, D_0 through D_7 of both devices, in parallel. There will be 10 address lines ($A_0 - A_9$) required to access any one of the 1024 possible locations. However, each 512 × 8 ROM has only 9 such lines ($2^9 = 512$), A_0 through A_8, which are connected in parallel. The tenth address line (A_9), which is the most-significant address bit, is connected in complementary fashion to the $\overline{CE}$ inputs of both ROMs.

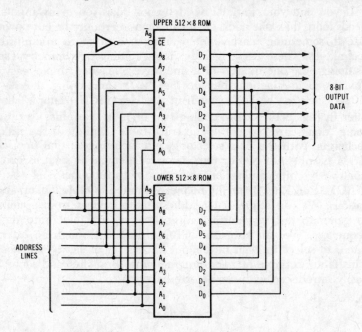

Fig. 9-6. Combining two 512 × 8 ROMs to form a 1K × 8 ROM system.

For the "lower" 512 × 8 ROM, $\overline{CE}_1$ is connected to the most-significant address line (A_9), while the $\overline{CE}_2$ input of the "upper" 512 × 8 ROM is connected to its complement (A_0). As long as A_9 is 0, only the lower 512 × 8 ROM is enabled, so that the maximum 10-bit binary address possible is 0111111111_2, which equals 511. Thus, there are 0 through 511, or 512 possible 8-bit memory locations. Once the lower, or first, 512 8-bit locations have been addressed, address line A_9 must be a 1 in order to address those locations greater than 512, which then disables the "lower" ROM. At the same time, it enables the "upper" ROM, which can also store 512 8-bit locations. Consequently, we are then able to select either ROM, giving a total capacity of 1024 8-bit locations, although only 512 8-bit locations are available at any given time.

Another possible scheme is shown in Fig. 9-7, which organizes a 1K × 8 ROM system by using eight 1K × 1 ROMs. This time, using bit-organized ROMs, the ten address lines of all eight ROMs are connected in parallel while all $\overline{CS}$ inputs are permanently connected to ground. All eight ROMs are then enabled simultaneously.

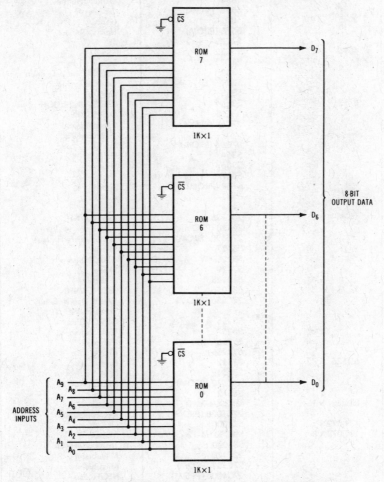

Fig. 9-7. Combining eight 1K × 1 ROMs to form a 1K × 8 ROM system.

Currently there are CMOS ROMs that span the range from 512 × 8 to 524,288 × 8. A select few are listed in Table 9-1.

Table 9-1. CMOS ROMs

Organization	Device	Manufacturer
512 × 8	1831	GE/RCA/Hughes
	1832	GE/RCA/Hughes
	6654 (PROM)	GE/Intersil
1024 × 4	6653 (PROM)	GE/Intersil
1024 × 8	1833	GE/RCA/Hughes
	1834	GE/RCA/Hughes
	6658 (PROM)	GE/Intersil
2048 × 8	1835	GE/RCA
	6616 (PROM)	Harris
	27C291	TI
4096 × 8	23C32	SSS
	23C33	SSS
	5332	GE/RCA/Sharp/Toshiba
	5333	GE/RCA/Toshiba
	5335	Toshiba
8192 × 8	23C64	NEC/SSS/NCR/GE/Hughes
	23C65	NCR/SSS
	27C64 (PROM)	Intersil/Fujitsu/ National/Ti
	87C64 (PROM)	Intersil
	5364	GE/RCA
	5366	Sharp
	5367	Sharp
	5764 (PROM)	Sharp
	6165	Hitachi
	6166	Hitachi
16,384 × 8	23C128	SSS/NCR/NEC
	24128 (PROM)	Toshiba
	27C128 (PROM)	TI/Mitsubishi/Fujitsu
	53127	Sharp
	53128	GE/RCA/Sharp
	53129	Sharp
	613128	Hatachi
32,768 × 8	23C256	SSS/NCR/NEC
	27C256 (PROM)	Intel/Fujitsu/National/ Hitachi/NEC/TI
	53256	GE/RCA/Sharp
	53257	Toshiba/Sharp
	65256	Motorola
	613256	Hitachi
65,536 × 8	83512	Fujitsu
	23C512	NCR
	27C512 (PROM)	NEC/Fujitsu/TI
	53512	Sharp
65,536 × 16	MK3901M-10	Mostek
	27C1028 (PROM)	NEC/Fujitsu
131,072 × 1	531000	OKI
131,072 × 8	MK3901M-12	Mostek
	531000	Toshiba/GE/RCA
	23C100	Mitsubishi
	27C101 (PROM)	Mitsubishi
	27C1000 (PROM)	NEC/Fujitsu
	27C1028 (PROM)	Fujitsu
	62301	Hitachi
	23C1000	NCR/NEC
	831124	Fujitsu
131,072 × 16	62402	Hitachi
262,144 × 8	62302	Hitachi
524,288 × 8	23C400	Mitsubishi

RANDOM-ACCESS MEMORIES

The *random-access memory*, or *RAM*, gets its name from the fact that any given memory-cell address can be selected in any sequence (i.e., at random). Furthermore, RAMs have the ability to change the data contents of any memory address at any time. This is referred to as a Write operation. (Like a ROM, a RAM also has the ability to output, or Read, the data stored at a given address.) Because of this two-fold function, RAMs are referred to as *read/write memories*. RAMs are classified as *volatile memories*, as they will lose their data contents if power to the device is briefly removed.

RAMs can be expanded in the same way as ROMs (Figs. 9-6 and 9-7). Since RAMs are able to be written to as well as read from, a Read/Write ($R/\overline{W}$), or Write Enable ($\overline{WE}$), line is used to select either operation.

RAMs are generally divided into two major categories: *static* and *dynamic*. A static RAM, or SRAM, retains its data indefinitely as long as power is supplied to it. On the other hand, the data stored by a dynamic RAM (DRAM) must be periodically "refreshed," or updated. This is usually accomplished by performing either a repetitive Read or Write operation.

Fig. 9-8 shows a 16K × 1 DRAM, a type which is found in many of the earlier home microcomputers. For a 16K device, there must be 14 address lines, which can be organized in a 7-row, 7-column matrix. The 1-bit data word is stored in the memory cell selected partially by the 7-row address bits $A_0 - A_6$, when a *row address strobe* ($\overline{RAS}$) signal is a 0. The remainder of the 14-bit address is comprised of the

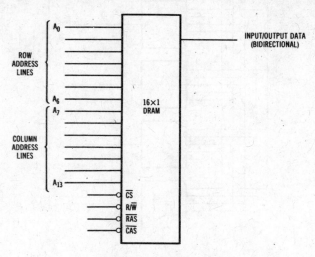

Fig. 9-8. A 16K × 1 DRAM.

7-column address bits, $A_7 - A_{13}$, and is selected when the *column address strobe* ($\overline{CAS}$) signal is also a 0. Since the RAM is organized into 16K 1-bit words, it has only a single bidirectional data I/O line, which is frequently referred to as the *I/O bus*. Depending on the logic level of the $R/\overline{W}$ input at any given time, the data I/O line either receives this 1-bit data and "Writes" it into one of the 16,384 possible memory cells, or it outputs it (i.e., "Reads" it) over the same line.

It is possible to construct a 16K × 8 DRAM system by combining eight 16K × 1 DRAMs. To do this, there must be eight 16K RAMs,

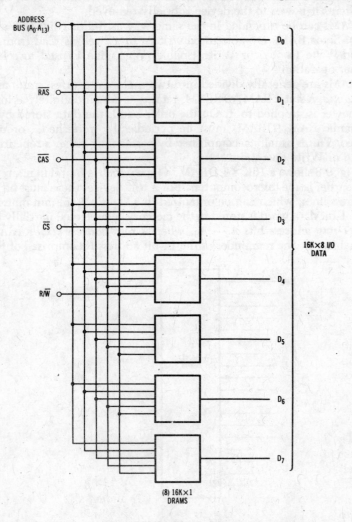

Fig. 9-9. Combining eight 16K × 1 DRAMs to form a 16K × 8 DRAM system.

connected as shown in Fig. 9-9. The 14 address lines are connected in parallel as an address bus. The refresh lines ($\overline{RAS}$, $\overline{CAS}$), as well as the $\overline{CS}$ and R/$\overline{W}$ lines, are also connected in parallel.

Where only 16K × 1 RAMs used to be common, it is now possible to buy 256K × 4 and 1 MEG × 1 DRAMs for under $30.00. The pin diagram of the Hitachi HM511000 CMOS 1 MEG × 1 DRAM is shown in Fig. 9-10. Notice that there are only ten address lines (A_0 through A_9), whereas a 1-MEG memory requires 20 address lines. For DRAMs, 10 bits of the memory address are first entered when the $\overline{RAS}$ is brought low. The final 10 bits of the memory address are activated by bringing the $\overline{CAS}$ line low. As a result, a total of 20 address lines are present, giving $2^{20} = 1,048,576$ memory locations. For smaller memory organizations, the same arrangement also applies. This reduces the overall number of pins required for a given integrated-circuit package.

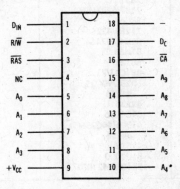

Fig. 9-10. Pin diagram of 1 MEG × 1 CMOS DRAM.

Table 9-2 lists a few of the many CMOS static and dynamic RAMs that are available. In addition, many non-CMOS RAMS (NMOS types) have inputs and/or outputs that are directly CMOS compatible.

PROGRAMMABLE LOGIC ARRAYS

The *programmable logic array* (PLA) is an alternate to a PROM that has some special uses. It has more inputs but does not exhaustively decode them. The Intersil IM5200 programmable logic array of Fig. 9-11 is a typical example. There are 14 inputs. The complements of these 14 inputs are internally generated, and these 28 lines go to a programmable 28 × 48 array. Up to 48 different input combinations, including "don't-care" conditions, can be AND-logic-gen-

Table 9-2. CMOS RAMs

Organization		Device	RAM Type
64	×4	74C98	SRAM
32	×8	1824	SRAM
128	×8	1823	SRAM
256	×1	4720/74C200	SRAM
256	×4	6561/43501	SRAM
512	×4	5102	SRAM
1024	×1	6508	SRAM
1024	×4	6148/2148	SRAM
2048	×8	1433/6116	SRAM
4096	×1	6147/2147/6504	SRAM
4096	×4	51C68	SRAM
8192	×8	1630/6264	SRAM
16,384	×1	1403/4311/6177	SRAM
16,384	×4	1620	SRAM
16,384	×8	8816	SRAM
16,384	×16	92560/51C66	SRAM
32,768	×8	8832/71256	SRAM
		2833	DRAM
65,536	×1	7187/6287/4361	SRAM
		51C64	DRAM
65,384	×4	71258/43254	SRAM
		81C466/2464	DRAM
65,384	×16	81H64	SRAM
131,072	×8	88128	SRAM
262,144	×1	881C81/71257	SRAM
		50256/90C256/ 41256	DRAM
262,144	×2	42256	DRAM
262,144	×4	81C4256	SRAM
		41005/44C256/ 64256	DRAM
524,288	×1	41512	DRAM
1,048,576	×1	81C1000	SRAM
		511000/421001	DRAM

erated by this array. Each of these 48 *product terms* can then be encoded onto eight output lines, just like a ROM, in a 48 × 8 array. Finally, we have the programmable option of individually inverting or not inverting each output.

Since we can encode "don't-care" states into our 48 product terms, we can get more than one possible input combination into each product term. For instance, only one product-term decoding would be needed for the entire second half of the truth table in Fig. 9-3 if we use a PLA. Since some of our decodings are more easily done in negative-logic form, the output-inverting option can be a handy way of simplifying the number of products used.

A PLA will work only if there are lots of "don't-care" states in your truth table and only if there are lots of times when one or more inputs are ignored. While some interesting applications exist in microprocessor programming, sequence generators, and coding systems, the PLA at present is not a particularly popular or economical way of doing logic design.

456

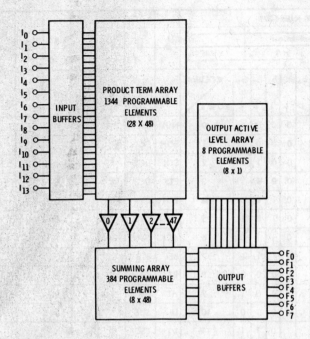

Fig. 9-11. Circuit arrangement of the IM5200 programmable logic array (PLA).

SOME EXAMPLES

Let's wind up this chapter by looking at three examples of computer-peripheral logic design:

A. Build a positive-logic decoder for the ASCII machine commands Carriage Return (CR) and Cancel (CAN).

B. Build a hexadecimal-to-ASCII converter that will automatically display 8-bit microprocessor op code as television characters.

C. Build an ASCII keyboard encoder that includes a choice of key-pressed polarity and a choice of case.

Our ASCII code is shown in Fig. 9-12. It is the standard microprocessor and data-communications code. It is used between computers, terminals, modems, tv typewriters, printers, and other peripheral devices. ASCII code consists of a 7-bit (optionally, 8-bit) code, arranged into groupings of uppercase alphabet, lowercase alphabet, numbers, and the transparent or control commands.

b7	b6	b5	b4	b3	b2	b1	COLUMN → ROW ↓	0	1	2	3	4	5	6	7
								0 0 0	0 0 1	0 1 0	0 1 1	1 0 0	1 0 1	1 1 0	1 1 1
			0	0	0	0	0	NUL	DLE	SP	0	@	P	`	p
			0	0	0	1	1	SOH	DC1	!	1	A	Q	a	q
			0	0	1	0	2	STX	DC2	‖	2	B	R	b	r
			0	0	1	1	3	ETX	DC3	#	3	C	S	c	s
			0	1	0	0	4	EOT	DC4	$	4	D	T	d	t
			0	1	0	1	5	ENQ	NAK	%	5	E	U	e	u
			0	1	1	0	6	ACK	SYN	&	6	F	V	f	v
			0	1	1	1	7	BEL	ETB	'	7	G	W	g	w
			1	0	0	0	8	BS	CAN	(	8	H	X	h	x
			1	0	0	1	9	HT	EM	)	9	I	Y	i	y
			1	0	1	0	10	LF	SUB	*	:	J	Z	j	z
			1	0	1	1	11	VT	ESC	+	;	K	[	k	{
			1	1	0	0	12	FF	FS	,	<	L	\	l	¦
			1	1	0	1	13	CR	GS	–	=	M	]	m	}
			1	1	1	0	14	SO	RS	.	>	N	^	n	~
			1	1	1	1	15	SI	US	/	?	O	–	o	DEL

Fig. 9-12. ASCII code is the computer and data-communications standard.

A. An ASCII Decoder

This looks like a simple gates-only problem even though there are seven logic inputs. Since we need only two of the possible 32 machine-command decodings, chances are we won't need a full pair of MSI chips like the 4515 or the 74C154 for a complete and total decode. So, gates should work.

The obvious way is to invert all the zeros and route them to a seven-input AND gate along with the ones. But, as usual, obvious isn't best. Instead, we use our deMorgan equivalence to negative-logic-NAND the zeros (with a NOR gate) and positive-logic-AND the ones with the NAND output. As Fig. 9-13 shows, the Carriage Return (CR) decoding is very simple.

The Cancel (CAN) decoding would be equally simple if it weren't for the extra (fifth) input to the four-input 4002/74C25 gate. One

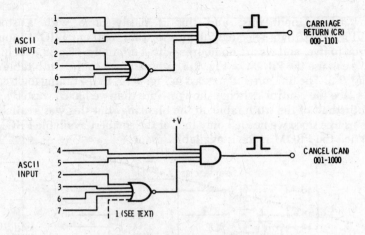

Fig. 9-13. Decoding ASCII control functions.

possible solution is to simply ignore the ASCII "1" input. This gives us a possible "don't care" state, so the output will respond to the little-used ASCII machine command End-of-Medium (EM) as well as to CAN. For many systems, this is acceptable.

But, what if we had to have a total decoding? Three possible solutions are (1) to OR ASCII inputs 1 and 2 with Mickey-Mouse logic consisting of two diodes and a resistor, (2) to use one-fourth of a 4071 or 74C32 OR gate to do the same thing, or (3) to somehow invert the 1 input and shove 1 into the extra AND gate input that goes to +V. How many more ways can you think of?

B. A Hexadecimal-to-ASCII Converter

The *hexadecimal*, or base-16, code consists of the four-bit binary digits for 0 through 9 and the letters A through F for binary 10 through 15, respectively. An 8-bit microprocessor having an op-code word of D6 is binarily equal to 1101-0110. For a tv-style display of a group of op-code words, we would like to be able to look at the upper bits ("D") or the lower bits ("6") in turn. Or, if we really wanted to get fancy, we could also make things automatic to display upper bits (short pause), lower bits (long pause), upper bits (short pause), and so on. So, we would also need a way to get an ASCII Space output or a 010-0000. Our converter would accept hexadecimal op-code on the input and output tv-typewriter-compatible or video-display-compatible ASCII.

A 10-input, 8-output 8K PROM will do the job nicely, using eight bits for the hex input, one for upper/lower selection, and one for blanking. But, these are big and expensive, so we will go to a PROM minimization route. In this case, we want to pick the upper-four or lower-four input bits, and this is nicely done with a stock 4019/74C157 data selector. Put this in front, and our PROM inputs drop to five, and we are home free with a minimum-size 32 × 8 chip. We can use the PROM of Fig. 9-5, programmed to the truth table of Fig. 9-3. The final circuit appears in Fig. 9-14. Extra timing logic can replace the control switches shown. Note that we have "wasted" an entire half of the truth table on the blanking. But this was available to us free since we needed only half of the smallest available PROM. When the PROM space is available, use it.

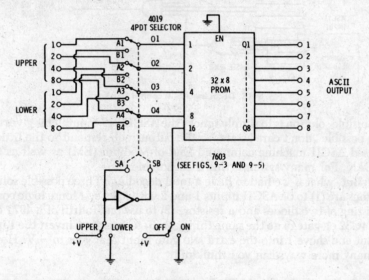

Fig. 9-14. A complete hexadecimal-ASCII converter circuit.

C. A Keyboard Encoder

Two 4051 analog switches, used as a keyswitch data-selector matrix, do our key scanning and provide us with an output that is the binary equivalent of the selected key, along with a key-pressed output.

The add-on "gates-only" logic of Fig. 9-15 converts the raw binary output and the raw keypressed output into a complete ASCII code. The lower EXCLUSIVE-OR gate shortens the keypressed command to a brief strobe and gives us a choice of polarity by working as a controllable complementer. The NAND output gates let us change any code into a CTRL, or control shift, by using a CTRL key. The shift logic on pin 5 makes sure that only number and punctuation are shiftable, and that they are in the right direction for right-side-up keys. Finally, our case-select logic lets us either provide or not provide a

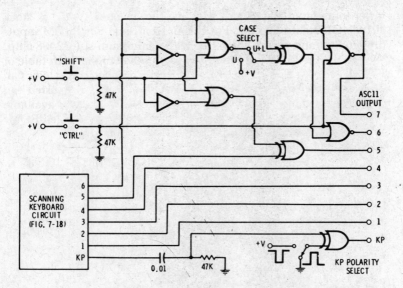

Fig. 9-15. Output logic for ASCII keyboard encoder.

lowercase alphabet as needed. While a lot of "loose ends" are picked up by this logic, note than only 2¼ simple gate packages are needed.

Chapter 10

Getting It All Together

By now, you should have just about all the information you need to design and use CMOS intelligently at the one- or two-IC level. What remains is to interconnect various bits and pieces to build entire systems that will do various useful things. In a sense, this entire book is about applications. Several whole-system examples have already been discussed and many more have been suggested. In the previously mentioned *TTL Cookbook*, published by Howard W. Sams & Company, you will find dozens more applications, circuits, and system-level ideas. Practically any of these circuits can be done cheaper, better, simpler, and with far less supply power if you upgrade to CMOS.

Rather than repeat ourselves, this chapter will look at some whole-system circuits that are new, vastly better than, or unique to CMOS. These newer applications are of two types. One type uses a single MSI or LSI dedicated circuit to do most or all of the job. We will check into some of these first. The second approach is the more tradational one of using a handful of CMOS devices that work together to get a desired result.

Generally, those things that everybody wants to do are already available as CMOS single chips or as circuits using one big chip with some other components. These applications include calculators, digital clocks, test instruments, digital meters, timers, and video games. And we can expect many more one-chip systems to appear soon. It is only when you are trying to build something new, different, or specialized that the "bits-and-pieces" approach of using individual CMOS gates, counters, and registers should be tried.

Before going the "bits-and-pieces" route, always check for the low-cost availability of a single-chip approach. Otherwise, you will end up spending a lot of time, effort, and money reinventing the wheel. And even if a one-chip approach seems expensive, ask yourself what the price might be later on and whether you will still be using this sort of thing when the cost finally does drop. If you find a single chip "almost but not quite" does your job, ask yourself if you can bend what you want to do to fit what is available, rather than going the "bits-and-pieces" route.

If you must go the bits-and-pieces route, don't be afraid to mix logic. One system can easily accommodate CMOS, Mickey-Mouse logic, LS TTL, and a big old PMOS or NMOS chip. CMOS, of course, is the best choice for low-power, low-frequency uses. Mickey-Mouse logic should be reserved for corner cutting in noncritical circuit areas. If part of your circuit has to reliably run in the 3- to 10-MHz range, do this part with low-power Schottky or LS TTL, and the rest with CMOS. Remember that one B-series CMOS output will drive one LS input, and that an LS output will drive any amount of CMOS with a 2.2K pull-up resistor added.

Be sure to consider using a PROM (programmable read-only memory) if there is some messy or irrational logic involved. This is particularly true for circuits that will be used only one time and circuits where changes are possible or likely.

Finally, if your IC count gets too high, be sure to look at the possibility of doing as much of it as you can with software and a microprocessor chip or a microcomputer chip system. Our "break-even" point is very much a moving target. At this writing, chances are that a microprocessor is a better choice if you have over 10 MSI circuits in your system. And, it is virtually certain to become the best choice in the future.

Will microprocessors obsolete CMOS? Not really. There are two reasons why they won't: first, because many microprocessors *are* CMOS, and second, because the input and output circuits, the interface, and much microprocessor housekeeping can be done simply and easily with CMOS. We will continue to see fewer and fewer chips used to do the same task, more and more use of software to replace hardware, and much more use of redundant, fixed architecture (as in a PROM or RAM-ROM-CPU-I/O microprocessor setup) rather than custom, hard-wire logic.

MULTIPLEXED DIGITAL DISPLAYS

Many CMOS applications have a numeric display of some sort. This is true of calculators, clocks, and digital test instruments. Chapter 6 explained how we could build a counter/decoder/display sys-

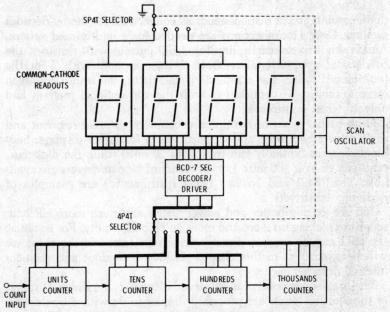

COMMON-CATHODE
READOUTS

SCAN
OSCILLATOR

BCD-7 SEG
DECODER/
DRIVER

4P4T
SELECTOR

COUNT
INPUT

| UNITS COUNTER | TENS COUNTER | HUNDREDS COUNTER | THOUSANDS COUNTER |

Fig. 10-1. Multiplexed digital display system.

tem using one or more ICs per decade. For more decades, you simply repeat the circuitry as often as you need to.

It turns out that there is a much better way to handle many digits at once. This is called *multiplexing* and is shown in Fig. 10-1.

In a multiplexed display, all of the digits share *one* decoder/driver. The digits are scanned rapidly and in sequence. Only one digit is powered or grounded at a time. At the same time that a digit receives power, the input to the decoder/driver is connected to the appropriate counter or latch output for that digit.

The scanning rate is set faster than the eye can follow. A typical scanning rate ranges from 400 hertz to 1 kilohertz. Each digit is operated at higher current than usual to make up for its being off most of the time. The visual result is identical to having all the digits on all of the time at normal brightness.

We have shown common-cathode readouts in Fig. 10-1. For commond-anode readouts and a common-anode decoder/driver, the upper sp4t selector would go to a positive supply source rather than to ground. In either system, the selector, readout, and decoder/driver must be matched to work together.

Multiplexing results in an obvious simplification of circuitry since we need only one decoder/driver. More importantly, it cuts down the number of display-to-circuit connections. For instance, a four-digit, seven-segment display needs 29 connections for the by-decades method. Only 11 connections are needed for a multiplexed system. And if you are externally limiting LED current with resistors, the by-decades approach takes 28 of them, compared with 7 for the multiplexed route. For a nine-digit calculator, the results are even more dramatic—64 connections with the conventional method and only 16 when multiplexed.

With earlier types of multiplexed counters, external segment and digit drivers had to be added, particularly for larger displays. Segment currents normally run from 5 to 20 mils, while the digit currents can go over 100 mils. Lots of standard bipolar drivers are available. The 75491 and 75492 (Texas Instruments) are examples of early bipolar drivers.

ICs are getting better and newer LEDs are much more efficient, so we are picking up more and more on-chip capability. For instance, the 4511 can be used to directly drive segments, and the chips we will be looking at in this chapter include *both* digit and segment internal drivers.

The break-even point of multiplexed displays is 3½ digits. Displays of three or less digits are probably simpler to do with a by-decades approach, while multiplexing simplifies anything higher. The more digits in use, the greater the benefits of multiplexing.

There are several limits to multiplexing and several areas where multiplexing should *not* be used. For multiplexing to work at all, the display elements must be nonlinear or else must have a well-defined threshold. If not, you will get "sneak paths" backwards through supposedly off segments that will light unwanted portions of the display at all times. The segments must be capable of handling high pulse currents as well. LEDs are ideal for multiplexing as they are diodes with a good threshold, and their efficiency often goes *up* with high current pulses. Similarly, seven-segment gas-discharge displays have a good threshold and multiplex well. However, they do take high-voltage drivers, translation, and a high-voltage power supply which makes them less attractive than LEDs for many uses. On the other hand, liquid-crystal displays do not have this well-defined threshold, nor do they pulse well. In general, multiplexing of a liquid-crystal display ranges from diffcult to impossible.

Another place where you should avoid multiplexing is in clock radios, communications receivers, or other systems that involve very close and sensitive rf amplifiers. The scanning circuitry and high-current switching can induce noise into nearby circuits and cause all sorts of problems. Most often, the channel and frequency displays associated with receivers are only two or three digits anyway and you wouldn't gain anything by multiplexing, even without noise problems.

466

A DIGITAL WRISTWATCH

As a simple example of a multiplexed display with internal drivers, consider the digital wristwatch of Fig. 10-2. This circuit gives you a choice of an hours and minutes display, a seconds display, or a day/date display. It also gives you an option of two brightness levels for day and night use.

The circuit starts with a 32,768-Hz crystal that is adjustable to an exact frequency with the trimmer capacitor shown. The CMOS chip divides this frequency down with an internal long binary divider to produce a 1-second clock rate. The basic 1-second clock rate is then divided down by sixes, tens, twelves, twos, sevens, and thirty-ones to pick up minutes, hours, am/pm, and finally the parallel-derived day and date. These values are internally stored and routed to a multiplexed decoder and driver. All of the drive circuitry is on the chip, both for segment and digit outputs.

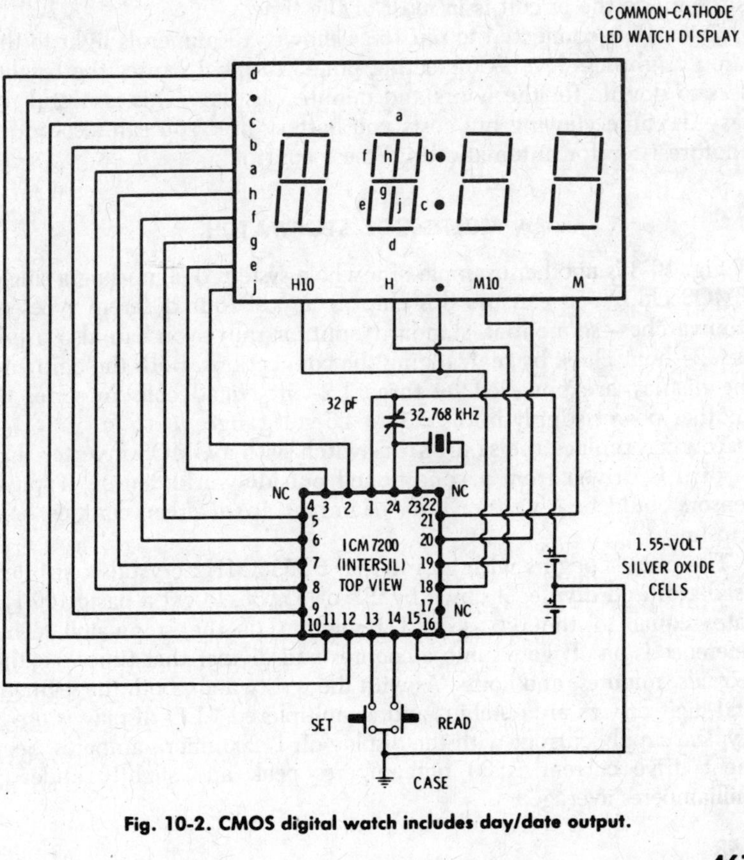

Fig. 10-2. CMOS digital watch includes day/date output.

467

The display is normally off. Pressing the READ button once gives you the hours and minutes. Pressing it again gives you the day and date, while pressing it a third time gives you seconds. To set the time, you push the SET button. This puts you in the date-set mode. You then hit the READ button as often as needed in order to set the date. Hitting SET twice puts you in the hours-set mode. Hit SET three times for days, four times for minutes, and five times for seconds.

With the display off, only four microamperes of current are needed. Up to seven milliamperes of peak segment current are available when the display is on.

There are several options with this chip. Various chips in the series offer 12- and 24-hour operation (the colon shows you pm for the 12-hour operation), and some chips also offer a day output. Note that the day output requires nine-segment readouts in the two left-most positions. You need these to properly shape the day letters.

A 1-second output appears at pin 22. This is useful for calibration of the crystal. You normally calibrate with the display *off* since this is the mode the circuit is in most of the time.

If pin 15 is connected to pin 16 as shown, all numerals light to the same brightness level. Connecting pin 15 to pin 1 causes the brightness to double for the hours and minutes display. This is useful for easy daytime viewing but costs you battery life. You can also add a photoresistor for automatic brightness control.

A MULTIMODE STOPWATCH

Fig. 10-3 is another example of a whole-system design using a single CMOS chip. You can use this chip as any of four different types of stopwatches—sequential, standard, split, or rally; you can also use it as a 24-hour clock by rearranging the connections. Both the chip and the display are powered by three 1.2-volt Nicad cells in series or another power supply in the 2.5- to 4.5-volt range.

You can replace the START-STOP switch with a CMOS inverter that in turn is driven from a conditioned outside-world sensor. Typical sensors could be a light or laser beam, a trap, a screen contact, or a trip line.

The circuit begins with an internal 6.5536-MHz crystal-controlled oscillator and divides it down by 2^{16}, or 65,536, to get a basic 100-Hz rate, equal to the 0.01-second resolution of the stopwatch. This reference is on-off gated into a counter and divider that times out the seconds, minutes, and hours. As with the wristwatch, both the segment and digit drivers are on-chip, and a multiplexed LED display is used. Typical supply current with the display off is 200 micro-amperes. Segment drive current is 20 milliamperes peak and slightly under 3 milliamperes average.

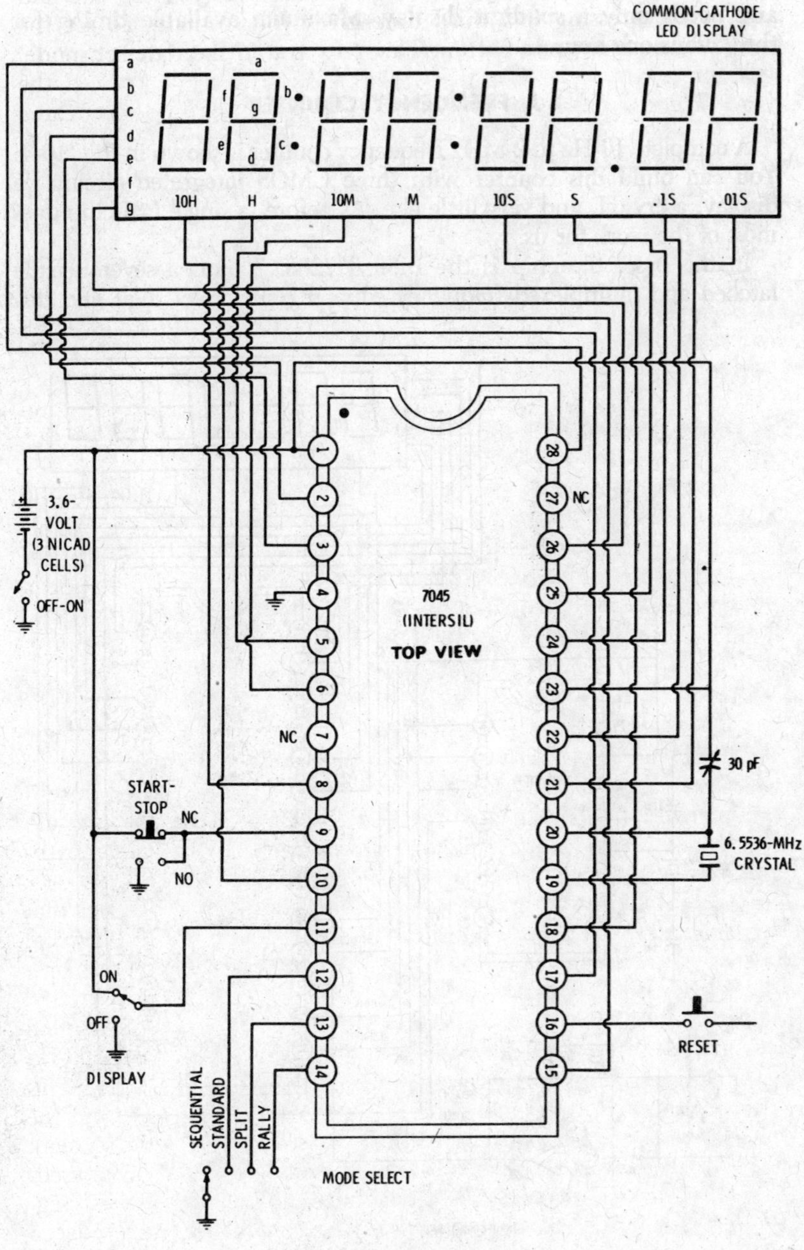

Fig. 10-3. Multimode stop watch.

The 7205 is a somewhat similar chip. It is cheaper and packaged in a smaller, 24-pin package. It provides only two stopwatch modes and needs only a six-digit display. Maximum available timing for the 7205 is one hour.

A FREQUENCY COUNTER

A complete 10-Hz to 2-MHz frequency counter is shown in Fig. 10-4. You can build this counter with three CMOS integrated circuits, a display, a crystal, and very little else. As before, a single LSI chip does most of the work for us.

In this case, the chip is the Intersil 7208. This is a seven-decade latched and multiplexed frequency counter with direct digit and dis-

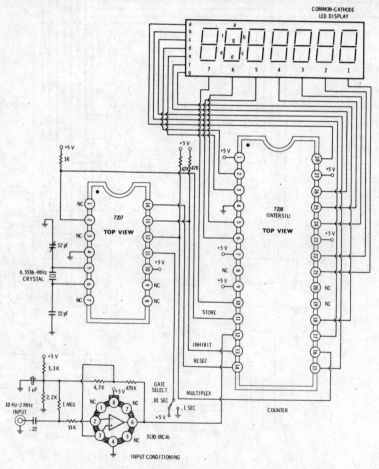

Fig. 10-4. A complete frequency counter using only 3 CMOS packages.

play drive. It needs 210 microamperes from a 5-volt supply with the display off.

The key timing waveforms are generated by a smaller companion chip, the 7207. The waveforms involved are shown in Fig. 10-5. The 7207 starts with a 6.5536-MHz crystal and divides it by 2^{12} to produce a 1600-Hz square wave. This is used to multiplex the 7208 counter and display. The 1600-Hz square wave is further divided down to produce counting gates of 0.01 second or 0.1 second, along with suitable reset and update commands.

The 0.01-second gate is handy with higher frequencies and shorter-length displays, but the 0.1-second gate gives us more digits for a lower frequency. It also has a hidden advantage that is even greater. One-tenth of a second is exactly six power-line cycles. Any frequency-modulated hum on the input or any hum that gets into the front end of a v/f or a/d converter will average to zero six times during a 0.1-second gate. With a 0.01-second gate, it will assume erratic random values and possible jitter.

Our 0.1-second gating sequence actually takes twice that long to complete. We start by holding the inhibit input high for a tenth of a second. The previously reset counter does nothing during this time and ignores input count pulses derived from the 3130 input-conditioning circuit. Our inhibit is then released for 0.1 second, and input counts are accepted by the counter for one-tenth of a second and counted. At the end of the tenth of a second, a brief low update command transfers the count to internal storage latches in the 7208. These latches in turn update the display. Immediately following update, a brief low on the reset line clears the counter to all zeros to start a new sequence. Resetting the counter does not change the count value stored in the latch and appearing on the display.

The 7208 has some options not shown in Fig. 10-4. We can force-feed a reset by grounding pin 14 with a push-button switch, even

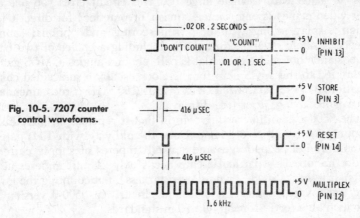

Fig. 10-5. 7207 counter control waveforms.

471

with the 7207 still connected. Pin 9 is a display enable; grounding it turns off the display. This is handy for brightness control (use pulse-width modulation), to extend battery life in portable "push-to-read" instruments, and to flash the display to indicate overrange or a negative value. For stand-alone uses, the 7208 chip has an internal multiplexer. The two resistors and one capacitor needed for self-oscillation are shown in Chapter 2.

Note that the 7207 can't produce a 1-second time gate by itself. If you start with a lower-frequency crystal or an input divide-by-ten, you will get out of the range of the dynamic first-stage dividers in the chip unless you use a very low supply voltage. Worse yet, the multiplex frequency may drop down low enough to cause objectional display flicker.

You can add a divide-by-10 or divide-by-100 prescaler to the input to extend the frequency range to 20 or 200 MHz. A different, higher-frequency, input-conditioning circuit will be needed. You can also extend the counter's range on the low end with a phase-locked-loop multiplier (see Chapter 7) to allow rapid measurement of very low frequencies and for tachometer displays. A voltage-to-frequency converter on the front end lets you convert analog voltages to digital values. You can measure time periods rather than frequencies by interchanging the count and enable inputs. A high-frequency reference is routed to the count input. The enable input is held low for the duration of the event being timed. For instance, with a 1-MHz reference, an enable-low reading of 7643 would mean that the event lasted 7643 microseconds, or 7.643 milliseconds.

FOUR-IN-ONE VIDEO GAME

The video-game circuit of Fig. 10-6 is yet another example of doing most of a job with a single large IC. This circuit gives you one of four video games (pong, soccer, squash, or practice) for direct television display, along with on-screen scoring and optional sound effects. It offers several skill options, including a choice of two ball speeds and choice of two available ball rebound angles. CMOS gates are added to this MOS chip for the clock oscillator and video combination. A crystal-stabilized clock of 2.012160 MHz is recommended for a nonweaving, jitter-free display.

The 8500-1 is an interesting chip in that it replaces an incredible amount of bits-and-pieces logic. Done the old way with TTL gates, it took around 50 packages and several amperes of supply current to do the job. This particular chip needs only 24 milliamperes at 9 volts (if it was CMOS, it would need far less), not counting the low-duty-cycle speaker current for sound effects. An 8500-0 version is also available for European television standards.

472

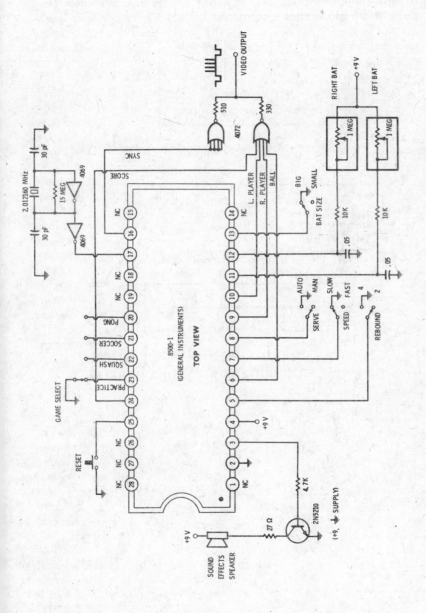

Fig. 10-6. Four-in-one tv game offers pong, soccer, squash, and practice.

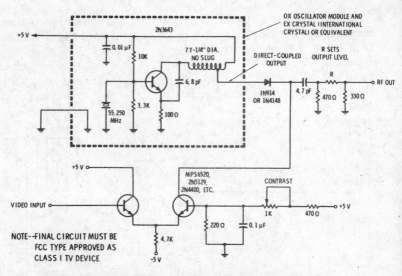

(A) Using "ox" crystal oscillator module.

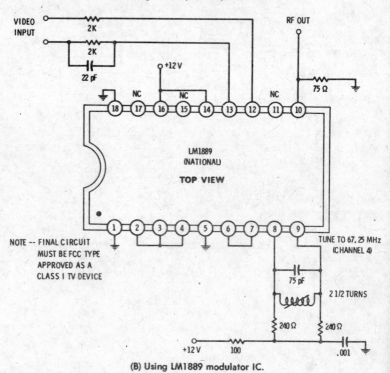

(B) Using LM1889 modulator IC.

Fig. 10-7. Rf modulators for video games or tv typewriters.

Games like the 8500 and tv typewriters can either be connected directly into the video circuitry or be clipped onto the rf terminals of a tv set. Direct video is simpler and cheaper, and offers wider bandwidth and better overall performance. However, it means you have to modify the set, which can introduce a severe shock hazard on a hot-chassis tv set. This shock hazard is present on about half the tv sets in use today and can be eliminated by optical coupling or by isolation of the ac power line.

Clip-on rf entry input is the other choice. It doesn't require any set modifications. However, circuits of this type are called Class-1 tv devices by the FCC, and their final circuits require FCC type approval. Three *approaches* to Class-1 tv devices appear in Figs. 10-7 and 10-8.

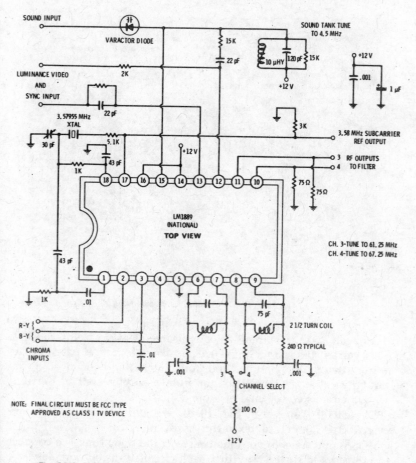

Fig. 10-8. Deluxe rf modulator includes color, sound, and choice of channels.

THE CMOS MICROLAB

The CMOS microlab in Fig. 10-9 is a good example of bits-and-pieces CMOS circuit design. The Microlab is a self-contained logic demonstrator. It is particularly useful for basic logic training, especially for survey courses, software people, and other places where heavy emphasis on device-dependent hardware isn't needed.

The Microlab is made possible by a low-cost clip-on connector system. It is internally powered by four D cells. It is reasonably student-proof since no combination of front-panel connections, however wrong, can damage the device. And thanks to self-indicating logic blocks and some internal resistors that "anticipate" what the student is up to, it is particularly easy to use and clearly demonstrates logic action without hardware hassles.

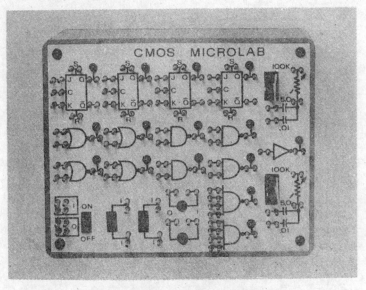

Fig. 10-9. CMOS Microlab for student lab use.

You can make the circuit as complicated or as simple as you like. This version offers four universal flip-flops, four NOR gates, an inverter, two pulse circuits, and four internally debounced push-button or slide switches. Bought from a surplus source, the nine CMOS circuits cost a total of around $3.50.

The circuit is shown in Fig. 10-10. We add resistors to each and every CMOS input to obey the usage rule which states that all CMOS inputs have to go somewhere. At the same time, we connect the resistors in such a way that each circuit turns off completely if

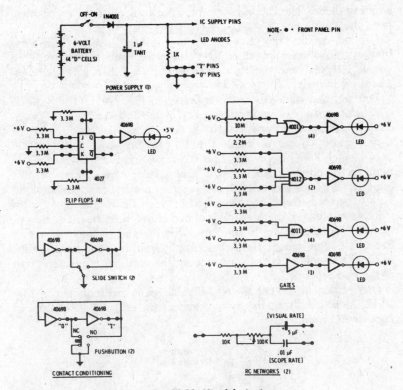

Fig. 10-10. Microlab circuits.

unused, and attempts to anticipate what the student is trying to do if something is forgotten.

For instance, the resistors on the flip-flops disable both direct inputs. The J and K inputs are made high so that you get a binary divider if you connect only the clock. The resistors are very high values that are completely swamped when connected to another CMOS package or to a conditioned contact. The resistor connections to the 4001 NOR gate are particularly interesting. Do nothing to the inputs, and the output goes low, turning off the LED. This eliminates the confusion of a lit, but unused, LED and also conserves the battery current. Connect either input and you get an inverter. Connect both inputs and you get the NOR logic. In the inverter (one-connection) setup, the one input drags the other input along with it, either directly or with a slight voltage-divider action.

Our NAND gates simply have each input resistor going to +6 volts. This disables the unused blocks and lets you operate with one or more input pins unconnected. All of the logic blocks also go to a 4069B inverter whose current-limited output directly drives a light-

477

emitting diode. The diode lights for a logic block high, or positive-logic one, condition.

Contact conditioning is by way of force-fed, set-reset flip-flops built with cascaded inverters. Our RC values are chosen for both visual (1-10 Hz) and oscilloscope or midaudio (0.5–5 kHz) frequencies of operation.

You will find some construction details in Fig. 10-11. The push-on connector system is very cheap. You make it from Molex "bead-chain" posts and crimped push-on connectors. A short length of shrink tubing covers the connector and provides stress relief for the 6- or 10-inch-long jumper lead. Note that two posts are used at every point in the circuit. This lets you "daisy-chain" any number of pins together. You can use a rigid vinyl or other plastic front panel, along with a printed-circuit board spaced ¼ inch below the panel with suitable spacers. A double-sided, plated-through pc board is recommended for this project to simplify soldering and reduce the size to fit a small, high-impact plastic case. Light-emitting diodes with built-in shoulders will be far easier to mount than ordinary ones. The cost of the two types of LEDs is about the same.

One version of the completed Microlab pc board is shown in Fig. 10-12. More details on this system appear in the June, 1974 issue of *Popular Electronics.*

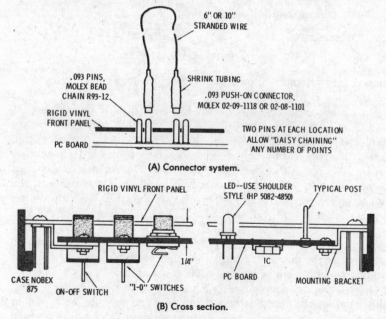

(A) Connector system.

(B) Cross section.

Fig. 10-11. Microlab construction details.

Fig. 10-12. Printed-circuit board for the CMOS Microlab.

BIT-BOFFER DIGITAL CASSETTE RECORDER

You can use a CMOS bits-and-pieces approach to build an effective and reliable digital data recorder using ordinary, unmodified, medium-quality audio cassette recorders. The system is called the *Bit Boffer,* and it meets the "Kansas City" standard for 300-baud hobbyist interchange.

Two essential keys to using an unmodified audio recorder for digital data are to record only constant-amplitude sine waves switched at their zero crossings and to provide speed independence by slowing down or speeding up the receiver circuitry as the tape speed changes. Speed independence is important because of the speed variations in an individual recorder, but it is crucial for recording data on one machine and playing it back on a second machine. This latter capability is essential to mass-distributed and interchangeable hobbyist computer software.

The Kansas City hobbyist 300-baud standard is summarized in Fig. 10-13. A logic one is eight sine-wave cycles of 2400 Hz, while a zero is four sine-wave cycles of 1200 Hz. Both frequency standards are coherently switched at zero crossing with continuous phasing between any transitions. The data format closely resembles the ASR-33 Teletype format, except that it is about three times faster. We also have two format options, one for recording ASCII characters with parity and a second for recording 8-bit computer op code.

The bit-boffer circuit is shown in Fig. 10-14. You should recognize several parts of the circuit as design examples that we have used previously. Our bit boffer works with a UART or microprocessor input/output (I/O). Whatever you choose, it is essential that it have

479

separate 16 × receiver and transmitter clocks. The CMOS 6402 or most standard UARTs may be used.

The transmitter starts with a 64 × , 300-baud reference or 19,200 Hz. This reference, or half of it, is routed to a 4018 sine-wave synthesizer followed by an active filter. A 1-volt-peak-to-peak clean sine wave of either frequency is routed to the tape recorder during the record process. The UART automatically synchronizes things for us and outputs the correct number of cycles since its own 16 × transmitter clock is also derived from the same reference.

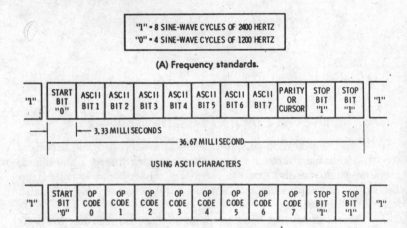

"1" = 8 SINE-WAVE CYCLES OF 2400 HERTZ
"0" = 4 SINE-WAVE CYCLES OF 1200 HERTZ

(A) Frequency standards.

| "1" | START BIT "0" | ASCII BIT 1 | ASCII BIT 2 | ASCII BIT 3 | ASCII BIT 4 | ASCII BIT 5 | ASCII BIT 6 | ASCII BIT 7 | PARITY OR CURSOR | STOP BIT "1" | STOP BIT "1" | "1" |

USING ASCII CHARACTERS

| "1" | START BIT "0" | OP CODE 0 | OP CODE 1 | OP CODE 2 | OP CODE 3 | OP CODE 4 | OP CODE 5 | OP CODE 6 | OP CODE 7 | STOP BIT "1" | STOP BIT "1" | "1" |

USING 8-BIT MICROPROCESSOR OP CODE

(B) Format and time sequence.

Fig. 10-13. 300-baud bit-boffer standards.

The receiver starts with a filter and limiter followed by an edge detector. The edge detector trips a retriggerable monostable set to *two-thirds* the low frequency half-period. A type-D flip-flop recovers the one/zero information for us. Gating reconstitutes a 16× receiver clock whose speed varies with that of the tape. Of the remaining type-D flip-flops, one is used for a receiver data-reset generator, while the last two flip-flops form a tuning aid that facilitates the setting of the baud-rate control.

You will get best operation with a recorder that has EAR and AUX inputs, tone and volume controls set to "6," an automatic level control or agc, and exceptionally clean tape heads. High-quality tape is essential, and it should be checked for dropouts before use.

While the circuit seems rather complicated, it easily fits on a compact, single-sided pc board, and the cost is very low. One version of the bit boffer is shown in Fig. 10-15.

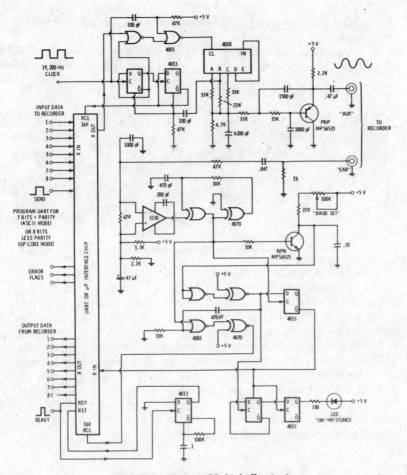

Fig. 10-14. One possible bit-boffer circuit.

Can we do better? Right now, it doesn't seem to make too much sense to use a microprocessor and I/O to replace a UART and about $6 worth of other parts for *stand-alone* systems. However, it is another story entirely if we already have a microcomputer somewhere else in our system. Just a little bit of software and perhaps part of a read-only memory can greatly simplify the circuit and also pick up some other benefits, such as freedom from adjustment.

For instance, the parallel outputs of the microprocessor could directly drive the sine-wave synthesizer. Or, with fancier software, we could even synthesize sine waves with more steps and perhaps get by with nothing but a capacitor as a filter. The receiver will probably still need its CMOS op-amp comparator. But the compara-

481

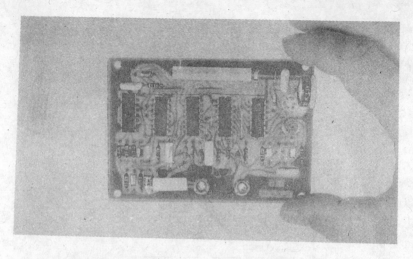

Fig. 10-15. One version of a bit-boffer digital cassette system.

tor output could directly interrupt a microprocessor and be measured by suitable software to extract speed information as well as data. This could eliminate the need for a calibration potentiometer. Optionally, a phase-locked loop could be added to the receiver for even more noise protection.

Right now, your best policy is probably "hardware today, software tomorrow." Use the bits-and-pieces circuit approach when it is the simplest way and when it saves money. However, be on the lookout for software techniques and general-purpose redundant-hardware-architecture (ROM or microprocessor-RAM-I/O) approaches for any problem, because that is the direction in which almost all circuit design is headed.

THE CMOS MUSIC MODULES

The June, 1976 issue of *Popular Electronics* describes in detail a series of music modules that make heavy use of a bits-and-pieces CMOS design approach. These modules are fully *polytonic* (all the notes available all of the time) and are more than flexible enough to adapt to organ, synthesizer, or direct microprocessor-controlled architecture. A typical power-supply chassis with its plug-in music modules is shown in Fig. 10-16.

There are three key modules. One is the top-octave generator, which is similar to the circuit shown in Fig. 4-30B. The second module is a triple divider, which is three of the circuits shown in Fig. 6-2. The third module is called a *dual hex vca*. It contains two groups of six voltage-controlled amplifiers. A schematic of one of the vca

Fig. 10-16. The CMOS music modules include a top-octave generator, dividers, and vca keyers.

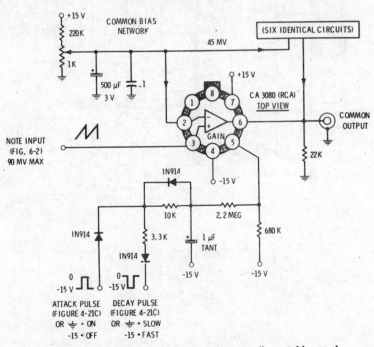

Fig. 10-17. Keyer or vca circuit offers electronically variable attack, sustain, and delay.

channels appears in Fig. 10-17. When the vca is compared with the variable-pulse, attack-sustain-decay circuit of Fig. 4-21C, you simultaneously get complete electronic control of attack, sustain, and fallback of all notes.

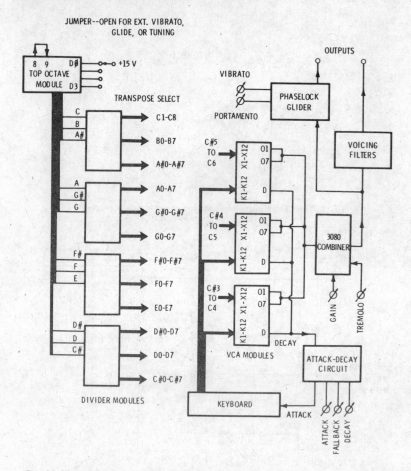

Fig. 10-18. A 72-note, 36-key polytonic music system using three music modules.

Fig. 10-18 shows how one top-octave module, four divider modules, and three vca modules can be combined to give us a 72-note, 36-key polytonic system. Use of the octave translators in the top-octave module lets our short keyboard move wherever we like out of the possible 72 notes. Modules are rearranged for simpler or more complex systems (Fig. 10-19). The top-octave generator is easily adaptable to direct microprocessor control.

484

Fig. 10-19. Experimental polytonic synthesizer using CMOS music modules.

TVT-4 TV TYPEWRITER

The TVT-4 is a circuit that will display the contents of a 512-character ASCII memory as a video display of 16 lines of 32 characters each on an ordinary tv set. An early prototype version is shown in Fig. 10-20. While this system was designed back when CMOS was expensive and microprocessors were nonexistent, you may still find it handy for ultralow-cost and stand-alone tvt applications.

A tv typewriter must be associated with a memory. Fig. 10-21 shows a typical 1K × 8 NMOS memory that you can use with a stand-alone tvt. Similar memory areas inside a microcomputer may be used on a timesharing or *direct-memory-access* basis as well.

The memory in Fig. 10-21 has ten address lines, eight input lines, eight output lines, a read/write control line, and a chip-enable pin. The latter is grounded to use the memory, since making it positive floats the outputs and ignores inputs. The address lines pick the particular group of cells to be written into or read out of. The contents of the selected cells appear on the output lines when the address is stable and the read/write control is high. Bringing the read/write control low will enter new data presented on the input lines.

Fig. 10-20. TVT-4 alphanumeric video display system.

Normal cycle times are 1 microsecond or less. It is particularly important that the address lines be stable immediately before, during, or after a memory-write cycle. This type of memory is called a *static, volatile* memory in that no minimum operating speeds exist, no special timing clocks are needed, but data will be lost with loss of supply power.

Our memory approach can also have a common input/output bus setup as is typical with microcomputers. This is handy for screen reading, editing, and retransmission, but we have to be sure that the memory can be read to the tvt circuitry anytime that it is needed. We can also use an all-CMOS 5101 memory if we like. This will give us a "nonvolatility" option with a backup battery but will cost extra.

The TVT-4 circuit is shown in Fig. 10-22. Since this was an early design, there is some corner cutting to get around things like the very expensive (at that time) counters that were used. Total IC cost today is around $6 plus the cost of the character generator. You will find that a few parts have to be hand picked and some debugging might be needed for your initial setup with this circuit.

The key to this circuit is the multiplying phase-locked loop that lockes the tvt timing to the power line. A master oscillator is run in a

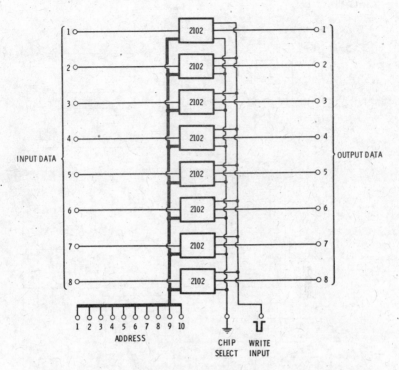

Fig. 10-21. TVT-4 main memory is arranged as 1K × 8; it stores two display pages.

burst mode. It clocks out 32 cycles of character time, lasting 1.3 microseconds each, and then waits for the rest of a 63-microsecond horizontal scan. Key waveforms are shown in Fig. 10-23. The off time of the burst mode is modulated by the phase-locked loop to lock the timing to the power line. At the same time, this setup eliminates the need for horizontal blanking and blanking skew.

The master oscillator is divided by 32 for the horizontal addresses and the dead-time-delay generation. It is then divided by ten for the character addresses. Our divide-by-ten counter is really a divide-by-eight counter that reaches around and inhibits two counts out of each ten that are input. This eliminates the need for between-line blanking.

A further division by 16 generates the four vertical address lines. There is yet another binary divide-by-two counter beyond the last vertical stage, but its count is shortened to give us a 160-horizontal-line live scan and a 102-line retrace. This is done with the three-diode decode-and-reset circuit. The final output frequency is 60 Hz.

The 60-Hz timing signal is compared with a power-line reference and then filtered and used to modulate the off time of the master

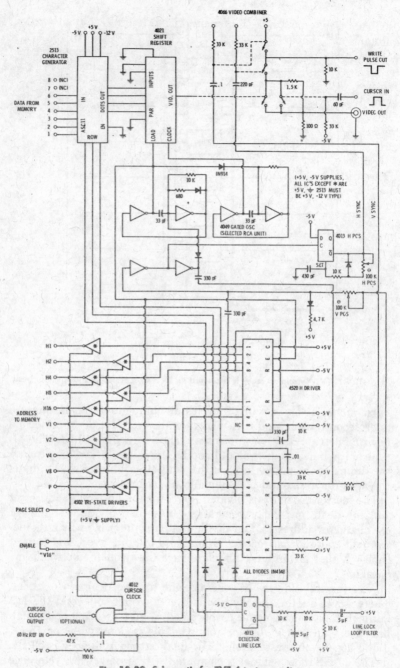

Fig. 10-22. Schematic for TVT-4 tv typewriter.

oscillator. This gives us a simple and cheap line-lock circuit with a minimum of components.

Meanwhile, our addressed memory outputs are routed to a character generator which in turn drives a parallel-to-serial video-shift-register converter. The register is clocked and loaded by a second gated oscillator which is driven and locked to the master oscillator. Our video output signal is combined with delayed and pulse-shaped vertical and horizontal sync pulses by an analog switch in the output circuit.

All address lines are three-state. This lets you share a microprocessor's memory or lets you use a simple cursor system that addresses and writes during vertical retrace. A cursor clock output produces one pulse per character per frame for cursor logic.

You may have to hand-pick a 4049 to get proper operation of the gated oscillator. All RCA units seem to work fine, but others may not. Some capacitor and resistor values may need final trimming during your initial setup, particularly the horizontal dead-time capacitor and the capacitor that is used to skip two of the ten character counts. Incidentally, start the debugging by getting the counters to count and then close the loop to get line lock. The rest should fall in place easily after that.

Note that all the ICs run off a split +5-volt, −5-volt supply except for the three-state drivers that work off of a TTL-compatible +5-volt, and ground supply. The 2513 character generator has to be the old +5-volt, −5-volt, −12-volt type (Signetics, etc.) to work in this circuit. If you are into larger memories, or if your memory is far from your circuit, chances are you will want to use heavier TTL three-state drivers following the 4502s.

3½-DIGIT A/D CONVERTERS

All digital panel meters use some form of analog-to-digital (a/d) conversion. With the increasing use of computers for data acquisition, a/d converters are used to process analog data. Among the most popular CMOS devices available are the MC14433 from Motorola and Intersil's 7106 (LCD drive) and 7107 (LED drive).

The MC14433 is a single-chip 3½-digit a/d converter that uses a modified dual-ramp technique of a/d conversion, with a high input impedance, auto-polarity, and auto-zero. The digital output is a multiplexed 3½-digit BCD code, with the most-significant digit containing over- and underrange and polarity information, as well as the half digit (blank or "1").

Fig. 10-24 shows a simple, but effective, digital voltmeter that uses the MC14433. We can externally set the full-scale voltage to either ± 199.9 millivolts by using a 27K integrator timing resistor between pins 4 and 5 and an external reference voltage of 200 millivolts at pin 2. For a full-scale voltage of ± 1.999 volts, the external resistor is 470K and the reference voltage must be 2 volts.

A pair of $0.1\text{-}\mu\text{F}$ capacitors are used for the integrator (pins 5 and 6) and for the offset correction (pins 7 and 8). Although an external clock may be used at pin 10, the MC14433 contains its own clock so that a single resistor can be used between pins 10 and 11 to set the frequency of the conversion cycle.

A 4511 decodes the 4-bit BCD output to the 3½-digit common-cathode display. At pin 15, the overrange blanking signal (normally high) blanks the display whenever the input voltage exceeds the

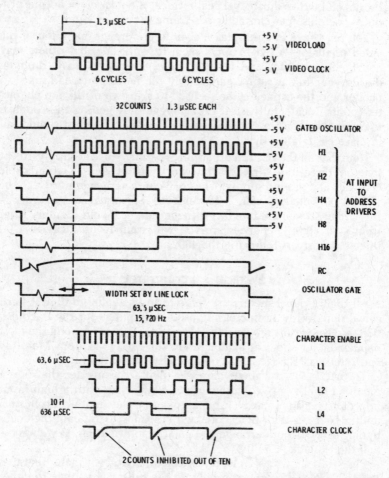

Fig. 10-23. Key waveforms

reference voltage at pin 2. The 75492 interfaces the four display enable outputs to the display during the strobing or multiplexing process. This basic circuit can be enhanced by the addition of the resistor-divider string shown in Fig. 10-25; the voltmeter will then be able to measure voltages up to ± 199.9 volts.

The 7106 and 7107 are 3½-digit a/d converters similar in function to the MC14433. The 7106 is used with a LCD and includes the necessary backplane drive signal at pin 21. The 7107 will directly drive 0.3- or 0.43-inch common-anode LED displays. The general

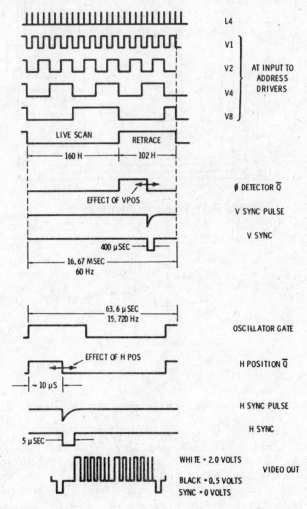

L4

V1

V2

V4

V8

AT INPUT TO
ADDRESS
DRIVERS

LIVE SCAN   RETRACE

160 H   102 H

Ø DETECTOR Q̄

EFFECT OF VPOS

V SYNC PULSE

V SYNC

400 µSEC

16. 67 MSEC
60 Hz

63. 6 µSEC
15. 720 Hz

OSCILLATOR GATE

EFFECT OF H POS

H POSITION Q̄

~ 10 µS

H SYNC PULSE

H SYNC

5 µSEC

WHITE - 2.0 VOLTS   VIDEO OUT
BLACK - 0.5 VOLTS
SYNC - 0 VOLTS

for the TVT-4 tv typewriter.

491

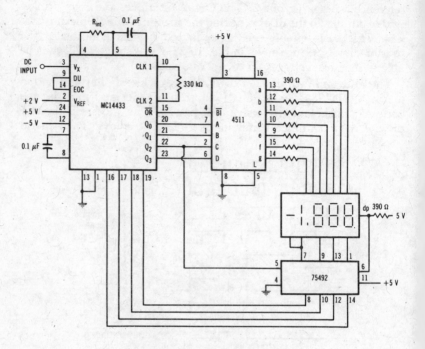

Fig. 10-24. Circuit for a 3½-digit DVM.

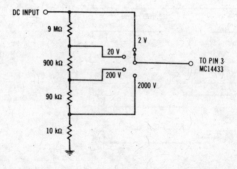

Fig. 10-25. Resistor-divider network for increasing DVM full-scale voltage range.

circuits of Figs. 10-26 and 10-27 are for a full-scale range of ± 199.9 millivolts. Like the MC14433, either the 7106 or 7107 may be wired to permit a ± 1.999-volt full-scale range, by changing the values of the reference resistor (R1), the integrator resistor (R2), and the auto-zero capacitor (C1) to 1.5K, 470K, and 0.047 μF, respectively.

The MC14433 and 7106/7107 are designed to operate from ± 5-volt supplies. However, when a separate negative 5-volt supply

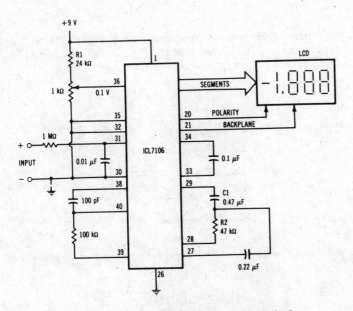

Fig. 10-26. Circuit for a 3½-digit DVM, with LCD display.

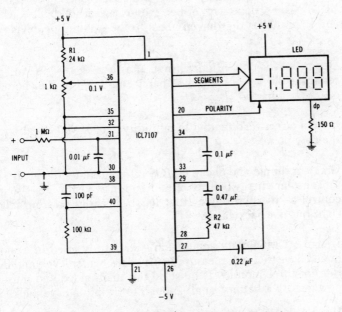

Fig. 10-27. Circuit for a 3½-digit DVM, with LED display.

493

is not available or convenient, as in portable applications, a suitable substitute, using a 4049 buffer and the circuit shown in Fig. 10-28, can be constructed.

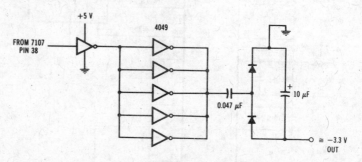

Fig. 10-28. Circuit for generating a negative supply voltage from a positive supply.

SOME CHALLENGES

Now it's your turn. What can *you* do with CMOS? How can you use this energy-saving, low-cost logic family for do-more-with-less, soft approaches to genuinely useful projects? To get you started, here is a baker's dozen of offbeat, unusual, or oddball CMOS-compatible problems that haven't been fully explored or developed yet:

- Design a *Kilowatt Minder*, that is a large, easy-to-read, and real-time display of the power being consumed in your home or office, and its cost.

- Build a direct-reading digital humidity gauge. Make it also read the dew point.

- Create a simple and cheap CMOS differential temperature sensor for solar-energy control of panels, tanks, vents, etc. Use it to control a "thermal diode" that lets heat easily go from left to right but not backwards.

- Design a CMOS data-acquisition system to be used in a snow-bound cave that will measure hourly date, water level, and flow rate for three months. It must do this unattended and must be powered by a battery small enough to fit into a cave pack.

- Use a WWVB receiver and CMOS logic to build an always accurate, self-resetting digital clock. Some design details of earlier systems appear in various issues of *Radio Electronics* and *Wireless World*.

- Use CMOS and a low-power helium-neon laser to build an accurate cave-mapping device.

- Apply CMOS to remote meter-reading and billing functions.

- Design a proton precession magnetometer using a bottle of water and CMOS logic. See *IEEE Transactions* GE9, No. 2, and several (around 1964) issues of the *Review of Scientific Instruments*.

- Class-D (switching-mode) audio amplifiers are theoretically 100 percent efficient. Build a very efficient amplifier using CMOS switching techniques.

- Design a simple and cheap solar-tracking system using CMOS. Make it also optimize power transfer from solar cell to load.

- Work up an electronic-music scanning keyboard that continuously scans the keys and keeps track of at least four notes simultaneously.

- Design a simple yet powerful CMOS system to maximize power transfer from a Winston collector concentrated bank of solar cells to a variable, 120-volt, 60-Hz ac load.

- Build a CMOS *super spirograph* that displays on a color tv, using those spirograph toys with the gears and colored pens as your model. Use thumbwheel switches to set the number of gear teeth and the trace diameters. Use an actual rotating cycloid in your synthesis.

Appendixes

TTL to CMOS Conversion Table

As an aid in selecting a given CMOS device for those having experience with TTL devices, the following chart summarizes the many CMOS devices that are either functional or pin-for-pin equivalents of their TTL counterparts.

The CMOS devices are listed primarily by using the 4000-series numbering system introduced by RCA. Other manufacturers generally follow the same numbering system, but with different prefixes. Motorola's CMOS devices belong to the MC14000 series, while those manufactured by Fairchild Semiconductor belong to the 34000 series.

Those devices listed as being available in the 74C-, 74HC-, and 74HCT-series class of devices are pin-for-pin equivalents of the 7400 series of TTL devices. For example, a 7402 TTL quad NOR gate performs the same function as a 4001, an MC14001 (Motorola), and a 34001 (Fairchild), and is pin-for-pin compatible with either the 74C02, the 74HC02, or the 74HCT02.

Table A-1. TTL-to-CMOS Conversion

TTL	4000 Series	7400 C	HC	HCT
7400	4011	X	X	X
7401	40107			
7402	4001	X	X	X
7403			X	(open drain)
7404	4009, 4049	X	X	X
7406	4009, 4049			
7407	4010,4050			
7408	4081	X	X	X
7410	4023	X	X	X
7411			X	
7414	MC14584, 40106	X	X	
7420	4012	X	X	
7425	4002			
7427	4025		X	
7428	4001			
7430	4068	X	X	
7432	4071	X	X	X
7437	4011			
7440	4012			
7442	4028	X	X	
7445	4028			
7446	4055, 4511			
7447	4055, 4511			
7448	4511	X		
7449	4511			
7450	4085			
7451			X	
7453	4086			
7454	4086			
7470	4096			
7472	4095			
7473	4027	X	X	
7474	4013	X	X	X
7475	4042		X	
7476	4027	X	X	
7477	4042		X	
7478	4027			
7483	4008	X		
7485	4063	X	X	
7486	4030	X	X	X
7489		X		
7490	4510	X		
7491	4015, 4094			
7493	4520	X		
7494	4035			
7495	40104, 40194	X		
7499	40104, 40194			

Table A-1. TTL-to-CMOS Conversion (Continued)

TTL	4000 Series	7400 C	7400 HC	7400 HCT
74100	4034			
74104	4095			
74105	4095			
74107	4027	X		
74110	4095			
74111	4027			
74112			X	
74113			X	
74121	4047, 4098			
74122	4047, 4098			
74123	4098		X	
74125	4502		X	
74126	4502		X	
74132	4093		X	
74133			X	
74136	4030, 4070			
74137			X	
74138			X	X
74139			X	X
74141	4028			
74145	4028			
74147			X	
74148	4532		X	
74150	4067	X		
74151	4051, 4097	X	X	
74152	4051, 4097			
74153	4052		X	
74154	4514, 4515	X	X	
74155	4555, 4556			
74156	4555, 4556			
74157	4019	X	X	X
74158			X	
74160		X	X	
74161		X	X	
74162		X	X	
74163		X	X	
74164	4015	X	X	
74165	4021	X	X	
74166	4014			
74167	4527			
74173	4076	X	X	
74174		X	X	X
74175		X	X	X
74178	4035			
74179	4035			
74180	40101			
74181	40181		X	

Table A-1. TTL-to-CMOS Conversion (Continued)

TTL	4000 Series	7400 C	HC	HCT
74182	40182			
74190	4510		X	
74191	4516		X	
74192	40192	X	X	
74193	40193	X	X	
74194	40104, 40194	X	X	
74195	4035, 40195	X	X	
74198	4034			
74200	4061	X		
74221		X	X	
74365			X	
74367			X	
74368			X	
74373		X	X	X
74374		X	X	X
74375			X	
8095		X		
8097		X		

Some Product Sources

AFS Precision Tool
Box 354
Cupertino, CA 95014

Analog Devices
Box 280
Norwood, MA 02062

AP Products
72 Corwin Drive
Painesville, OH 44077

Applied Solderwrapper
1 Main Place
Dallas, TX 75250

Augat
33 Perry Avenue
Attleboro, MA 02703

Bead Chain
110 Mountain Grove Street
Bridgeport, CT 06605

Bishop Graphics
20450 Plummer Street
Chatsworth, CA 91311

Brady
2221 West Camden Road
Milwaukee, WI 53201

Calectro
400 South Wyman Street
Rockford, IL 61101

Cambion
445 Concord Avenue
Cambridge, MA 02138

Christiansen Radio
1950 San Remo
Laguna Beach, CA 92657

Circuit Aides
Bergenfield, NJ 07631

Circuit Stik
24015 Garnier Street
Torrance, CA 90510

Continental Specialties
44 Kendall Street
New Haven, CT 06512

Corning TSR
550 High Street
Bradford, PA 16701

Datak
65 Seventy First Street
Gutenberg, NJ 07493

Dynachem
13000 So. Firestone Blvd.
Santa Fe Springs, CA 90670

Eldre Components
1500 Jefferson Road
Rochester, NY 14623

Gardner-Denver
1333 Fulton Street
Grand Haven, MI 49417

Hewlett Packard
620 Page Mill Road
Palo Alto, CA 94304

Kodak PMT
Eastman Kodak
Rochester, NY 14650

Micro Electronic Systems
8 Kevin Drive
Danbury, CT 06810

Molex
2222 Wellington Court
Lisle, IL 60532

Nobex
1027 California Drive
Burlingame, CA 94010

Panavise
10107 Adella Avenue
Southgate, CA 90280

Phillips
400 Crossways Parkway
Woodbury, NY 11797

Pomona Electronics
1500 East Ninth Street
Pomona, CA 91766

Rodgers Corp.
Rodgers, CT 02063

Roper-Whitney
2833 Huffman Blvd.
Rockford, IL 61101

Scotch Color Key
3M Center
St. Paul, MI 55701

Sensorex PH
9713 Bolsa Avenue
Westminster, CA 92683

Stamp It (M-Tech)
Box C
Springfield, VA 22571

Telequipment & Tektronix
Box 500
Beaverton, OR 97077

Teletype Corp.
5555 Touhy Avenue
Skokie, IL 60076

Ungar
233 E. Manville Street
Compton, CA 90220

Vector Electronics
12460 Gladstone Avenue
Sylmar, CA 91342

Velcro Corp.
5347 N. Spaulding Drive
Westlake Village, CA 91891

Index